DATA MINING FOR DESIGN AND MANUFACTURING

T0189468

Data Mining for Design and Manufacturing

Methods and Applications

Edited by

Dan Braha
Ben-Gurion University

KLUWER ACADEMIC PUBLISHERS
DORDRECHT / BOSTON / LONDON

A C.I.P. Catalogue record for this book is available from the Library of Congress.

ISBN 978-1-4419-5205-9

Published by Kluwer Academic Publishers,
P.O. Box 17, 3300 AA Dordrecht, The Netherlands.

Sold and distributed in North, Central and South America
by Kluwer Academic Publishers,
101 Philip Drive, Norwell, MA 02061, U.S.A.

In all other countries, sold and distributed
by Kluwer Academic Publishers,
P.O. Box 322, 3300 AH Dordrecht, The Netherlands.

Printed on acid-free paper

Printed in the Netherlands.

PREFACE

PART I: OVERVIEW OF DATA MINING

PART II: DATA MINING IN PRODUCT DESIGN

PART III: DATA MINING IN MANUFACTURING

Data Mining for Design and Manufacturing

The productivity of individual companies as well as the efficiency of the global economy can be dramatically affected by Engineering Design and Manufacturing decisions and processes. Powerful data acquisition systems (such as minicomputers, microprocessors, transducers, and analog-to-digital converters) that collect, analyze, and transfer data are in use in virtually all mid-range and large companies. Over time, more and more current, detailed, and accurate data are accumulated and stored in databases at various stages of design and production. This data may be related to designs, products, machines, materials, processes, inventories, sales, marketing, and performance data and may include patterns, trends, associations, and dependencies. There is valuable information in the data. For instance, understanding the data and the quantitative relationships among product design, product geometry and materials, manufacturing process, equipment capabilities, and related activities could be considered strategic information. Extracting, organizing, and analyzing such useful information could be utilized to improve and optimize company planning and operations.

The large amount of data, which was generated and collected during daily operations and which contain hundreds of attributes, needs to be simultaneously considered in order to accurately model the system's behavior. It is the abundance of data, however, that has impeded the ability to extract useful knowledge. Moreover, the large amount of data in many design and manufacturing databases make it impractical to manually analyze for valuable decision-making information. This complexity calls for new techniques and tools that can intelligently and (semi)automatically turn low-level data into high-level and useful knowledge.

The need for automated analysis and discovery tools for extracting useful knowledge from huge amounts of raw data suggests that Knowledge Discovery in Databases (KDD) and Data Mining methodologies may become extremely important tools in realizing the above objectives. Data mining is primarily used in business retail. Applications to design and manufacturing are still under utilized and infrequently used on a large scale. Data Mining is often defined as the process of extracting valid, previously unknown, comprehensible information from large databases in order to improve and optimize business decisions[1]. Some researchers use the term KDD to denote the entire process of turning low-level data into high-level knowledge.

---

[1] Fayyad, U.M., Piatetsky-Shapiro, G., Smyth, P., and Uthurusamy, R. (Eds.), Advances in Knowledge Discovery and Data Mining. Cambridge, MA: AAAI Press/MIT Press, 1996.

Although data mining algorithms are at the core of the data mining process, they constitute just one step that usually takes about 15% to 25% of the overall effort in the overall data mining process. A collaborative effort of domain expert(s) (e.g., designer, production manager), data expert(s) (e.g., IT professionals) and data mining expert(s) is essential to the success of the data mining integration within design and manufacturing environments. A successful implementation of the data mining process often includes the following important stages[1]. The first step involves understanding the application domain to which the data mining is applied and the goals and tasks of the data mining process; e.g., understanding the factors that might affect the yield of a Silicon wafer in the semiconductor industry. The second step includes selecting, integrating, and checking the target data set that may be stored in various databases and computer-aided systems (such as CAD, CAM, MRP or ERP). The target data set may be defined in terms of the records as well as the attributes of interest to the decision-maker. The third step is data preprocessing. This includes data transformation, handling missing or unknown values, and data cleaning. In the fourth step, data mining takes place for extracting patterns from data. This involves model and hypothesis development, selection of appropriate data mining algorithms, and extraction of desired data. In the fifth step, the extracted patterns are interpreted and presented in a user-readable manner; e.g., using visualization techniques, and the results are tested and verified. Finally, the discovered knowledge may be used and a knowledge maintenance mechanism can be set up. The data mining process may be refined and some of its steps may be iterated several times before the extracted knowledge can be used for productive decision making.

Data Mining techniques are at the core of the data mining process, and can have different goals depending on the intended outcome of the overall data mining process. Most data mining goals fall under the following main categories[1]:

- **Data Processing** is concerned with selection, integration, filtration, sampling, cleaning and/or transformation of data.
- **Verification** focuses mainly on testing preconceived hypotheses (generated by the decision-maker) and on fitting models to data.
- **Regression** is concerned with the analysis of the relationships between attribute values within the same record, and the automatic production of a model that can predict attribute values for future records. For example, multivariate linear regression analysis may be used to identify the most significant factors affecting process capability.
- **Classification** involves assigning a record to predetermined classes. For example, wafer-cleaning processes (that include parameters set at various levels) may be classified according to the quality of the cleaning process outcome; thus, the outcome of new cleaning processes can be identified.
- **Clustering** focuses on partitioning a set of data items with similar characteristics together. For example, identifying subgroups of silicon

wafers that have a similar yield.

♦ **Association** (link analysis) involves the discovery of rules that associate one set of attributes in a data set to other attribute sets. An example for a relation between attributes of the same item is "if the Lot Size > 100 ∧ Tool Change = Manual → (Inventory = High)."

♦ **Sequential Pattern Analysis** is concerned with the discovery of causal relationships among temporally oriented events. For example, the event of environmental attack and high service stresses can lead to a stress-corrosion breaking within the next 2 hours.

♦ **Model Visualization** focuses on the decision makers' attempt to convey the discovered knowledge in an understandable and interpretable manner. Examples include histograms, scatter plots, and outliers identification.

♦ **Deviation Analysis** is used to detect deviation over time, deviation from the mean, and deviation between an observed value and a reference value as applied, for instance, in quality control.

A variety of techniques are available to enable the above goals[2]. Different data mining techniques serve different purposes, each offering its own advantages and disadvantages. The most commonly used techniques can be categorized in the following groups: Statistical methods, Artificial Neural Networks, Decision Trees, Rule Induction, Case-Based Reasoning, Bayesian Belief Networks, and Genetic Algorithms and Evolutionary Programming. An introductory overview of data mining is provided in Chapters 1 and 2.

Several techniques with different goals can be applied successively to achieve a desired result. For example, in order to identify the attributes that are significant in a photolithography process, clustering can be used first to segment the wafer-test database into a given predefined number of categorical classes, then classification can be used to determine to which group a new data item belongs.

Over the last decade, data mining mechanisms have been applied in various organizations, and have led to a wide range of research and development activities. It is primarily used today by companies with a strong customer focus such as in retail, insurance, finance, banking, communication, and direct marketing. Although data mining is widely used in many such organizations, the interest in data mining reveals an astute awareness among manufacturing companies across many industry sectors regarding the potential of data mining for changing business performance. For example, data mining techniques have been used by Texas Instruments (fault diagnosis), Caterpillar (effluent quality control and warranty claims analysis), Ford (harshness, noise, and vibration analysis), Motorola (CDMA Base Station Placement), Boeing (Post-flight Diagnostics), and Kodak (data visualization). Still, the application

---

[2] Berry, M. J., and Linoff, G., Data Mining Techniques. New York: John Wiley & Sons, 1997.

of data mining to design and manufacturing is not broadly integrated within companies. Decision makers are hampered from fully exploiting data mining techniques by the complexity of broad-based integration. The objective of this book is to help clarify the potential integration and to present a wide range of possibilities that are available by bringing together the latest research and application of data mining within design and manufacturing environments. In addition, the book demonstrates the essential need for the symbiotic collaboration of expertise in design and manufacturing, data mining and information technology.

Data Mining in Product Design and Development

The integration of data mining to design and manufacturing should be based on goals and capabilities of data mining as well as goals and weaknesses of current design and manufacturing environments. To broaden our understanding of how data mining can overcome a variety of problems in design and manufacturing we consider a wide range of activities within manufacturing companies. The first important activity is the product design and development process. A product development process is the sequence of activities that a manufacturing company employs in order to turn opportunities and ideas into successful products. As can be seen in Figure 1, product development goes through several stages, starting with identifying customer needs and ending with production and then delivery to market[3].

The nature of the product development process can be viewed as a sequential process. The design process evolves from concept through realization. For instance, a part cannot be assembled until the components are machined; the components cannot be machined until the NC code is created; the NC code cannot be created without a dimensioned part model; the part model cannot be dimensioned without a set of requirements and a general notion of what the part looks like; and presumably the last two items come from a need that must first be identified. All this points to the seemingly undeniable truth that there is an inherent, sequential order to most design processes. One can reason equally effectively, however, that product development is an iterative process. First, product development teams are only human and have a bounded rationality. They cannot simultaneously consider every relevant aspect of any given product design. As the product development process progresses, new information, ideas, and technologies become available that require modifying the product design. Second, design systems are limited; there is no known system that can directly input a set of requirements and yield the optimum design. Rather, the designer must

---

[3] Braha, D, and Maimon, O., A Mathematical Theory of Design: Foundations, Algorithms and Applications. Boston, MA: Kluwer Academic Publishers, 1998.

iteratively break down the set of requirements into dimensions, constraints, and features and then test the resulting design to see if the remaining requirements were satisfied. Finally, the real world often responds differently than is imagined. The real world is full of chaotic reactions that are only superficially modeled in any design system. All this points to the seemingly undeniable truth that there is an inherent, iterative nature to the product development process[3].

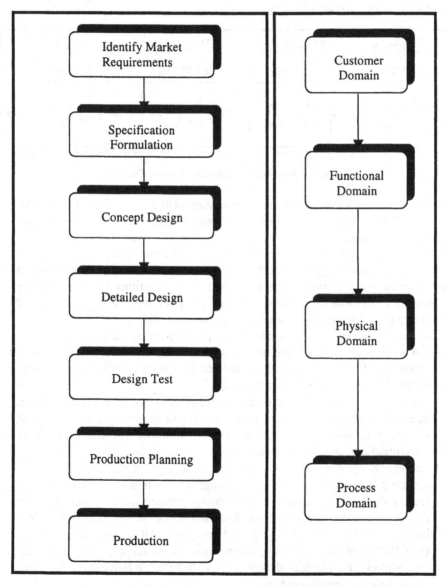

Figure 1 Traditional View of Product Design and Development

In order to reconcile these two disparate visions of the product development process, we categorize product development iteration is categorized as occurring either *between* stages (*inter-stage* iteration) or *within* a stage (*intra-stage* iteration)[3]. In this model (see Figure 2), product development still flows sequentially from initial concept through realization, each process stage providing the data and requirements for the subsequent stage. Within each process stage, however, the designer iteratively creates a design that meets the given requirements.

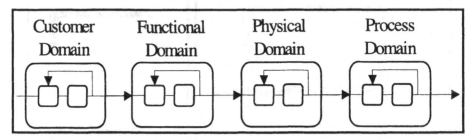

Figure 2 Combining Sequential and Iterative Design

This model largely represents the current state-of-the-art in CAD/CAM/CAE systems. While there are numerous software modules to assist the designer during intra-stage design iteration (e.g., QFD software to help identify customer needs and CAE software to analyze a current design), the tools are generally not well integrated at the inter-stage level. It has been recognized that minimizing the inter-stage and intra-stage iterations tends to reduce product development lead times and cost. Data mining has the potential of becoming one of the key components in achieving these goals.

During the product design and development process data mining can be used in order to determine relationships among "internal" factors at each stage and "external" factors at consecutive and previous stages. Following are some examples of how data mining can be utilized.

- ♦ Data mining can be used to extract patterns from customer needs, to learn interrelationships between customer needs and design specifications, and to group products based on functional similarity for the purpose of benchmarking, modular design, and mass customization.
- ♦ At the concept design stage, data mining can support concept selection by dynamic indexing and retrieval of design information in knowledge bases (e.g., patents and benchmarking products), clustering of design cases for design reuse, extracting design knowledge for supporting knowledge-based systems, extracting guidelines and rules for design-for-X (manufacturability, assembly, economics, environment), and exploring interactively conceptual designs by visualizing relationships in large product development databases.

♦ During system-level design, data mining can aid in extracting the relationships between product architecture, product portfolio, and customer needs data.

♦ At the detailed design stage, data mining can support material selection and cost evaluation systems.

♦ In industrial design, information about the complex relationships between tangible product features (such as color and ergonomics) and intangible aspects of a product that relate to the user (such as aesthetics, comfort, enthusiasm, and feelings) can be extracted with data mining and used in redesign.

♦ When testing the design, product characteristics can be extracted from prototypes. This may be used for determining the best design practice (e.g., design for reuse).

♦ During product development planning, data mining can be beneficial to activities such as the prediction of product development span time and cost, effectiveness of cross-functional teams, and exploration of tradeoffs between overlapping activities and coordination costs. Data mining may also be used for identifying dependencies among design tasks, which can be used to develop an effective product development plan.

♦ On an organizational level, data mining can be seen as a supportive vehicle for organizational learning, e.g., based on past projects, the factors (and their interdependencies) that affect a project's success/failure may be identified. This may include the intricate interaction between the project and the company, market, and macro environment.

In summary, the utilization of data mining in understanding the relationships among internal and external factors facilitates the inter-stage and intra-stage iterations. This will lead to improved product design and development.

Data Mining in Manufacturing

At the end of the product design and development process, after a stage of production ramp-up, the ongoing manufacturing system operation begins. There are a wide range of domains within manufacturing environments to which data mining techniques can be applied. On an abstract level, data mining can be seen as a supportive vehicle for determining causal relationships among "internal" factors associated with the manufacturing processes and "external" elements related to the competitiveness of the manufacturing company (e.g., production indices, performance parameters, yield, company goals). Since manufacturing environments have an inherently temporal or spatial context, time and/or space factors may be taken into account in the mining process in order to correctly interpret the collected data (e.g., from a certain production date the number of defects is much higher than normal). For instance, the process of wafer fabrication is a series of 16-24 loops, each adding a layer to the device. Each loop comprises some or all of the major steps of photolithography, etching, stripping, diffusion, ion implantation, deposition, and chemical mechanical planarization. At each stage, there are various inspections and measurements performed to monitor the equipment and process. The large number of parameters that are measured after each operation could identify causal interrelationships between processing steps and various test data. This information could then be utilized to eliminate faults during the manufacturing process, and thus enhance yield levels.

Additional examples where data mining can be applied successfully include:

♦ fault diagnosis such as predicting assembly errors and defects, which may be used to improve the performance of the manufacturing quality control activity;

♦ preventive machine maintenance, which is concerned with deciding the point in time and type of maintenance of tools and instruments. For instance, cutting tool-state may be classified and used for tool condition monitoring;

♦ manufacturing knowledge acquisition by examining relevant data, which implicitly contains most of the required expert knowledge. The extracted knowledge rules can then be incorporated by expert systems for decision support such as fuzzy controllers, diagnosis systems, and intelligent scheduling systems;

♦ operational manufacturing control such as intelligent scheduling systems that learn the effect of local dynamic behavior on global outcomes, and use the extracted knowledge to generate control policies. These operational systems are inherently adaptive, since data that is accumulated in real-time can change the baseline policies generated by the data mining algorithms;

♦ learning in the context of robotics (e.g., navigation and exploration, mapping, feature recognition, and extracting knowledge from numerical and graphical sensor data;

♦ quality and process control, which is concerned with monitoring standards; taking measurements; and taking corrective actions in case deviation from the norm is detected and/or discernible patterns of data over time are present. Extracted knowledge may include classification to predetermined types of deviation from the norm, and causal relationships among temporally oriented events;

♦ adaptive human-machine interface for machine operation;

♦ summarization and abstraction of large and high-dimensional manufacturing data;

♦ enabling supply and delivery forecasting, e.g., by classifying types of suppliers involved in transportation and distribution of the product.

Enabling Technology for Data Mining

In order too successfully implement data mining to design and manufacturing environments, several key issues such as the selection, integration, cleansing, and preparation of data should be addressed. Thus, enabling or supportive technologies that help carry out these processes are valuable.

♦ One of the most important supportive technologies is data warehousing, which is defined as the process of integrating legacy operational systems (storing data related to product, process, assembly, inventory, purchasing, etc.) within a company to provide centralized data management and retrieval for decision support purposes. Thus, the preprocessing, including cleaning and transforming data with the intent of analysis and discovery, can be facilitated by a data warehouse.

♦ Report generators, which are used to present the extracted patterns in a user-readable way, are another type of supportive technology. If discovered knowledge is further used by various computers (e.g., CNC systems, industrial robots, etc.), it is imperative that computers be able to interpret the output.

♦ Computationally massive data mining operations can be enabled through parallel-computing platforms and distributed computation. The parallelism aspect is especially important when data mining is deployed proactively and systematically throughout the manufacturing environment, and is used for continuous tasks such as preventive machine maintenance and real-time monitoring of the overall manufacturing processes.

Data Mining for Design and Manufacturing: Methods and Applications brings together for the first time the research and application of data mining within design and manufacturing environments. The contributors include researchers and practitioners from academia and industry. The book provides an explanation of how data mining technology can be exploited beyond prediction and modeling, and how to overcome several central problems in design and manufacturing environments. The book also presents the formal tools required to extract valuable information from design and manufacturing data (e.g., patterns, trends, associations, and dependencies), and thus facilitates interdisciplinary problem solving and optimizes design and manufacturing decisions.

The book includes aspects of topics such as: data warehouses and marts, data mining process, data mining tasks (e.g., association, clustering, classification), data mining methods (e.g., decision trees and rules, neural networks, self-organizing feature maps, wavelets, fuzzy learning, and case-based reasoning), machine learning in design (e.g., knowledge acquisition, learning in analogical design, conceptual design, and learning for design reuse), data mining for product development and concurrent engineering, design and manufacturing warehousing, Computer-aided Design (CAD) and data mining, data mining for Computer-aided Process Planning (CAPP), data mining for Material Requirements Planning (MRP), manufacturing data management, process and quality control, process analysis, data representation/visualization, fault diagnosis, adaptive schedulers, and learning in robotics.

Chapters are arranged in four sections: Overview of Data Mining; Data Mining in Product Design; Data Mining in Manufacturing; and Enabling Technologies for Data Mining in design and manufacturing.

Dan Braha
2001

Acknowledgements

I wish to acknowledge the authors for their fine contributions and cooperation during the book preparation. Chapters underwent a rigorous selection process. Many thanks to the reviewers. I appreciate the friendly assistance of Armin Shmilovici and Arieh Gavious. The technical support and suggestions of Sageet Braha were invaluable. I am grateful to John Martindale of Kluwer Academic Publishers for his forthcoming support of this important endeavor. Finally, to little Neoreet who provided entertainment along the way.

CHAPTER 1

Data Mining: An Introduction

Ishwar K. Sethi
isethi@oakland.edu
Intelligent Information Engineering Laboratory, Department of Computer Science and Engineering, Oakland University, Rochester, MI 48309

ABSTRACT

This chapter provides an introductory overview of data mining. Data mining, also referred to as knowledge discovery in databases, is concerned with nontrivial extraction of implicit, previously unknown and potentially useful information from data in databases. The main focus of the chapter is on different data mining methodologies and their relative strengths and weaknesses.

D. Braha (ed.), Data Mining for Design and Manufacturing, 1–40.
© 2001 *Kluwer Academic Publishers. Printed in the Netherlands.*

INTRODUCTION

In order to understand data mining let us first consider the distinction between data, information, and knowledge through an example. Suppose having completed weekly grocery shopping you are at your local grocery store in one of the checkout lanes. When your grocery goes past the checkout scanner, data is being captured. When your grocery store looks at the accumulated data for different customers at some point in time and finds certain grocery shopping patterns, the captured data is transformed into *information*. It is important to note that whatever it is that converts data into information resides external to the data and in the interpretation. When we have a very high degree of certainty or validity about information, we refer to it as *knowledge*. Continuing with our example, if the grocery shopping patterns discovered by our local grocery store are found to hold at many other grocery stores also, we have a situation where data is finally transformed into knowledge. Thus, we see that information and knowledge are both derived from data.

The modern technology of computers, networks, and sensors has made data collection an almost effortless task. Consequently, data is being captured and stored at a phenomenal pace. However, the captured data needs to be converted into information and knowledge to become useful. Traditionally, analysts have performed the task of extracting information and knowledge from recorded data; however, the increasing volume of data in modern industrial and business enterprises calls for computer-based methods for this task. Such methods have come to be known as *data mining* methods and the entire process of applying computer-based methodology is known as *knowledge discovery*. Note that this data mining viewpoint does not impose any restriction on the nature of the underlying computer data analysis tools. This is the viewpoint that is held by most of the vendors of data mining products. However, some people, especially those belonging to the artificial intelligence community, have a slightly narrower definition for data mining. According to their viewpoint, the underlying data analysis tools must be based on one or more sub-technologies of artificial intelligence, for example machine learning, neural networks, or pattern recognition, to qualify as the data mining method.

Importance in Business Decision Making

Data mining technology is currently a hot favorite in the hands of decision-makers as it can provide valuable hidden business intelligence from historical corporate data. It should be remembered, however, that fundamentally, data mining is not a new technology. The concept of extracting information and knowledge discovery from recorded data is a well-established concept in scientific and medical studies. What is new is the convergence of several factors that have created a unique opportunity for data mining in the corporate world. Businesses are suddenly realizing that the data that they have been collecting for the past 15-20 years can give them an immense competitive edge. Due to the client-server paradigm, data warehousing technology, and the currently available immense desktop computing power, it has become very easy for an end-user to look at stored data from all sorts of perspectives and extract valuable business intelligence. Data mining is being used to perform market segmentation to launch new products and services as well as to match existing products and services to customers' needs. In the banking, healthcare, and insurance industries, data mining is being used to detect fraudulent behavior by tracking spending and claims patterns.

Data Classification

One can classify data into three classes: (1) *structured data*, (2) *semi-structured data*, and (3) *unstructured data*. Most business databases contain structured data consisting of well-defined fields of numeric or alphanumeric values. Semi-structured data has partial structure. Examples of semi-structured data are electronic images of business and technical documents, medical reports, executive summaries, and repair manuals. The majority of web documents also fall in this category. An example of unstructured data is a video recorded by a surveillance camera in a departmental store. Such visual or multimedia recordings of events or processes of interests are currently gaining widespread popularity due to reducing hardware costs. This form of data generally requires an extensive amount of processing to extract contained information. Structured data is often referred to as *traditional data* while the semi and unstructured data are lumped together as *non-traditional data*. Because of the presence of structure in it, traditional data has no ambiguity. On the other hand, non-traditional data is difficult to interpret and often has multiple interpretations. Most of the current data mining methods and commercial tools are meant for traditional data; however, development of data mining tools for non-traditional data is growing at a rapid rate.

Another way of looking at data, especially traditional data, is to look at the behavior of recorded attributes with respect to time. Certain attributes, for example a customer's social security number, do not change with time. A database containing only such kinds of records is considered to have *static data*. On the other hand, there are attributes, for example a customer's monthly utility consumption, that change with time. A database containing such records is considered to have *dynamic* or *temporal data* as well. The majority of the data mining methods are more suitable for static data and special consideration is often required to mine dynamic data.

DATA MINING AND DATA WAREHOUSING

In this section, the relationship between data warehousing and data mining is addressed. Although the existence of a data warehouse is not a prerequisite for data mining, in practice the task of data mining is made a lot easier by having access to a data warehouse.

A data warehouse can be viewed as a data repository for an organization set up to support strategic decision-making. The architecture and the features of a data warehouse are very different from other databases in an organization, which are operational databases designed to support day-to-day operations of an organization. The function of a data warehouse is to store historical data of an organization in an integrated manner to reflect the various facets of the organization's business. The data in a warehouse is never updated but used only to respond to queries from end-users, who are generally the decision-makers. This is in contrast with the users of operational databases, whose interaction with the database consists of either reading some records from it or updating them. Unlike operational databases, data warehouses are huge in size storing billions of records. In many instances, an organization may have several local or departmental data warehouses. Such data warehouses (see Figure 1) are often called *data marts* due to their smaller size.

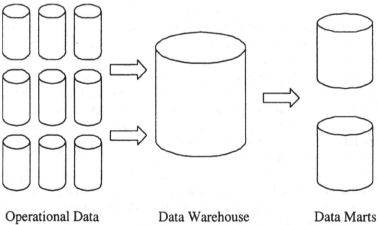

Operational Data   Data Warehouse   Data Marts

Figure 1. Operational databases, a data warehouse, and data marts

The link between data mining and data warehousing is a mutually reinforcing link. It is difficult to say whether the prospects of an informed and focused decision-making through data mining are responsible for a surge in industry-wide interest in data warehousing; or whether the availability of clean, well-formatted historical data in warehouses is the cause of recent boom in data mining. In any case, data mining is one of the major applications for data warehousing since the sole function of a data warehouse is to provide information to end-users for decision support. Unlike other query tools explained later, the data mining tools provide an end-user with a capability to extract hidden information. Such information, although more difficult to extract, can provide a bigger business advantage and yield higher returns on data warehousing investments for a company.

DATA MINING AND QUERY TOOLS

All databases provide a variety of query tools for users to access information stored in the database. For ease of operation, these query tools generally provide a graphical interface to users to express their queries. In the case of relational databases, the query tools are known as *SQL* tools because of the use of *Structured Query Language* to query the database. In the case of dimensional databases, the query tools are popularly known as the *on-line analytical processing (OLAP)* tools. Following the viewpoint that data mining is concerned with the extraction of information and knowledge from databases, one obvious question to raise is "How is data mining different from structured query language (SQL) and on-line analytical processing (OLAP) tools?" In this section, we try to provide an answer to this question.

Data Mining and SQL

SQL is the standard language for relational database management systems. SQL is good for queries that impose some kind of constraint on data in the database to extract an answer. In contrast, data mining is good for queries that are exploratory in nature. For example, if we want to obtain a list of all utility customers whose monthly utility bill is greater than some specified dollar amount, we can get this information from our database using SQL. However, SQL is not a very convenient tool if we want to obtain the differences in customers whose monthly bills are always paid on time with those customers who are usually late in their payments. It is not that this question cannot be possibly answered by SQL. By several trial and error steps, perhaps one can arrive at the answer through SQL. Data mining is, on the other hand, very good at finding answers to the latter types of questions. Thus, we can say that SQL is useful for extracting obvious information, i.e. shallow knowledge, from a database but data mining is needed to extract not so obvious, i.e. hidden, information from a database. In other words, SQL is useful when we know exactly what we are looking for; but we need data mining when we know only vaguely what we are looking for. Thus, SQL and data mining are complementary and both are needed to extract information from databases.

Data Mining and OLAP

OLAP tools have become very popular in recent years as they let users play with data stored in a warehouse by providing multiple views of the data. In these views, different dimensions correspond to different business characteristics, e.g. sales, geographic locations, product types etc. OLAP tools make it very easy to look at dimensional data from any angle or to "slice-and-dice" it. For example, it is easy to answer questions like "How have increased advertising expenditures impacted sales in a particular territory?" with OLAP tools. To provide such answers, OLAP tools store data in a special format which corresponds to a multi-dimensional hyper-box structure. Although OLAP tools like data mining tools provide answers that are derived from data, the similarity between the two sets of tools ends here. The derivation of answers from data in OLAP is analogous to calculations in a spreadsheet; OLAP tools do not learn from data; nor do they create new knowledge. They are simply special-purpose visualization tools that can help an end-user learn patterns in the data. In contrast, the data mining tools obtain their answers via learning the relationships between different attributes of database records. Often, these discovered relationships lead to creation of new knowledge by providing new insights to business. Thus, OLAP tools are useful for data mining because of their capabilities to visualize relationships between different data dimensions; however, they are not a substitute for data mining.

THE DATA MINING PROCESS

The data mining process consists of several stages and the overall process is inherently interactive and iterative [Fayyad et al, 1996]. The main stages of the data mining process are: (1) domain understanding; (2) data selection; (3)

cleaning and preprocessing; (4) discovering patterns; (5) interpretation; and (6) reporting and using discovered knowledge.

Domain Understanding

The domain understanding stage requires learning the business goals of the data mining application as well as gathering relevant prior knowledge. Blind application of data mining techniques without the requisite domain knowledge often leads to the discovery of irrelevant or meaningless patterns. This stage is best executed by a team of business and information technology persons to develop an all-around understanding of the data mining task being undertaken.

Data Selection

The data selection stage requires the user to target a database or select a subset of fields or data records to be used for data mining. Having a proper domain understanding at this stage helps in the identification of useful data. Sometimes, a business may not have all the requisite data in-house. In such instances, data is purchased from outside sources. Examples of data often purchased from outside vendors include demographic data and life-style data. Some applications of data mining also require data to be obtained via surveys.

Cleaning and Preprocessing

This is the most time-consuming stage of the entire data mining process. Data is never clean and in the form suitable for data mining. The following are typical of data corruption problems in business databases:

- Duplication - This kind of data corruption occurs when a record, for example a customer's purchases, appears several times in a database. It is one of the most common data corruption problems found in databases of businesses, such as direct mailers and credit card companies, dealing with individual customers. Misspelling due to typing/entry errors generally causes this kind of corruption. Sometimes customers are known to misspell deliberately to avoid linkage with their own past records.
- Missing Data Fields - Missing fields are present in a database due to a variety of reasons. For example, a customer may simply get tired of filling in the desired information; or a missing field may be caused by a data

entry error with an improper entry for a field. Filling in the missing values is generally a non-trivial task. Often, the records with missing fields are ignored for further processing.

□   Outliers - An outlier is a data value in a field, which is very different from the rest of the data values in the same field. The presence of outliers in a database is generally due to incorrect recordings of outlier fields. In many instances, outliers are easy to spot by considering domain consistency constraints. Sometimes, an outlier may be present due to exceptional circumstances such as a stolen credit card, when considering the monthly expenditure field in a credit card database. Detection of such outliers requires a considerable effort and often such a detection step itself is called the *data discovery step*.

Preprocessing

The preprocessing step of the third stage of data mining involves integrating data from different sources and making choices about representing or coding certain data fields that serve as inputs to the data discovery stage. Such representation choices are needed because certain fields may contain data at levels of details not considered suitable for the data discovery stage. For example, it may be counter-productive to represent the actual date of birth of each customer to the data discovery stage. Instead, it may be better to group customers into different age groups. Similarly, the data discovery stage may get overwhelmed by looking at each customer's address and may not generate useful patterns. On the other hand, grouping customers on a geographical basis may produce better results. It is important to remember that the preprocessing step is a crucial step. The representation choices made at this stage have a great bearing on the kinds of the patterns that will be discovered by the next stage of data discovery.

Discovering Patterns

The data-pattern-discovery stage is the heart of the entire data mining process. It is the stage where the hidden patterns and trends in the data are actually uncovered. In the academic or research literature, it is only this stage that is referred to as data mining. There are several approaches to the pattern discovery stage. These include association, classification, clustering, regression, sequence analysis, and visualization. Each of these approaches can be implemented through one of several competing methodologies, such as statistical data analysis, machine learning, neural networks, and pattern recognition. It is because of the use of methodologies from several disciplines that data mining is often viewed as a multidisciplinary field (see Figure 2). Here, we will provide details about different approaches to pattern discovery. The details about different methodologies will be presented in the next section.

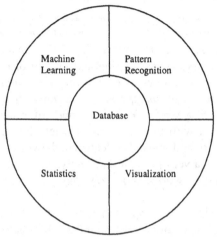

Figure 2. Core technologies for data mining

☐ Association - This approach to data discovery seeks to establish associative relationships between different items in database records. The association approach is very popular among marketing managers and retailers who find associative patterns like "90% of customers who buy product X also buy product Y" extremely helpful for market targeting and product placement in stores. The association approach to data discovery is successful when one has an idea of different associations that are being sought out. This is because one can find all kinds of correlations in a large database. Statistics, machine learning, and neural networks are popular methodologies for the association approach to data discovery.

☐ Classification - The classification approach to data discovery is perhaps the most widely used approach. It consists of classifying records into two or more pre-determined classes. As an example, consider a utility company planning to offer an appliance service plan to its customers. To get maximum response for a planned telemarketing effort, the utility may want to classify its customers into two classes – customers likely to respond and customers not likely to respond. Once such a classification is done, the telemarketers can concentrate on only those customers that fall in the first category. The application of the classification approach requires a classification rule, which is generally extracted from an existing set of pre-classified records. Such a set of records is often termed as a *training set* and the process of extracting the classification rule is commonly known as *learning*. Decision tree classifiers and neural network classifiers are two of the most popular methodologies for implementing the classification approach to data discovery.

☐ Clustering - Clustering implies data grouping or partitioning. This approach to data discovery is used in those situations where a training set of pre-classified records is unavailable. The major applications of clustering are in market segmentation and mining of customers' response data. Clustering is also known as *exploratory data analysis* (EDA). Other terms for clustering are *unsupervised learning* and *self-organization*. Performing clustering with a known number of groupings is relatively easy in comparison with those situations when the number of groups is not known a-priori and must be determined by the clustering process itself. Clustering is a very popular data analysis tool and statistical pattern recognition, neural networks, and fuzzy logic offer a variety of clustering algorithms.

☐ Sequence Analysis - This approach is used for discovering patterns in time-series data. For example, we could use this approach to determine the buying patterns of credit-card customers to predict their future purchases. Such predictive information can be used for identifying stolen credit cards. Sequence analysis is also used to establish associations over time. For example, it can be used to find patterns like "80% of customers who buy product X are likely to buy product Y in the next six months." This allows marketers to target specific products and services that the customers are more likely to buy. The popular methodologies for sequence analysis are rooted in statistics and neural networks.

☐ Visualization - The visualization approach to data mining is based on an assumption that human beings are very good at perceiving structure in visual forms. This approach thus consists of providing the user with a set of visualization tools to display data in various forms. The user while viewing the visualized data makes the actual discovery of the patterns in the data. An extreme of the visualization approach to data mining consists of creating an immersive virtual reality (VR) environment so that a user can move through this environment discovering hidden relations.

Interpretation

The interpretation stage of the data mining process is used by the user to evaluate the quality of discovery and its value to determine whether previous stages should be revisited or not. Proper domain understanding is crucial at this stage to put a value on discovered patterns.

Reporting

The final stage of the data mining process consists of reporting and putting to use the discovered knowledge to generate new actions or products and services or marketing strategies. Without this step, the full benefits from data mining cannot be realized. Reporting can take many forms, including detailed graphical presentation of the patterns discovered and the transfer of the mined knowledge or model to the appropriate business application.

DATA MINING METHODOLOGIES

This section presents basic concepts of different data mining methodologies. We present a few selected techniques from different methodologies. From statistical data analysis methods, we describe linear regression, logistic regression, linear discriminant analysis, and clustering techniques. From pattern recognition, we focus mainly on nearest neighbor classification. After presenting the basic neuron model, we describe briefly multiple-layer feed-forward network and self-organization feature map methods from neural network methodology of data mining. From machine learning, we describe decision tree methods and genetic algorithms.

Many different types of variables or attributes, i.e. fields in a database record, are common in data mining. Not all of the data mining methods are equally good at dealing with different types of variables. Therefore, we first explain the different types of variables and attributes to help readers determine the most suitable methodology for a given data mining application.

Types of Variables

There are several ways of characterizing variables. One way of looking at a variable is to see whether it is an *independent* variable or a *dependent* variable, i.e. a variable whose value depends upon values of other variables.

Another way of looking at a variable is to see whether it is a *discrete* variable or a *continuous* variable. Discrete variables are also called *qualitative* variables. Such variables are measured or defined using two kinds of non-metric scales – *nominal* and *ordinal*. A nominal scale is an order-less scale, which uses different symbols, characters or numbers, to represent the different states of the variable being measured. An example of a nominal variable is the customer type identifier, which might represent three types of utility customers – residential, commercial, and industrial, using digits 1, 2, and 3, respectively. Another example of a nominal attribute is the zip-code field of a customer's record. In each of these two examples, numbers designating different attribute values have no particular order and no necessary relation to one another. An ordinal scale consists of ordered discrete gradations, e.g. rankings. An example of an ordinal attribute is the preference ordering by customers; say of their favorite pizza. An ordered scale need not be necessarily linear, e.g. the difference in rank orders 3 and 4 need not be identical to the difference in rank orders 6 and 7. All that can be established from an ordered scale is the greater-than or less-than relations. The continuous variables are also known as *quantitative* or *metric* variables. These variables are measured using either an *interval* scale or a *ratio* scale. Both of these scales allow the underlying variable to be defined or measured with infinite precision. The difference between the interval and ratio scales lies in how the zero point is defined in the scale. The zero point in the interval scale is placed arbitrarily and thus it does not indicate the complete absence of whatever is being measured. The best example of an interval scale is the temperature scale, where zero degrees Fahrenheit does not mean total absence of temperature. Because of the arbitrary placement of the zero point, the ratio relation does not hold true for variables measured using interval scales. For example, 80 degrees Fahrenheit does not imply twice as much heat as 40 degrees Fahrenheit. In contrast, a ratio scale has an absolute zero point and consequently the ratio relation holds true for variables measured using this scale. This is the scale that we use to measure such quantities as height, length, energy consumption, and salary.

Statistical Data Analysis

Statistical data analysis is the most well established methodology for data mining. Ranging from 1-dimensional analysis, e.g. mean, median, and mode of a qualitative variable, to multivariate data analysis simultaneously using many variables in analysis, statistics offers a variety of data analysis methods [Hair et al, 1987]. These data analysis methods can be grouped into two categories. The methods in the first category are known as *dependence* methods. These methods use one or more independent variables to predict one or more dependent variables. Examples of this category of methods include multiple regression and discriminant analysis. The second category of statistical data analysis methods is known as *interdependence* methods. These methods are used when all of the variables involved are independent variables. Examples of interdependence methods are different types of clustering methods and multidimensional scaling.

Dependence Methods

Multiple Linear Regression

Multiple regression method is used when a single dependent quantitative variable (also called the *outcome variable*) is considered related to one or more quantitative independent variables, also known as *predictors*. The objective of regression analysis is to determine the best model that can relate the dependent variable to various independent variables. Linear regression implies the use of a general linear statistical model of the following form

$$y = a_0 + a_1 x_1 + a_2 x_2 + \cdots + a_k x_k + \varepsilon$$

where y is the dependent variable and $x_1, x_2 \ldots, x_k$ are the independent variables. The quantities, $a_0, a_1 \ldots, a_k$, are called unknown parameters and ε represents the random error. The unknown parameters are determined by minimizing the sum of squared error (SSE). It is defined as

$$SSE = \sum_{i=1}^{n} (y_i - \hat{y}_i)^2$$

where y_i and $\hat{y}_i$, respectively, are the observed and predicted values of the dependent variable for the i-th record in the database of n records. The process of model fitting, when only one independent variable is involved is equivalent to best straight-line fitting in least square sense. The term *simple regression* is often used in that situation. It should be noted that the term *linear* in the general linear model applies to the dependent variable being a linear function of the unknown parameters. Thus, a general linear model might also include some higher order terms of independent variables, e.g. terms such as $x_1^2, x_1 x_2$, or x_2^3. The major effort on the part of a user in using the multiple regression technique lies in identifying the relevant independent variables and in selecting the regression model terms. Two approaches are common for this task: (1) sequential search approach, and (2) combinatorial approach. The sequential search approach consists primarily of building a regression model with a set of variables, and then selectively adding or deleting variables until some overall criterion is satisfied. The combinatorial approach is a brute force approach, which searches across all possible combinations of independent variables to determine the overall best regression model. Irrespective of whether the sequential or combinatorial approach is used, the most benefit to model building occurs from a proper understanding of the application domain.

Logistic Regression

In many applications, the dependent variable is a qualitative variable, e.g. the credit rating of a customer, which can be good or bad. In such cases, either logistic regression or discriminant analysis is used for prediction. Rather than predicting the state of the dependent variable, the logistic regression method tries to estimate the probability p that the dependent variable will be in a given state. Thus, in place of predicting whether a customer has a good or bad credit rating, the logistic regression approach tries to estimate the probability of a good credit rating. The actual state of the dependent variable is determined by looking at the estimated probability. If the estimated probability is greater than 0.50, then the prediction is yes (good credit rating), otherwise no (bad credit rating). In logistic regression, the probability p is called the *success probability* and is estimated using the following model:

$$\log(p / (1 - p)) = a_0 + a_1 x_1 + a_2 x_2 + \cdots + a_k x_k$$

where $a_0, a_1 \ldots, a_k$ are unknown parameters. This model is known as the *linear logistic model* and $\log(p / (1 - p))$ is called the *logistic* or *logit* transformation of a probability. Unlike the multiple linear regression where the unknown parameters are estimated using the least squares method, the logistic regression procedure determines unknown parameters by the likelihood maximization method. With respect to the credit rating example, this means maximizing the likelihood of a good credit rating.

Linear Discriminant Analysis

Linear discriminant analysis (LDA) is concerned with problems that are characterized as classification problems. In such problems, the dependent variable is categorical (nominal or ordinal) and the independent variables are metric. The objective of discriminant analysis is to construct a discriminant function that yields different scores when computed with data from different classes. A linear discriminant function has the following form:

$$z = w_1 x_1 + w_2 x_2 + \cdots w_k x_k$$

where $x_1, x_2 \ldots, x_k$ are the independent variables. The quantity z is called the *discriminant score*. The variables $w_1, w_2 \ldots, w_k$ are called *weights*. A geometric interpretation of the discriminant score is shown in Figure 3. As this figure shows, the discriminant score for a data record represents its projection onto a line defined by the set of weight values.

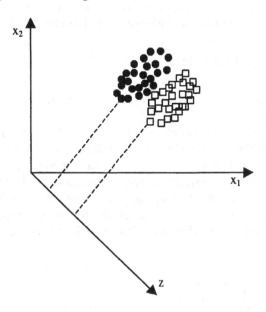

Figure 3. Geometric interpretation of the discriminant score

The construction of a discriminant function involves finding a set of weight values that maximizes the ratio of the between-groups to the within-group variance of the discriminant score for pre-classified records (training examples) from the database. Once constructed, the discriminant function is used to predict the class of a given data record, i.e. the state of the dependent variable from the independent variables. The classification is performed by the following classification rule. Assign the *i-th* data record to class A (e.g. good credit rating) if its discriminant score z_i is greater than or equal to the cutting score; otherwise assign it to class B (i.e. bad credit rating). The cutting score thus serves as a criterion against which each individual record's discriminant score is judged. The choice of cutting score depends upon whether both classes of records are present in equal proportions or not, as well as the underlying distributions. It is common to assume the underlying distributions to be normal. Letting $\tilde{z}_A$ and $\tilde{z}_B$ be the mean discriminant scores of pre-classified data records from class A and B, respectively, the optimal choice for the cutting score, z_{cut}, is given as

$$z_{cut} = \frac{\tilde{z}_A + \tilde{z}_B}{2}$$

when the two groups of records are of equal size and are normally distributed with uniform variance in all directions. A weighted average of mean discriminant scores, calculated as follows, is used as an optimal cutting score when the groups are not of equal size:

$$z_{cut} = \frac{n_A \tilde{z}_A + n_B \tilde{z}_B}{n_A + n_B}$$

The quantities n_A and n_B in above, respectively, represent the number of records in each group.

While a single discriminant function is constructed for two-way classification, multiple discriminant functions are required when dealing with more than two classes. The term *multiple discriminant analysis* is used in such situations. For an *M*-way classification problem, i.e. a dependent variable with *M* possible outcomes, *M* discriminant functions are constructed. The classification rule in such situations takes the following form: "Decide in favor of the class whose discriminant score is highest." This is illustrated in Figure 4.

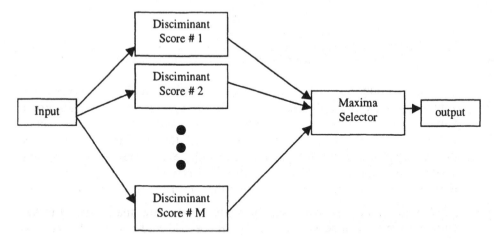

Figure 4. Illustration of classification in multiple discriminant analysis

Interdependence Methods

There are many applications where there is no dependent variable. In such applications, interdependence methods such as clustering and multidimensional scaling are used.

Clustering

Clustering or cluster analysis refers to methods for grouping objects – individuals, products, and services, in such a way that each object is more similar to objects in its own group than to objects in other groups. Since clustering methods are used in a wide range of disciplines, there exist a variety of names for clustering such as *unsupervised classification, Q analysis, typology,* and *numerical taxonomy.* A clustering method is characterized by how it measures similarity and what kind of method is employed to perform grouping. There are several ways of measuring similarity, with Euclidean distance being the most commonly used measure. Given two objects, A and B, the *Euclidean distance* between them is defined as the length of the line joining them. Other distance measures are the *absolute* and *maximum* distance functions. Irrespective of the distance measure being used, a small distance value between two objects implies high similarity and vice-versa a large distance value implies low similarity. Since objects in any application will have many attributes, each measured with a different scale, it is very common to use a normalized distance function in clustering. A normalized distance function incorporates a raw data normalization step so that each raw value is converted into a standard variate with a zero mean and a unit variance. The most commonly used normalized distance measure is the *Mahalanobis distance,* which not only takes care of different scales for different attributes but also accounts for inter-correlations among the attributes.

Most of the common clustering methods can be classified into two general categories: (1) *hierarchical* and (2) *partitional.* A hierarchical clustering procedure achieves its clustering through a nested sequence of partitions, which can be represented in a treelike structure. On the other hand, a partitional clustering method performs clustering in one shot. Figure 5 shows these differences in two general types of clustering procedures.

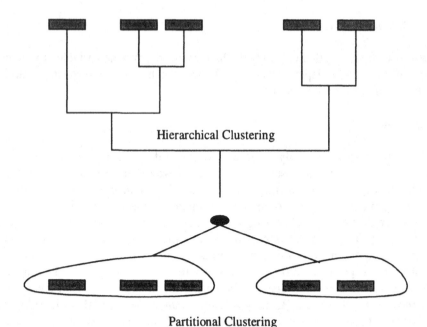

Partitional Clustering

Figure 5. Differences in hierarchical and partitional clustering

Hierarchical Clustering

Hierarchical clustering procedures can be further divided into two basic types – *agglomerative* and *divisive*. In agglomerative clustering, each object starts out as its own cluster. The subsequent stages involve pair-wise merging of two most similar clusters or objects. This process continues until the number of clusters is reduced to the desired number, or eventually all objects are grouped into a single cluster as shown in Figure 5. The treelike structure of Figure 5 is often referred to as a *dendrogram*. Divisive approach to hierarchical clustering is exactly opposite to the agglomerative approach. Here, we begin with one large cluster of all objects. In subsequent steps, objects most dissimilar are split off to yield smaller clusters. The process is continued until each object becomes a cluster by itself. Agglomerative procedures are much more popular and are provided by most statistical software packages. Some of the popular agglomerative clustering procedures are *single linkage*, *complete linkage*, and *average linkage*. These methods differ in how the similarity is computed between clusters. Figure 6 illustrates these differences.

Partitional Clustering

Partitional clustering procedures can be classified as *sequential* or *simultaneous* procedures. In a sequential partitional clustering procedure, objects to be clustered are handled one by one and the ordering of presentation of objects usually has an influence on the final clusters. Simultaneous methods, in contrast, look at all objects at the same time and thus generally produce better results. Many partitional clustering methods achieve clustering via optimization. Such procedures are known as *indirect methods* in contrast with direct partitional clustering methods, which do not use any optimization method. The most well known partitional clustering method is the *K-means* method, which iteratively refines the clustering solution once the user specifies an initial partition or from a random initial partition.

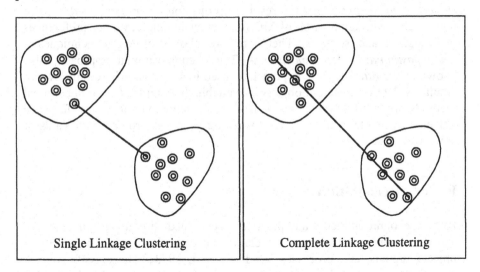

| Single Linkage Clustering | Complete Linkage Clustering |

Figure 6. Similarity computation in single linkage and complete linkage clustering

Comparison of Hierarchical and Partitional Clustering

Historically, the hierarchical clustering techniques have been more popular in biological, social, and behavioral sciences whereas partitional methods are more frequent in engineering applications. The main advantage of hierarchical procedures is their speed. The other advantage is that no knowledge of the number of clusters is required. The disadvantage of hierarchical methods is that they sometime lead to artificial groupings because grouping mistakes cannot be reversed due to the hierarchical nature of the grouping process. In contrast, the partitional methods are relatively slow but tend to produce better results. However, these methods rely on the user to provide good initial seed points for clusters and thus demand a better domain understanding on the part of the user. Irrespective of the selected clustering procedure, it is generally advisable to compute several solutions before settling on one as the final solution.

Multidimensional Scaling

Multidimensional scaling (MDS) relies on a projection from a high dimensional space to a low dimensional space (two or three dimensions) to uncover similarities among objects. For the mapping from a high-dimensional space to a low-dimensional one to be useful in MDS, it is required that the mapping preserve inter-point distances as far as possible. MDS is also known as *perceptual mapping* and the resulting projection as the *perceived relative image*. The main application of MDS lies in evaluating customer preferences for products and services. There are two classes of MDS techniques – *decompositional* and *compositional*. The decompositional approach, also known as the *attribute-free* approach, is used in situations where only overall similarity data for different objects is available. In contrast, the compositional methods are used when detailed data across numerous attributes for each object is available. Most statistical software packages provide both kinds of MDS methods.

Pattern Recognition

Pattern recognition theory and practice is concerned with the design, analysis, and development of methods for classification or description of patterns – objects, signals, and processes [Duda & Hart, 1973]. The classification is performed using such physical properties of patterns as height, width, thickness, and color. These properties are called *features* or *attributes* and the process of obtaining feature measurements for patterns is called *feature extraction*. Pattern recognition systems are used in two kinds of applications. The first kind of applications are those where a pattern recognition system provides cost and speed benefits. Examples of such applications are part location and identification in manufacturing, handwriting recognition in banking and offices, and speech recognition. The second kinds of applications are those where a pattern recognition system is used to perform a complex identification task either to assist or replace an expert. Examples of this kind of application include fingerprint classification, sonar signal classification, and flaw and crack detection in structures.

Pattern Recognition Approaches

There are three basic approaches to pattern recognition (PR) – *statistical*, *structural*, and *neural*. In the context of data mining, statistical and neural approaches are useful and will be discussed. The neural approach is discussed under neural networks. We shall limit ourselves here to the statistical pattern recognition (SPR) approach. The statistical approach is rooted in statistical decision theory. This approach to pattern recognition treats each pattern (object or data record) as a point in an appropriate feature space. Similar patterns tend to lie close to each other, whereas dissimilar patterns, those from different classes, lie far apart in the feature space. The taxonomy of statistical pattern recognition techniques is shown in Figure 7.

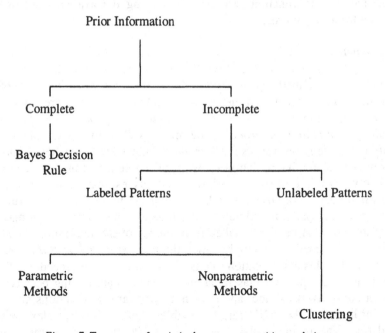

Figure 7. Taxonomy of statistical pattern recognition techniques

When complete information, i.e. a-priori probabilities and distribution parameters, about a pattern recognition task is available, the preferred PR approach is the *Bayes decision rule*, which provides optimal recognition performance. However, availability of complete information is rare and invariably a PR system is designed using a set of training or example patterns. This is analogous to data mining. When the example patterns are already classified, we say that we have *labeled patterns*. In such situations, *parametric* or *nonparametric* classification approaches are used. When example patterns do not have class labels, classification is achieved via clustering. The clustering methods in SPR are the same as those discussed earlier under statistical data analysis and, hence, will not be discussed any further.

Parametric Methods

The parametric methods are used when the form of class conditional densities is known. In practice, these densities are commonly assumed to be multivariate Gaussian. The Gaussian assumption leads to linear or quadratic classifiers. To implement these classifiers, the parameters of the class-conditional density functions are estimated using the available set of pre-classified training patterns.

Nonparametric Methods

The nonparametric methods are used in those situations where the form of the underlying class conditional densities is unknown. There are two basic nonparametric approaches – *density estimation approach* and *posteriori probability estimation approach*. The most well known example of the density estimation approach is the *Parzen window* technique where a moving window is used for interpolation to estimate the density function. The most well known example of the posteriori-probability estimation approach is the *k-nearest neighbor* (k-NN) method, which leads to the following rule for classification. Classify an unknown pattern to that class which is in majority among its k-nearest neighbors taken from the set of labeled training patterns. When k=1, this method is simply called the *nearest neighbor classifier*. It is easy to see that this classification rule is like a table look up procedure. The k-NN rule is a very popular classification method in data mining because it is purely a data-driven method, and does not imply any assumptions about the data. Furthermore, the k-NN rule is capable of producing complex decision boundaries. The main disadvantage of the k-NN rule is its computational burden, as an unknown pattern must be compared against every pattern in the training set, which can be exceedingly large. Many efficient implementation schemes are available in the literature to offset this drawback of the k-NN classifier.

All of the classification approaches – parametric and nonparametric, discussed so far are single shot approaches, i.e. a classification decision is made in one step. Decision tree classifiers in contrast offer a multistage decision methodology where stepwise exclusion of alternatives takes place to reach a final decision. Furthermore, the decision tree methodology does not require any assumption about class conditional densities. Consequently, tree classifiers are very popular in statistics, pattern recognition, machine learning, and neural networks. We shall discuss them later under machine learning.

Neural Networks

Artificial neural networks (ANNs) are currently enjoying tremendous popularity with successful applications in many disciplines. The interest in artificial neural networks is not new; it dates back to the work of McCulloch and Pitts, who about fifty years ago proposed an abstract model of living nerve cells or neurons [Zurada, 1992]. Since then, a very diverse set of researchers has been interested in ANNs because of a variety of different reasons. A main reason that has led many to look at ANNs is their non-algorithmic learning capability to solve complex classification, regression, and clustering problems.

Neuron Model

A neural network consists of a number of elementary processing units, called *neurons*. Figure 8 shows a typical neuron model. A neuron receives a number of inputs $x_1, x_2 \ldots, x_k$. Each input line has connection strength, known as *weight*. The connection strength of a line can be *excitatory* (positive weight) or *inhibitory* (negative weight). In addition, a neuron is given a constant bias input of unity through the bias weight w_0. Two operations are performed by a neuron —summation and output computation. The summation operation generates a weighted linear sum of inputs and the bias according to the following equation:

$$net = \sum_{i=0}^{k} w_i x_i$$

The output computation is done by mapping the weighted linear sum through an activation function:

$$y = f(net)$$

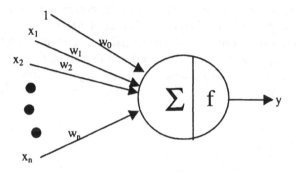

Figure 8. A typical neuron model

There are two basic types of activation functions – hard and soft. Hard activation function implies that the output of a neuron can exist in only one of the two possible states as shown below.

$$y = \text{sgn}(net - w_0) = \begin{cases} 1, & net > 0 \\ 0, & net < 0 \end{cases}$$

Such neurons are generally called *discrete* neurons or *perceptrons*. Sometimes the two allowed states are 1 and -1. Such neurons are called *bipolar discrete* neurons. Neurons with soft activation functions are called *soft* neurons or *continuous* neurons. An example of a soft activation function is the *sigmoidal* activation function:

$$y = \frac{1}{(1 + \exp(-\alpha(net - w_0)))}$$

which produces a continuously varying output in the range [0 1]. The quantity α in the above equations determines the slope of the activation function.

Neural Network Models

A neural network is a collection of interconnected neurons. Such interconnections could form a single layer or multiple layers. Furthermore, the interconnections could be unidirectional or bi-directional. The arrangement of neurons and their interconnections is called the *architecture* of the network. Different neural network models correspond to different architectures. Different neural network architectures use different learning procedures for finding the strengths (weights) of interconnections. Learning is performed using a set of training examples. When a training example specifies what output(s) should be produced for a given set of input values, the learning procedure is said to be a *supervised* learning procedure. This is the same as using a set of pre-classified examples in statistical data analysis and pattern recognition. In contrast, a network is said to be using an *unsupervised* learning procedure when a training example does not specify the output that should be produced by the network. While most neural network models rely on either a supervised or an unsupervised learning procedure, a few models use a combination of supervised and unsupervised learning.

There are a large number of neural network models, as shown in Figure 9, which have been studied in the literature [Looney, 1997]. Each model has its own strengths and weaknesses as well as a class of problems for which it is most suitable. We will briefly discuss only two models here that are common in data mining applications. These are (1) *multiple-layer feed-forward network*; and (2) *self-organizing feature map*. For information on other models, the reader should refer to books on neural networks.

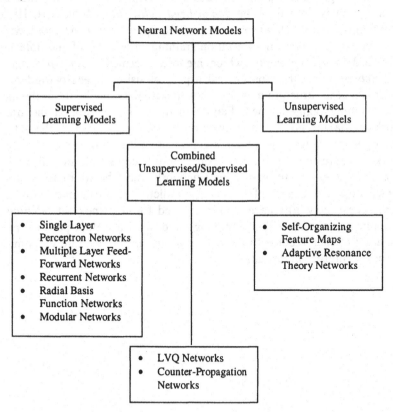

Figure 9. A classification of neural network models

Multiple-Layer Feedforward Network Model

The multiple-layer feedforward neural network model is perhaps the most widely used neural network model. This model consists of two or more layers of interconnected neurons, as shown in Figure 10. Generally, all neurons in a layer are connected to all neurons in the adjacent layers through unidirectional links. The leftmost layer is called the *input* layer, the rightmost the *output* layer. The rest of the layers are known as *intermediate* or *hidden* layers. It is known that a three-layer feed-forward network is capable of producing an arbitrarily complex relationship between inputs and outputs. To force a feedforward network to produce a desired input-output relationship requires training the network in an incremental manner by presenting pairs of input-output mapping. This training is done following an error minimization process, which is known as the *generalized delta rule* [Rumelhart, Hinton, and Williams, 1986]. However, the term *backpropagation* is more widely used to denote the error-minimization training procedure of multiple layer feedforward neural networks, which are often termed as *backpropagation neural networks* (BPN). One critical aspect of using a feedforward neural network is that its structure must match well with the complexity of the input-output mapping being learned. Failure to do so may result in the trained network not having good performance on future inputs. That is, the network may not generalize well. To obtain a good match, it is common to try several network structures before settling on one. Despite certain training difficulties, the multiple layer feedforward neural networks have been employed for an extremely diverse range of practical predictive applications with great success. These networks have also been used for sequence data mining. In such applications, data within a moving window of a certain pre-determined size is used at each recorded time instant as an input to capture the temporal relationships present in the data.

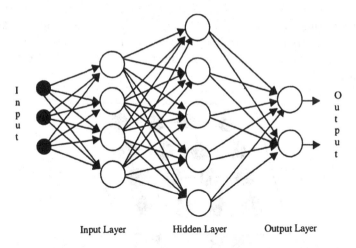

I
n
p
u
t

O
u
t
p
u
t

Input Layer   Hidden Layer   Output Layer

Figure 10. An example of a multiple-layer feedforward neural network

Self-Organizing Feature Map Model

The backpropagation training procedure is applicable only when each example in the training set has an associated output response. As mentioned before, however, many important applications do not involve any dependent variable. In order to perform data analysis in such instances, a neural network must be able to self-organize, i.e. it should be able to analyze and organize data using its intrinsic features without any external guidance. Kohonen's self-organizing feature map (SOFM) is one such neural network model that has received large attention because of its simplicity and the neuro-physiological evidence of the similar self-organization of sensory pathways in the brain [Kohonen, 1982]. The basic structure for a SOFM neural network is shown in Figure 11. It consists of a single layer of neurons with limited lateral interconnections. Each neuron receives an identical input. The network training is done following the *winner-takes-all* paradigm. Under this paradigm, each neuron competes with others to claim the input pattern. The neuron producing the highest (or smallest) output is declared the winner. The winning neuron and its neighboring neurons then adjust their weights to respond more strongly, when the same input is presented again. This training procedure is similar to k-means clustering of statistical data analysis, and suffers from the same merits and de-merits. Next to the feedforward neural networks, the SOFM networks are the most widely used neural networks. In addition to performing clustering, these networks have been used for feature extraction from raw data such as images and audio signals. A variation of SOFM is the *learning vector quantization* (LVQ) model that seeks to combine supervised and unsupervised modes of learning. In training LVQ, rough classification rules are first learned without making use of the known classification information for training examples. These rough rules are refined next using the known classification information to obtain finely tuned classification rules.

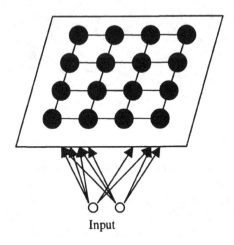

Input

Figure 11. Kohonen's SOFM model

In addition to the above three models, other important models gaining widespread popularity include the *radial-basis function* (RBF) network model, and the *neural tree* model. The RBF network model consists of three layers and exhibits performance similar to multiple layer feedforward neural networks. However, the training time for RBF networks is much shorter. The neural tree is another class of feedforward neural networks. Such networks have limited interconnectivity, similar to a decision tree structure, but are more powerful than tree classifiers in terms of predictive capability. We will discuss neural trees under tree-based methods.

Applying Neural Networks

A typical neural network application requires consideration of the following issues: model selection; input-output encoding; and learning rate. The choice of the model depends upon the nature of the predictive problem, its expected complexity, and the nature of the training examples. Since inputs and outputs in neural networks are limited to either [0-1] or [-1-1], the encoding of inputs and outputs requires careful considerations. Often, the encoding scheme for input-output has a large influence on the resulting predictive accuracy and training time. Another important factor in neural networks is the choice of learning rate. The learning rate determines the magnitude of weight changes at each training step. An improper learning rate (too small or too large) can cause an inordinately long training time, or can lead to sub-optimal predictive performance. Consequence, the use of neural networks is often called as an art.

Machine Learning

Machine learning is a sub-discipline of artificial intelligence. The goal of machine learning is to impart computers with capabilities to autonomously learn from data or the environment. The machine learning community has been a major contributor of several new approaches to data mining. These include tree-based methods for learning, genetic algorithms, intelligent agents, and fuzzy and rough set-based approaches.

Tree-Based Methods

The tree-based methods for classification and regression are popular across several disciplines – statistical data analysis, pattern recognition, neural networks, and machine learning. Many similar tree-based algorithms have been developed independently in each of these disciplines. There are several reasons for the popularity of tree-based methods. First, a tree-based method allows a complex problem to be handled as a series of simpler problems. Second, the tree structure that results from successive decompositions of the problem usually provides a better understanding of the complex problem at hand. Third, the tree-based methods generally require a minimal number of assumptions about the problem at hand; and finally, there is usually some cost advantage in using a tree-based methodology in certain application domains.

Decision Tree Classifiers

The term *decision tree* is commonly used for the tree-based classification approach. As shown in Figure 12, a decision tree classifier uses a series of tests or decision functions to assign a classification label to a data record. The evaluation of these tests is organized in such a way that the outcome of each test reduces uncertainty about the record being classified. In addition to their capability to generate complex decision boundaries, it is the intuitive nature of decision tree classifiers, as evident from Figure 16 that is responsible for their popularity and numerous applications. Like any other data mining methodology for classification, the use of decision tree classification requires automatic extraction of a tree classifier from training data. Several automatic algorithms exist for this purpose in the pattern recognition and machine learning literature. Most of these decision tree induction algorithms follow the top-down, divide-and-conquer strategy wherein the collection of pre-classified examples is recursively split to create example subsets of increasing homogeneity in terms of classification labels until some terminating conditions are satisfied.

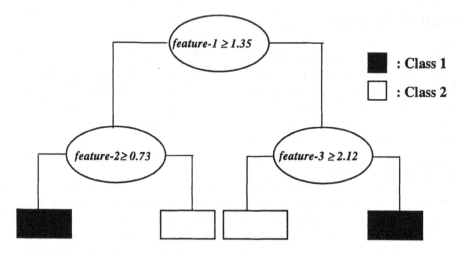

Figure 16. An example of a decision tree classifier. The left branches correspond to positive outcomes and right branches to negative outcomes of the tests at internal tree nodes.

The top-down, divide-and-conquer decision-tree-induction methodology consists of four components. First, it needs a splitting criterion to determine the effectiveness of a given split on training examples. Second, it requires a method to generate candidate splits. Third, a stopping rule is needed to decide when to stop growing the tree. Finally, it needs a method to set up a decision rule at each terminal node. The last component is the easiest part of the tree induction process. The majority rule is often used for this purpose. Different decision tree induction methods differ essentially in terms of the remaining three components. In fact, the differences are generally found only in the splitting criterion and the stopping rule.

The three most well known decision-tree-induction methodologies in pattern recognition, statistical data analysis, and machine learning literature are AMIG [Sethi and Sarvarayudu, 1982], CART [Breiman et al, 1984], and ID3 [Quinlan, 1986]. AMIG and ID3, both follow an information theory based measure, the *average-mutual information gain*, to select the desired partitioning or split of training examples. Given training examples from c classes, and a partitioning P that divides them into r mutually exclusive partitions, the average mutual information gain measure of partitioning, $I(P)$, is given as

$$I(P) = \sum_{i=1}^{r} \sum_{j=1}^{c} p(r_i, c_j) \log_2 \frac{p(c_j / r_i)}{p(c_j)}$$

where $p(r_i, c_j)$ and $p(c_j / r_i)$, respectively, are the joint and conditional probabilities and $p(c_j)$ is the class probability. Using the maximum likelihood estimates for probabilities, the above measure can be written as

$$I(P) = \sum_{i=1}^{r} \sum_{j=1}^{c} \frac{n_{ij}}{N} \log_2 \frac{n_{ij} N}{N_i n_j}$$

where n_j is the number of training examples from class c_j and n_{ij} is the number of examples of class c_j that lie in partition r_i. The quantity N is the total of all training examples of which N_i lie in partition r_i. The split of training examples providing the highest value of $I(P)$ is selected. The CART procedure uses the *Gini index of diversity* to measure the impurity of a collection of examples. It is given as

$$G = 1 - \sum_{j=1}^{c} p^2(c_j)$$

The split providing maximum reduction in the impurity measure is then selected. The advantage of this criterion is its simpler arithmetic.

   Determining when to stop top-down splitting of successive example subsets is the other important part of a decision-tree-induction procedure. The AMIG procedure relies for stopping on the following inequality that specifies the lower limit on the mutual information to be provided by the induced tree. The tree growing stops as soon as the accumulated mutual information due to successive splits exceeds the specified limit. CART and ID3 instead follow a more complex but a better approach of growing and pruning to determine the final induced decision tree. In this approach, the recursive splitting of training examples continues until 100% classification accuracy on them is achieved. At that point, the tree is selectively pruned upwards to find a best sub-tree according to some specified cost measure.

   The generation of candidate splits at any stage of the decision-tree-induction procedure is done by searching for splits due to a single feature. For example in AMIG, CART[*], and ID3, each top-down data split takes either the

---

[*] CART provides for limited Boolean and linear combination of features.

form of "Is xi ≥ t?" when the attributes are ordered variables or the form of "Is xi true?" when the attributes are binary in nature. The reason for using single feature splits is to reduce the size of the space of legal splits. For example with n binary features, a single feature split procedure has to evaluate only n different splits to determine the best split. On the other hand, a multi-feature-split procedure must search through a very large number of Boolean combinations, 2^{2^n} logical functions if searching for all possible Boolean functions, to find the best split. In recent years, many neural network-based methods, called *neural trees*, for finding decision tree splits have been developed [Sethi and Yoo, 1994; Sirat and Nadal, 1990]. These methods are able to generate efficiently multi-feature splits. Most of these neural tree methods make use of modified versions of the perceptron training procedure. Although neural trees do not provide predictive capabilities identical to those exhibited by multi-layer feedforward networks, their popularity rests on the intuitive appeal of step-wise decision making.

Regression Trees

While the majority of tree-based methods are concerned with the classification task, i.e. the data mining situations where the dependent variable is a discrete variable, the tree-based approach is equally suitable for regression. In such situations, the tree is called a *regression tree*. The well-known examples of the regression tree approach are CART (Classification and Regression Trees) and CHAID (Chi-Square Automatic Interaction Detection). CHAID is an extension of AID (Automatic Interaction Detection). The difference between the two is that AID is limited to applications where the dependent variable has a metric scale (interval or ratio). In contrast, CHAID is much broader in scope and can be applied even when the dependent variable happens to be a categorical variable; for example in market segmentation applications using brand preference data [Myers, 1996]. The steps involved in developing a regression tree from training examples are similar to the decision-tree-induction process discussed earlier. The major difference is in the splitting criterion. Unlike using the average-mutual information gain type measures suitable for classification tasks, measures such as the least square regression error are used in building a regression tree. There are two main differences between CART and CHAID. First, the trees produced by CART are binary trees, i.e. each node divides data into two groups only. In CHAID, the number of splits can be higher. The second difference is that CHAID is a pure top-down procedure, while CART methodology also uses bottom-up pruning to determine the proper tree size.

Genetic Algorithms

Genetic algorithms (GAs) belong to the broad area of evolutionary computing, which in recent years has gained widespread popularity in machine learning [Goldberg, 1989]. *Evolutionary computing* is concerned with problem solving by applying ideas of natural selection and evolution. Essentially, genetic algorithms are derivative-free search or optimization methods that use a metaphor based on evolution. In this approach, each possible solution is encoded into a binary bit string, called a chromosome. Also associated with each possible solution is a fitness function, which determines the quality of the solution. The search process in genetic algorithms begins with a random collection of possible solutions, called the *gene pool* or *population*. At each search step, the GA constructs a new population, i.e. a new generation, using genetic operators such as crossover and mutation. Just like the natural evolution process, members of successive generations exhibit better fitness values. After several generations of genetic changes, the best solution among the surviving solutions is picked as the desired solution.

The application of the GA to a problem requires consideration of several factors, including the encoding scheme, choice of the fitness function, parent selection, and genetic operators. The selection of an encoding scheme often depends upon how well it captures the problem-specific knowledge. The selected encoding scheme also influences the design of the genetic operators. The choice of fitness function is tied with the nature of the problem being solved. For classification problems, the fitness function may be the classification accuracy on the training data. On the other hand, the fitness function may be related to the mean square error for regression problems. The parent selection component of the GA specifies how the members of the present population will be paired to create the next generation of possible solutions. The usual approach is to make the probability of mating dependent upon a member's value of the fitness function. This ensures that members with better fitness values reproduce, leading to survival of the fittest behavior. Crossover and mutation operators are the two genetic operators that determine the traits of the next generation solutions. The *crossover operator* consists of interchanging a part of the parent's genetic code. This ensures that good features of the present generation are retained for future generations. In *one-point crossover operation*, a point on the genetic code of a parent is randomly selected and two parent chromosomes are interchanged at that point. In *two-point crossover operation*, two crossover points are selected and the genetic code lying between those points is interchanged. While crossover operators with many more points can be defined, one and two-point crossover operators are typically used in GAs.

The *mutation operator* consists of randomly selecting a bit in the chromosome string and flipping its value with a probability equal to a specified mutation rate. The mutation operator is needed because crossover operation alone cannot necessarily provide a satisfactory solution, as it only involves exploiting the current genes. Without mutation, there is a danger of obtaining locally optimum solutions. It is common to use a very low mutation rate to preserve good genes. Furthermore, a high mutation rate produces behavior similar to random search, and is, therefore, not used. In addition to selection, crossover, and mutation, GAs often use additional rules to determine the members of the next generation. One popular rule in this regard is the *principle of elitism*, which requires a certain number of best members of the present population to be cloned.

While GAs are often employed on their own to solve problems, it is not uncommon to see GAs being used in conjunction with other data mining methods. For example, GAs have been used in the decision tree methodology to determine data partitions. In neural networks, these have been used for network pruning or for finding the optimal network configuration. In addition to the attractiveness of the evolution paradigm, there are several other reasons that have contributed to the overall popularity of GAs. First, GAs are inherently parallel, and thus can be implemented on parallel processing machines for massive speedups. Second, GAs are applicable to discrete as well as continuous optimization problems. Finally, GAs are less likely to get trapped in a local minimum. Despite these advantages, these algorithms have not yet been applied to very large-scale problems. One possible reason is that GAs require a significant computational effort with respect to other methods, when parallel processing is not employed.

Intelligent Agents

Intelligent-agents (IA) is another machine learning approach that is rapidly becoming very popular in several applications, including data mining. Like GAs, the IA approach is also a population-based approach. However, unlike GAs where a member of the population interacts with another member to generate a new solution, the intelligent agents approach consists of finding solutions through social interaction among population members, each known as an intelligent agent [Epstein and Axtell, 1996]. Each intelligent agent behaves like an "individual" with different intelligent agents having different habits and behavior patterns. The habits and behavior patterns of agents are assigned based on existing data. As these simulated agents interact with each other, they adjust and adapt to other agents. Monitoring of behavior of a large collection of agents yields valuable information hidden in the data. For example, letting each agent simulate the buying pattern behavior of a consumer and running IA simulation for a sufficiently long time, we can count how many units of the new products are sold and at what rate. The information thus coming out of simulation can tell us whether the new product will be successful or not.

Fuzzy and Rough Sets

All of the methods discussed thus far have no provision for incorporating the vagueness and imprecision that is common in everyday life. The concepts of fuzzy sets and rough sets provide this provision and are useful in many applications. The fuzzy set concept was introduced by Lotfi Zadeh in 1965 [Zadeh, 1965] and since then has been applied to numerous applications. The last few years have especially seen a rapid increase in interest in fuzzy sets. Unlike conventional sets where an object either belongs to a set or not, objects in fuzzy sets are permitted varying degrees of memberships. As an example, consider the set of "tall" persons. In the conventional approach, we would define a cutoff height, say 5'10'', such that every person with height equal to or greater than the cutoff height would be considered tall. In contrast, the fuzzy set of "tall" people includes everyone in it. However, each person belongs to the "tall" fuzzy set with a different grade of membership in the interval of 0-1. The main advantage of fuzzy sets is that it allows inference rules, e.g. classification rules, to be expressed in a more natural fashion, providing a way to deal with data containing imprecise information. Most of the fuzzy set-based methods for data mining are extensions of statistical pattern recognition methods. The application of fuzzy set-based methods, e.g. the fuzzy k-means clustering procedure, to clustering is especially appealing. The statistical clustering techniques assume that an item can belong to one and only one cluster. Often in practice, no clear boundaries exist between different clusters, as shown in the several examples of Figure 13. In such situations, the notion of fuzzy clustering offers many advantages by letting each object have a membership in each cluster.

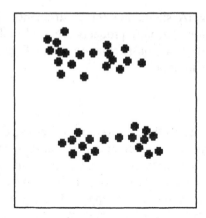

Figure 13. Examples of fuzzy clusters

The concept of rough sets is relatively new; Pawlak introduced it in 1982 [Pawlak, 1982]. The basic philosophy of rough set methodology is that lowering the degree of data precision makes data regularities easier to find and characterize in terms of rules. However, this lowering of precision is not without risk, as it can lead to loss of information and differentiation between different objects. The theory of rough sets provides tools to deal with this problem by letting roughness vary in a controlled manner to find a data precision level that ensures sufficient pattern discrimination. Often, rough sets are confused with fuzzy sets as they both deal with imperfect knowledge. However, both deal with different aspects of imperfection. The imperfection dealt with in fuzzy sets is with respect to objects within the same class. In contrast, rough sets deal with imperfections between groups of objects in different classes. The main application of rough sets so far has been in classification tasks. Furthermore, most of the reported applications have been for relatively small size problems. However, this is expected to change as rough sets gain more popularity and commercial software starts becoming available.

ACCURACY OF THE MINED MODEL

Irrespective of the data mining methodology selected for a particular task, it is important to assess the quality of the discovered knowledge. There are two components to this assessment – *predictive accuracy* and *domain consistency*. Predictive accuracy implies how well the discovered classification or regression model will perform on future records. Domain consistency means whether the discovered knowledge is consistent with other domain knowledge that the end user might have. Since the model assessment based on domain consistency is problem specific, we will discuss only the predictive assessment component in the following.

The goal of assessing predictive accuracy is to find the true error rate - an error rate that would be obtained on an asymptotically large number of future cases. The most common approach for assessing predictive accuracy is the train-and-test error rate estimation approach. In this approach, the entire data set is randomly divided into two groups. One group is called the training set and the other the testing set. The selected data mining methodology is applied to the test set to build the predictive model. Once the model is built, its performance is evaluated on the test set to determine the test-sample error rate, which provides a very good indication of the predictive performance on future records. When the test set consists of 5000 test cases, selected independent of the training set, then the test-sample error rate is considered virtually the same as the true error rate. For test sets of smaller size, the test-sample error rate is considered a slightly optimistic estimate of the true error rate. It is therefore a good practice to have as large a test set as possible. When this is not possible due to the small size of the total data set, a situation not likely to occur in data mining, various re-sampling techniques such as cross-validation and boot-strapping are used to obtain the predictive accuracy [Weiss and Kulikowski, 1991].

COMPARISON OF DATA MINING METHODOLOGIES

Having described a number of methodologies, a natural question to ask is "Which data mining methodology is best?" This is a difficult question to answer. All that can be said is that there is no universally best methodology. Each specific application has its own unique characteristics that must be carefully considered and matched with different methodologies to determine the best method for that specific application. However, there are a few general remarks that can be made with respect to different methodologies.

The strength of statistical methods is that these methods come from a mature field. The methods have been thoroughly studied and have a highly developed theoretical foundation. In consequence, the knowledge discovered using these methods is more likely to be accepted by others. However, the weakness of the statistical methods is that they require many assumptions about the data. Furthermore, these methods require a good statistical knowledge on the part of an end-user to properly interpret the results. Another weakness of the statistical methods is that they do not scale well for large amounts of non-numeric data.

Many of the statistical pattern recognition methods share the same strengths and weaknesses as those of the statistical methods. However, the case of the k-NN rule is entirely different. It is a true data-driven method, which does not require any assumption about the data. Its decision making model is also intuitively understandable. The weakness of this method is its predictive capability; it cannot provide accuracy at par with other methods. This weakness of the k-NN rule, however, is offset by its zero learning time.

The neural network and machine learning methods are relatively new. In many cases, these techniques have not been theoretically well studied. However, the main strength of neural network and machine learning methods is that they require relatively fewer assumptions about data, and are thus data-driven. The neural network methods are criticized mainly on two counts. First, the results are difficult to replicate, and second, the concern about interpretability of the behavior of a trained neural network. The concern about replicating the results arises due to the algorithmic nature of neural network training. Since several components, such as the random initial weights and the presentation order of training examples, of this algorithmic process are not necessarily duplicated every time a network is trained, it is generally difficult to replicate a network performance. The interpretability concern about neural network is associated with the *black box* image of neural networks. Since the explanations about a neural network behavior are stored in its weights, they are not easily comprehensible to a user. This has led to the black box view of neural networks. This view is further compounded by the fact that obtaining a proper neural network model for an application involves iteratively identifying proper network size, learning rate, and stopping criteria. This iterative process often overwhelms a new user. In consequence, the acceptance of neural networks is occasionally hard to come by. However, the black box image of neural networks is slowly changing as researchers are finding ways to uncover the knowledge captured by a neural network [Lu et al, 1995; Sethi and Yoo, 1996].

In comparison to other methods, the decision tree methodology appears to offer most advantages – a competitive predictive accuracy, minimal assumptions about data, better computational efficiency, and good scalability. Furthermore, the tree-based methods are able to deal with all types of variables. Consequently, this methodology is highly popular in data mining. The decision tree methodology also yields a mined model that is extremely easy to understand and use. On the negative side, the performance of decision trees is occasionally unsatisfactory because of single feature splits for data partitioning. Another criticism of decision tree methods is their greedy tree growing methodology, which often hides better models from discovery. It should be noted that many new decision tree induction methods combining machine and neural learning are being developed to address the above drawbacks of the decision tree methodology. Their availability in commercial software packages is expected to lead to more usage of the tree methodology in data mining.

SUMMARY

This chapter has provided an introduction to data mining and the core technologies behind it. Data mining is not a single technique but rather a collection of techniques that allows us to reach out to valuable hidden information present in large business databases. Although computer-based learning technologies - neural networks, pattern recognition, decision trees - form the core of data mining, there is more to data mining than simply using a neural network or decision tree algorithm. It is an interactive and an iterative process of several stages driven with the goal of generating useful business intelligence. The success of a data mining effort is not determined by the accuracy of the resultant predictive or classification model but by the value of the model to the business. Effective data mining not only requires a clear understanding of the business issues involved but also needs an inordinate amount of data preparation - identifying important variables, cleaning data, coding and analyzing data. Without proper data preparation, data mining is apt to generate useless information. The proverbial garbage-in, garbage-out is never more apt than in a data mining application without a proper understanding of the business aspects of the problem and careful data preparation.

Data mining has emerged as a strategic tool in the hands of decision-makers to gain valuable business intelligence from their corporate data. Such business intelligence is helping companies improve their operations in many critical areas including marketing, product development and customer services. In consequence, the applications of data mining continue to grow.

REFERENCES

Breiman, L., Friedman, J., Olshen, R. and Stone, C.J., Classification and Regression Trees, Belmont, CA: Wadsworth Int'l Group, 1984.

Duda R. and Hart, P., Pattern Classification and Scene Analysis. New York: John Wiley and Sons, 1973.

Epstein J.M. and Axtell, R., Growing Artificial Societies. Washington, DC: Brookerings Institution Press, 1996.

Fayyad U.M., Piatetsky-Shapiro, G., Smyth, P., and Uthurusamy, R. (Eds.), Advances in Knowledge Discovery and Data Mining. Cambridge, MA: AAAI Press/MIT Press, 1996.

Goldberg D.E., Genetic Algorithms in Search, Optimization, and Machine Learning. Reading, MA: Addison-Wesley, 1989.

Hair Jr. J.F., Anderson, R.E., Tatham, R.L. and Black, W.C., Multivariate Data Analysis. New York: Macmillan Publishing Company, 1987.

Kohonen T., "Self-Organized Formation of Topologically Correct Feature Maps," Biological Cybernetics, 43, pp. 59-69, 1982.

Lu, H., Setiono, R. and Liu, H., "NeuroRule: A Connectionist Approach to Data Mining," in Proceedings of the 21st VLDB Conference, pp. 478-489, 1995.

Looney C.G., Pattern Recognition Using Neural Networks. New York: Oxford University Press, 1997.

Myers J.H., Segmentation and Positioning for Strategic Marketing Decisions. Chicago: American Marketing Association, 1996.

Pawlak Z., Rough Sets: Theoretical Aspects of Reasoning About Data. Dordrecht, The Netherlands: Kluwer Academic Publishers, 1991.

Quinlan J.R., "Induction of Decision Trees," Machine Learning, 1, pp. 81-106, 1986.

Rumelhart, D.E., Hinton, G.E., and William, R.J., "Learning Internal Representation by Error Propagation," in Parallel Distributed Processing, MIT Press: Cambridge, MA, 1986.

Sethi I.K. and Yoo, J.H., "Design of Multicategory Multifeature Split Decision Trees Using Perceptron Learning," Pattern Recognition, 27(7), pp. 939-947, 1994.

Sethi I.K. and Yoo, J.H., "Symbolic Mapping of Neurons in Feedforward Networks," Pattern Recognition Letters, 17(10), pp. 1035-1046, 1996.

Sirat J.A. and Nadal, J.-P., "Neural Trees: A New Tool for Classification," Networks, 1, pp. 423-438, 1990.

Weiss S.M. and Kulikowski, C.A., Computer Systems That Learn. San Mateo, CA: Morgan Kaufmann Publishers, 1991.

Zadeh L.A., "Fuzzy Sets," Information and Control, 8, pp. 338-353, 1965.

Zurada J.M., Artificial Neural Systems. St. Paul, MN: West Publishing, 1992.

CHAPTER 2

A Survey of Methodologies and Techniques for Data Mining and Intelligent Data Discovery

Ricardo Gonzalez
rgonzal@umich.edu
Rapid Prototyping Laboratory, College of Engineering and Computer Science, The University of Michigan, Dearborn, Michigan 48128-1491

Ali Kamrani
kamkode@umich.edu
Rapid Prototyping Laboratory, College of Engineering and Computer Science, The University of Michigan, Dearborn, Michigan 48128-1491

ABSTRACT

This paper gives a description of data mining and its methodology. First, the definition of data mining along with the purposes and growing needs for such a technology are presented. A six-step methodology for data mining is then presented and discussed. The goals and methods of this process are then explained, coupled with a presentation of a number of techniques that are making the data mining process faster and more reliable. These techniques include the use of neural networks and genetic algorithms, which are presented and explained as a way to overcome several complexity problems that the data mining process possesses. A deep survey of the literature is done to show the various purposes and achievements that these techniques have brought to the study of data mining.

D. Braha (ed.), Data Mining for Design and Manufacturing, 41–59.

INTRODUCTION

During the last few years, data mining has received more and more attention from different fields, especially from the business community. This commercial interest has grown mainly because of the awareness of companies that the vast amounts of data collected from customers and their behaviors contain valuable information. If this information can be somehow made explicit, it will be available to improve various business processes.

Data mining deals with the discovery of hidden knowledge, unexpected patterns and new rules from large databases. It is regarded as the key element of a much more elaborate process called knowledge discovery in databases, or KDD, which is closely linked to data warehousing. According to Adriaans and Zantinge (1996), data mining can bring significant gains to organizations, for example through better-targeted marketing and enhanced internal performance. They stated in their book that the long-term goal of data mining is to create a self-learning organization that makes optimal use of the information it generates.

Recent publications on data mining concentrate on the construction and application of algorithms to extract knowledge from data. Skarmeta, et al (2000) developed a data-mining algorithm for text categorization. Andrade and Bork (2000) used a data-mining algorithm to extract valuable information on molecular biology form large amounts of literature. Lin, et al (2000) developed an efficient data-mining algorithm to measure proximity relationship measures between clusters of data. Delesie and Croes (2000) presented a data-mining approach to exploit a health insurance database to evaluate the performance of doctors in cardiovascular surgeries nationwide.

The emphasis given by most authors and researchers on data mining focuses on the analysis phase of data mining. When a company uses data mining, it is important to also see that there are other activities involved in the process. These activities are usually more time-consuming and have an important influence on the success of the data mining procedure.

This paper is organized as follows: The following section introduces the data mining concept as well as it outlines the advantages and disadvantages of its use in the knowledge extraction process from databases. This section also introduces some basic expertise requirements that any company should possess in order to use data mining effectively. The next section discusses the different stages involved in the data mining process. Some data mining techniques and methods used during the mining phase of the process are then discussed. Finally, some conclusions are presented to emphasize the importance of the techniques and methods presented in the previous sections.

DATA MINING

In the past, data mining has been referred to as knowledge management or knowledge engineering. Until recently, it has been an obscure and exotic technology, discussed more by theoreticians in the artificial intelligence fields. Fayyad, et al (1996) defined data mining as a step in the KDD process consisting of applying computational techniques that, under acceptable computational efficiency limitations, produce a particular enumeration of patterns or models over the data. Adriaans and Zantinge (1996) gave a more general definition used by many researchers. They stated that data mining is the process of searching through details of data for unknown patterns or trends. They stressed the importance of having an efficient method of searching in large amounts of data until a sequence of patterns emerges, whether complete or only within an allowable probability.

Many times, large databases are searched for relationships, trends, and patterns, which prior to the search, are not known to exist nor visible. These relationships or trends are usually assumed to be there by engineers and marketers, but need to be proven by the data itself. The new information or knowledge allows the user community to be better at what it does. Often, a problem that arises is that large databases are searched for very few facts that will give the desired information. Moreover, the algorithm and search criteria used in a single database may change when a new trend or pattern is to be studied. Also, each database may need a different search criterion as well as new algorithms that can adapt to the conditions and problems of the new data.

More often than not, humans find it difficult to understand and visualize large data sets. Furthermore, as Fayyad and Stolorz (1997) described, data can grow in two dimensions defined as the number of fields and the number of cases for each one of these fields. As they explained, human analysis and visualization abilities do not scale to high dimensions and massive volumes of data.

A second factor that is making data mining a necessity is the fact that the rate of growth of data sets completely exceeds the rates that traditional 'manual' analysis techniques can cope with. This means that if a company uses a regular technique for extracting knowledge from a database, vast amounts of data will be left unsearched, as the data growth surpasses the traditional mining procedures. These factors call for a need of a technology that will enable humans to tackle a problem using large amounts of data without disregarding or losing valuable information that may help solve any kind of problem involving large data sets.

Yevich (1997) stated that: "data mining is asking a processing engine to show answers to questions we do not know how to ask". He explained that instead of asking in normal query language a direct question about a single occurrence on a database, the purpose of data mining is to find similar patterns that will somehow answer the desired questions proposed by the engineers or marketers. If the questions or the relationships asked to be found on a database are too specific, the process will be harder and will take more time. Moreover, a lot of important relationships will be missed or disregarded.

The interest on data mining has risen in the past few years. During the 1980s, many organizations built infrastructural databases, containing data about their products, clients, and competitors. These databases were a potential gold mine, containing terabytes of data with much "hidden" information that was difficult to understand. With the great strides shown by artificial intelligence researchers, machine-learning techniques have grown rapidly. Neural networks, genetic algorithms, and other applicable learning techniques are making the extraction of knowledge form large databases easier and more productive than ever.

Data mining is being used widely in the United States, while in Europe, it has been used to a less extent. Large organizations such as American Express and AT&T are utilizing KDD to analyze their client files. In the United Kingdom, the BBC has applied data mining techniques to analyze viewing figures. However, it has been seen that the use of KDD brings a lot of problems. As much as 80% of KDD is about preparing data, and the remaining 20% is about mining. Part of these 80% is the topic that will be analyzed and discussed in the next section.

It is very difficult to introduce data mining into a whole organization. A lot of data mining projects are disregarded as options because of these additional problems summarized by Adriaans and Zantinge (1996) and Kusiak (1999):

- Lack of long-term vision: company needs to ask themselves "what do we want to get from our files in the future?"
- Struggle between departments: some departments do not want to give up their data.
- Not all files are up to date: data is missing or incorrect; files vary greatly in quality.
- Legal and privacy restrictions: some data cannot be used, for reasons of privacy.
- Poor cooperation from the electric data processing department.
- Files are hard to connect for technical reasons: there is a discrepancy between a hierarchical and a relation database, or data models are not up to date.
- Timing problems: files can be compiled centrally, but with a six-month delay.
- Interpretation problems: connections are found in the database, but no one knows their meaning or what they can be used for.

In addition to these common problems, the company needs to have a minimum level of expertise on the data mining processes. Scenarios, in which the area expert does not have a specific question and asks the analyst to come up with some interesting results, are sentenced to fail. The same holds true for situations where the expert provides the data analyst with a set of data and a question, expecting the analyst to return the exact answer to that question. According to Feelders, et al (2000), data mining requires knowledge of the processes behind the data, in order to:

- Determine useful questions for analysis;
- Select potentially relevant data to answer these questions;
- Help with the construction of useful features from the raw data; and
- Interpret results of the analysis, suggesting possible courses of action.

Knowing what to ask and what to expect from the information in a database is not enough. Knowledge of the available data from within is also required or at least desired. This will enable the expert and the data analyst to know where the data is and have it readily available depending on the problem being studied.

Finally, data analysis expertise is also desired. Hand (1998) discussed that phenomena such as population drift and selection bias should be taken into account when analyzing a database. Data mining expertise is required in order to select the appropriate algorithm for the data mining problem and the questions being raised.

DATA MINING METHODOLOGY

Data mining is an iterative process. As the process progresses, new knowledge and new hypothesis should be generated to adjust to the quality and content of the data. This means that the quality of the data being studied will determine the time and precision of any given data mining algorithm, and if the algorithm is flexible enough, important information about a problem will be found even if the central question is not fully answered.

Fayyad, et al (1996) developed a structured methodology outlining the different steps and stages of the data mining process. They outlined a six-step methodology that involved defining the problem to be solved, the acquisition of background knowledge regarding this problem, the selection of useful data, the pre-processing of the data, the analysis and interpretation of the results, and the actual use of these results. In their methodology, they stressed the importance of the constant "jump-backs" between stages. Feelders, et at (2000) then used this methodology to explain each step of the data mining process. As figure 1 shows, the mining stage of the process (Analysis and Interpretation) is just one of the basic stages of the data mining methodology. The discussion on the stages of this methodology will show that all phases play a major role in the process, especially those that come before the mining stage.

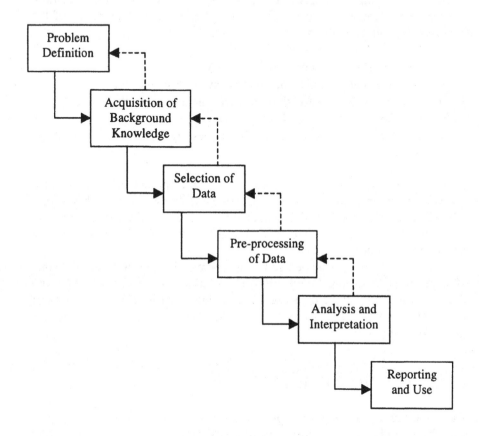

Figure 1. Major steps in the Data mining process

Problem Definition

The problem definition phase is the first stage of the data mining process. During this stage, the objectives of using data mining on the desired problem are identified. These are questions or assumptions that, as it was mentioned earlier, are known to exist by marketers and engineers, but that need to be proven by the data.

Most of the time, the initial question asked in a data mining project should be very vague. This will help the data mining process because it will disregard large amounts of data that are not useful to the problem. This way, the selection of data and the pre-processing of data will work with a set of data that has been initially 'dissected' to help solve the initial problem.

During the problem definition stage, it is also important to know how the results of the data mining process are going to be used. Glymour, et al (1997) discussed the different common uses of the results of a data mining process. Some of these include:

- Intervention and prediction: Results can lead to an intervention of the system being studied. Also, they can predict certain behaviors of the system.
- Description and insight: Results give an intelligent description an insight about the topic being studied.

Glymour, et al (1997) also stressed the fact that one should be cautious about the source of the data. Data may be bias, an issue that will directly affect the results of the data mining process. Biased descriptions and insights will lead to biased and possibly harmful predictions about a system. Another issue that may arise is the problem of causality. Feelders, et al (2000) suggested a closer look at the data before assuming the results given by the data mining process. This will ensure that the conclusions drawn by the process are not just the result of chance.

The problem definition stage sets the standards and expectations of the data mining process. To an extent, this stage helps the users know the quality of the data being studied. If many iterations are required, and the problem definition ends up being too vague without getting acceptable results, the problem may lie on the quality of the data and not in the definition of the problem.

Acquisition of Background Knowledge

As it was mentioned in the previous stage of the data mining process, possible bias and selection effects of the data being studied should be known. This knowledge will give the development team the possible limitations of the data under consideration.

Another important type of knowledge that is important to have before any selection of data is the typical causal relations found on data. Heckerman (1996) proposed Bayesian Networks as a solution to this problem. They allow the incorporation of prior knowledge, which may signal possible causality found on a given result from the data. The acquisition of prior knowledge will also prevent the occurrence of 'knowledge rediscovery', in essence, the data mining algorithm will tackle a problem with a number of assertions that will prevent it from having to relearn certain patterns that are known to exist. Feelders, et al (2000) proposed a method called rule induction, in which "the user would have to guide the analysis in order to take causal relations into account."

The acquisition of knowledge plays a critical role in the data mining process. This stage can help directly on the time it takes the process to give positive results. It also prevents the process to learn facts and rules that are already known to be true.

Selection of Data

After the background knowledge is known, the data mining process reaches the important stage of selecting the data that will be used and analyzed to give an answer to the problem under consideration.

This selection of the relevant data should be 'open-minded' because the purpose of data mining is not for the human to solve the problem, but rather to let the data speak for itself. With the background knowledge in place, the data mining process will prevent the human expert from introducing new unaccounted biases that could harm the conclusions made by the process.

Subramanian (1997) and Yevich (1997) proposed the use of a data warehouse as an ideal aid for selecting potential relevant data. However, a data warehouse is rarely available for use in the present time, so companies have to go through the process of 'creating' one before the selection of data to achieve an acceptable selection of data. If a data warehouse is not readily available at the selection stage, the process will be a long one, and the data mining process will suffer a big delay.

The selection of data is a crucial step in the data mining process. Assuming the previous steps are performed properly, data selection narrows down the range of the potential conclusions to be made in the following steps. It also sets the range where these conclusions may be applicable.

Pre-processing of Data

Even when a data warehouse that has all the relevant data of the problem is available, it is often required to pre-process the data before in can be analyzed.

This stage also gives the expert the freedom of adding new attributes to the process, which will in a way, help the data mining procedure. These additions constitute certain relations between data that may be difficult for the data mining algorithm to assert. Some algorithms, such as the classification tree algorithm discussed by Feelders, et al (2000), fail in the assertion of certain relations that may be important in the data mining process. Other algorithms take long amounts of time to make the assertions, so if the knowledge is readily available by the expert, it is better to add it directly as an attribute rather than letting the algorithm make the assertion.

Much of the pre-processing in the data mining process is due to the fact that many of the relations between entities in a database are one-to-many. Data mining algorithms often require that all data concerning one instance of the entity should be stored in one record, so that the analysis can be done in one big table that contains all the records regarding the possible instances of a single entity.

In order for the pre-processing to be successful, the expert should use domain knowledge and common sense to determine the possible attributes and the creation of records that will enable the data mining process to be successful over time.

Analysis and Interpretation

The analysis phase follows a long process of problem definition, selection of data and pre-processing of data. At this time of the process, at least 70% of the time used in data mining process has elapsed. During this phase, it is critical for the expert to have experience and knowledge on the subject area being studied, on data analysis, and of course, on data mining.

Knowledge on the subject being studied is required primarily to interpret results, and most importantly, to assert which results should be taken into account for further study for possible corrective actions if they are necessary at one point. Data analysis experience is principally required to explain certain strange patterns found on the data, and to give importance to more interesting parts of the data being studied. Finally, data mining experience and expertise are required for the technical interpretation of the results. This technical interpretation is in essence, as Feelders, et al (2000) mention in their paper, the translation of the results to the language of the domain and the data expert.

The analysis and interpretation stage of the data mining process is where the actual mining takes place. After the data has been selected, pre-processed, and the problem to be solved is known, this stage tries to find certain patterns, similarities, and other interesting relations between the available data. All these patters are usually translated into rules that are used in the last phase of the data mining process.

Reporting and Use

Results of a data mining process have a wide range of uses. Results can be used in a simple application, such as being input for a decision process, as well as in an important application, like a full integration into an end-user application.

The results of a data mining process can also be used in a decision support system or a knowledge-based system. This field of application enables the learning process of the knowledge-base system to be faster and more efficient since it will not have to wait or adjust to certain biases that a human expert may have. The knowledge obtained will be in some cases more accurate, and if the background knowledge obtained during the data mining process is reliable, then the results will have a higher reliability as well.

As it was mentioned in an earlier section of this paper, the data mining process has been used as a tool for various purposes. Its results can be used to predict patterns and behaviors, or just to organize, sort, and choose certain amounts of data to prove an assertion made by an expert in the field under consideration.

In almost any event, the results of the data mining process should be presented in an organized fashion. This implies the use and development of a well-defined user interface. A good and robust user interface is critical in the data mining process as well as in a wide range of problems. It enables the users to report and interact with the programs effectively, which improves the success chances of the process.

In summary, the data mining process is a long and iterative process. As figure 1 depicts, the process goes both ways, so for example, after selecting the data to be studied, the expert can "go back" to the previous stage of the process to add more background knowledge that may have been forgotten. Also, if the definition of the problem is too specific and no results are found, the expert has the advantage of going back to redefine the problem.

The selection stage and the pre-processing stage are the most critical steps in the data mining process. The two steps account for almost 80% of the time used in the data mining effort. Thus, special attention should be given two these two steps if the data mining process is to be a success.

The last two steps of the process are usually 'routine work' if the required data analysis and mining expertise are used properly. Many researchers have discussed these two steps, but the good performance and use of the analysis and the results of the process is completely dependent on the previous stages. Without a good definition of the problem, or, with an inadequate selection of data, the results are not going to be useful.

Companies have to realize that the data mining process is a long and sometimes complicated process. Expertise in many fields is required for the process to be successful. Workers have to be patient, as the process may take a long time to be able to come up with useful answers. Management has to support the procedure, and know that analyzing and reporting results are not the only two parts of the data mining process. In fact, they have to understand that these two steps are just the end work of an extensive and complex process.

DATA MINING GOALS AND TECHNIQUES

From the data mining methodology explained in the last section, it can be seen that the actual mining is only done at the end of the process, specifically at the analysis and interpretation stage. At this point of the process, various goals have been set and many innovative techniques can be used to reach these goals and improve the data mining stage of the process. These relatively new techniques have been widely used by researchers as a tool for solving data mining problems, as problems have turned harder, databases have grown larger, and the traditional methods for data mining have become obsolete. Each one of these techniques has some advantages, disadvantages, and limitations that should be accounted for, as the use of the techniques is very problem specific.

Before going into the details of these techniques, it is necessary to briefly present the basic goals of the data mining process. These goals were initially set with the thought in mind that a number of simple search techniques would be able to solve the mining problems, but, when problems became so big and complex, these traditional techniques became useless. However, it is important to note two things; first, that the goals of the data mining process remained the same, and second, that as in many areas of research, most of the new techniques are based on these 'old' search methods. They are just more improved and advanced methods that can handle more complex problems. Some of them also introduce the concept of machine learning into data mining, a concept that can only bring significant improvements to the data mining process.

Traditional Methods and Goals of Data Mining

Yevich (1997) summarized some of the traditional mining methods and goals and divided them into three categories: Association, clustering, and classification.

In the associations' method, associations usually refer to searching all details or transactions from the systems for patterns that have a high probability of repetition. An associative algorithm is required to find all the rules that will correlate one set of events with another set of events. This method is very effective whenever the size of the database being studied is relatively small. However, as the database grows, the method becomes unreliable because the searching space is too large for the algorithm to handle. In addition to the information needed in this method (probability of repetition of events), the algorithm will have to assume a lot of information in order to be able to cover all the search space.

The clustering method can be used when there is no known way to define a class of data to be analyzed. In essence, the goal of clustering is to come up with certain relations between data sets, by comparing data and finding relations from the initial information given by the data itself. Clustering algorithms are usually used to discover a previously unknown or suspected relation in a database. This methodology is also known as unsupervised learning and it has been used widely by many researchers in data mining and other areas of research. A problem that may be encountered with this methodology is the task of finding certain relations between data sets. As with associations, a lot of information has to be assumed in order to find relations between data points that seemingly have nothing in common. At this point is where these new techniques, especially those that involve machine learning, come into play.

The classification method identifies the process and must discover roles that define whether an item or event belongs to a particular set of data. Classification can become a very complex process because often many tables and thousands of attributes must be examined to find a relation between components. This creates a complexity problem similar to the one found on the last two methods. As databases grow in size, the time it takes for this method to reach its goal increases rapidly. Classification is also known as supervised learning.

These three methods of data mining and their goals are sometimes very similar in nature. In essence, clustering and classification are pursuing similar goals, and both are grouping together chunks of data in order to arrive at certain conclusions. However, on all three methods a complexity problem is encountered whenever conventional searching techniques are used and when databases grow in size. A number of techniques, old and new, are used to resolve these problems. As it was mentioned earlier, some involve methods such as machine learning while others involve the use of old techniques such as SPC.

Data Mining Techniques

Adriaans and Zantinge (1996) summarized the various techniques used for tackling and solving complex data mining problems, stressing that the selection of a technique should be very problem specific.

Some of the techniques available are query tools, statistical techniques, visualization, online analytical processing (OLAP), case-based learning, decision trees, neural networks, and genetic algorithms. The complexity of these techniques varies, and a good selection of the technique used is critical to arrive at good and significant results. Table 1 shows the techniques available for data mining and their basic advantages and limitations.

Table 1. Data mining techniques

DATA MINING TECHNIQUE	CHARACTERISTICS ADVANTAGES/DISADVANTAGES
Query tools	• Used for extraction of 'shallow' knowledge. • SQL is used to extract information.
Statistical techniques	• SPC tools are used to extract deeper knowledge from databases. • Limited but can outline basic relations between data.
Visualization	• Used to get rough feeling of the quality of the data being studied.
Online analytical processing (OLAP)	• Used for multi-dimensional problems. • Cannot acquire new knowledge. • Database cannot be updated.
Case-based learning	• Uses k-nearest neighbor algorithm. • A search technique rather than a learning algorithm. • Better suited for small problems.
Decision trees	• Good for most problem sizes. • Gives true insight into nature of the decision process. • Hard to create trees from a complex problem.
Neural networks	• Mimics human brain. • Algorithm has to be trained during encoding phase. • Complex methodology.
Genetic algorithms	• Use Darwin's evolution theory. • Robust and reliable method for data mining. • Requires a lot of computing power to achieve anything of significance.

The simplest technique available is the use of query tools for a rough analysis of the data. Just by applying simple structured query language (SQL) to a data set, one can obtain a lot of information. Before applying more advanced techniques, there is a need to know some basic aspects and structures of the data set. This was stressed in the previous section, and it was regarded as background knowledge. SQL is a good technique for discovering knowledge that is in the 'surface' of the data, which is information that is easily accessible from the data set.

Scott and Wilkins (1999), Adriaans and Zantinge (1996), and Yevich (1997) have stated that for the most part, about 80% of the interesting information can be abstracted from a database using SQL. However, as Adriaans and Zantinge (1996) stressed in their book, the remaining 20% of hidden information can only be extracted using more advanced techniques, and for most problems, this 20% can prove of vital importance when solving a problem or extracting valuable knowledge for future use.

Statistical techniques can be a good simple start for trying to extract this important 20% of knowledge from the data set. Patterns can be found in the data being studied with the help of histograms, Pareto diagrams, scatter diagrams, check sheets and other statistical tools. If important relations are not found using SPC tools, at least some information will be learned, and some relations that are definitely not in the data will be known, so that more advanced techniques do not have to look for weak relations in the data being studied.

Visualization techniques are very useful for discovering patterns in data sets, and are usually used at the beginning of a data mining process to get a feeling of the quality of the data being studied. This technique uses SPC as well to gain knowledge from databases based on simple but important patterns.

Online analytical processing (OLAP) tools are widely used by companies, they support multi-dimensional problems that involve many sorts of information requested by managers and workers at the same point in time. OLAP tools store the data being studied in a special multi-dimensional format, and managers can ask questions from all ranges. A drawback of OLAP tools is that data cannot be updated. Also, as Fayyad and Stolorz (1997) mentioned, OLAP is not a learning tool, so no new knowledge can be created, and new solutions cannot be found. Essentially, if the data has the solutions, OLAP will work well, but if the solutions are not there, then OLAP is useless and obsolete.

Case-based learning uses the k-nearest neighbor algorithm given by Adriaans and Zantinge (1996) to assert relations in a database. This algorithm however, is not really a leaning algorithm. Russell and Norvig (1995) used the algorithm as a search method rather than a learning technique. However, as they mention in their book, this search technique proves to be very useful since the data set itself is used as reference. In essence, the search technique only searches the data set space, and thus, it does not get 'corrupted' by outside data or knowledge.

A problem with the search technique used in case-based learning is its complexity. The algorithm searches and compares every single record or input from the database with each other, in order to find relations between records in a data set, so as the amount of data increases, the algorithm's complexity increases as well. According to Russell and Norvig (1995), this search technique leads to a quadratic complexity, which is obviously not desirable when searching large data sets. This technique has been used widely, but only in small problems that have small databases as input for data assertion.

Decision trees are usually used for classification purposes. The database is classified into certain fields that will enable the expert to assert certain behaviors and patterns found in the database. In essence, the data will be divided into categories and to make an assertion or find a pattern in the data, one has to follow a path in the decision tree to arrive at a conclusion. The path taken represents the assertions, facts, and other information used to make the desired conclusion.

Many algorithms have been proposed for the creation of such decision trees. Adriaans and Zantinge (1996) used the tree induction algorithm to create trees used by car companies to predict customer's behaviors. An advantage of this approach is the complexity of most of the algorithms. Most decision tree algorithms are very effective, and have an $n \ Log \ n$ complexity.

An important requirement for a decision tree to be successful is to have good knowledge of the problem at hand and of the data available. A lot of assertions and decisions regarding relations between data have to be made by the expert, and the technique relies heavily on these decisions.

Neural networks are an advanced technique that tries to depict the architecture and functionality of the human brain. By modeling neurons, the building block of the brain, neural networks try to mimic the human brain's learning process. This way, the data mining process will imitate the thinking process of a human, and thus, the results of the process will be more reliable than with the other techniques.

The architecture of a neural network has been discussed and defined by many researchers, as each researcher defines a neural network for the problem at hand. However, all authors have followed a basic architecture that consists of a set of input nodes, output nodes, and a potentially unlimited number of intermediate nodes. Input nodes receive input signals, output nodes give output signals, and the intermediate nodes are essentially a *black box* where the input is converted into output. Figure 2 shows an example of a neural network. Note that all intermediate nodes are connected with all input and output nodes. These connections emulate the existing links between neurons in the human brain.

Russell and Norvig (1995), among other researchers, stress the fact that a neural network is a two-phase process. The first phase is the encoding phase, in which the neural network is trained to perform a certain task. The second phase is the decoding phase, in which the network is used to classify examples, make predictions, or execute any learning task, involved in the data mining process.

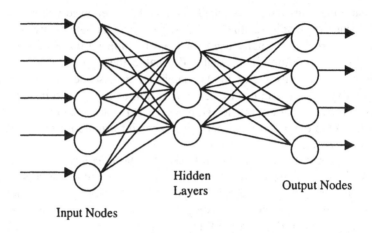

Input Nodes

Figure 2. A neural network architecture for data mining

Genetic algorithms follow the theory of evolution proposed by Darwin. His theory is based on the 'natural selection' process of evolution, which essentially states that each species has an over-production of individuals and in a tough struggle for life, only those individuals that are best adapted to the environment survive. The same principle can and has been adapted by many researchers that have used genetic algorithms as a learning tool on various data mining problems.

Holmes, et al (2000) developed and used a genetic algorithm to search and learn certain patterns for epidemic surveillance. By searching large databases, the algorithm creates a number of rules regarding possible causes, risks, and solutions for certain problems. Koonce, et al (1997) used genetic algorithms in data mining for learning manufacturing systems. Michalski (2000) developed the Learnable Evolution Model (LEM), a genetic approach that differs in many aspects to the Darwinian model used by most researchers. LEM uses machine learning to improve the evolutionary process. In data mining, this 'new' approach improves the performance and the ability of the algorithm to learn and find interesting patterns.

Vila, et al (1999) used genetic algorithms and neural networks to improve the quality of the data used in data mining, as they stressed that finding patterns and solutions on incomplete and/or unreliable data will result in useless conclusions. Fu (1999) compared a greedy search method with a genetic-based approach for two-dimensional problems involving large data sets. He explained that the improvements shown by the GA approach are vast. They include a more robust methodology, a faster method for finding patterns, and more reliable results. Yuanhui, et al (1998) combined a neural network with a genetic algorithm to mine classification rules. They showed that the genetic approach generates better rules than the decision tree approach.

The advantages and disadvantages of genetic algorithms follow those of natural selection in general. An important drawback is the large over-production of individuals. Genetic algorithms work with populations of chromosomes, and large amounts of data are used to solve even the easiest problems. Another problem with genetic algorithms is the random character of the searching process. As Adriaans and Zantinge (1996) discuss, the end-user of a genetic algorithm does not really see how the algorithm is creating and selecting individuals for finding patterns in large amounts of data.

An advantage of generic algorithms is their reliability. One can be sure that if a solution exists in a large database, the genetic algorithm will find it. This is of course, assuming that all the requirements of a genetic algorithm have been addressed and used properly. Another important advantage concerning reliability and robustness is that genetic algorithms do not need to have previous 'experience' on the problem at hand. This means that a genetic algorithm used for data mining will eventually find a pattern (if there is any) on the data even if the problem at hand is brand new and no past solutions have been found.

As databases expand, researchers have grown more and more dependent in artificial intelligence to solve these complicated data mining problems. AI techniques potentially have low complexities, and more importantly, they resemble the human way of thinking, which may make these techniques the most reliable of all, at least from the human's stand point.

CONCLUSIONS

This paper presented an introduction on data mining, the data mining methodology, and the goals and techniques available for solving all kinds of data mining problems. Why data mining is important and needed as a tool for improving the way business are being done is discussed first. This importance lies in the fact that in order to be successful; a company has to be aware of what the customers are thinking and saying about their product. This may seem like a simple task, but many times companies misunderstand customer responses, and at that time, data mining can present a better look at the available data.

As companies grow, in both time and size, the data collected by the company grows at an incredibly fast rate. Human understanding of this data can only go to a point, at which point data mining becomes a necessity.

The data mining process is an iterative one. As it was presented in a previous section of this paper, constant "jump-backs" can be made between stages of the process, until accurate results are found or until the database is shown to be unreliable. Also, it has to be noted that the most important stages of the process are the selection and the pre-processing of the data, instead of the mining stage itself.

Data mining can serve various purposes. It can be used for association, classification, clustering, and summarization, among other goals. However, the means of getting these results can vary from problem to problem. Query language, SPC, visualization and OLAP are techniques used to achieve a greater understanding on the data being studied, but as problems grow larger in size, and as data becomes more complex, new approaches have to be used. In essence, the computerization has to come back to what humans to best, that is, analyzing small portions of data using many resources that regular techniques do not possess, and that such techniques as neural networks and generic algorithms do.

REFERENCES

Adriaans, P. and D. Zantinge, Data Mining, Harlow: Addison-Wesley, 1996.

Andrade, M. and P. Bork, "Automated extraction of information in molecular biology," FEBS Letters, 476: 12-17, 2000.

Delesie, L. and L. Croes, "Operations research and knowledge discovery: a data mining method applied to health care management," International Transactions in Operational Research, 7: 159-170, 2000.

Fayyad, U., D. Madigan, G. Piatetsky-Shapiro and P. Smyth, "From data mining to knowledge discovery in databases," AI Magazine, 17: 37-54, 1996.

Fayyad, U. and P. Stolorz, "Data Mining and KDD: Promise and challenges," Future Generation Computer Systems, 13: 99-115, 1997.

Feelders, A., H. Daniels and M. Holsheimer, "Methodological and practical aspects of data mining," Information & Management, 37: 271-281, 2000.

Fu, Z., "Dimensionality optimization by heuristic greedy learning vs. genetic algorithms in knowledge discovery and data mining," Intelligent Data Analysis, 3: 211-225, 1999.

Glymour, C., D. Madigan, D. Pregibon and P. Smyth, "Statistical themes and lessons for data mining," Data Mining and Knowledge Discovery, 1: 11-28, 1997.

Hand, D.J., "Data mining: statistics and more?," The American Statistician, 52: 112-118, 1998.

Heckerman, D., "Bayesian Networks for Knowledge Discovery," Advances in Knowledge Discovery and Data Mining, AAAI Press, 273-305, 1996.

Holmes, J.H., D.R. Durbin and F.K. Winston, "The learning classifier system: an evolutionary computation approach to knowledge discovery in epidemiologic surveillance," Artificial Intelligence in Medicine, 19: 53-74, 2000.

Koonce, D.A., C. Fang and S. Tsai, "A data mining tool for learning from manufacturing systems," Computers & Industrial Engineering, 33: 27-30, 1997.

Kusiak, A., Computational Intelligence in Design and Manufacturing. Wiley-Interscience Publications, pp. 498-526, 1999.

Lin, X., X. Zhou and C. Liu, "Efficient computation of a proximity matching in spatial databases," Data & Knowledge Engineering, 33: 85-102, 2000.

Michalski, R.S., "Learnable evolution model: evolutionary processes guided by machine learning," Machine Learning, 38: 9-40, 2000.

Russell, S., P. Norvig, Artificial Intelligence: A Modern Approach. New Jersey: Prentice Hall, 1995.

Scott, P.D. and E. Wilkins, "Evaluating data mining procedures: techniques for generation artificial data sets," Information and software technology, 41: 579-587, 1999.

Skarmeta, A., A. Bensaid and N. Tazi, "Data mining for text categorization with semi-supervised agglomerative hierarchical clustering," International Journal of Intelligent Systems, 15: 633-646, 2000.

Subramanian, A., L.D. Smith, A.C. Nelson, J.F. Campbell and D.A. Bird, "Strategic planning for data warehousing," Information and Management, 33: 99-113, 1997.

Yevich, R., "Data Mining," in Data Warehouse: Practical Advice from the Experts, pp. 309-321, Prentice Hall, 1997.

Yuanhui, Z., L. Yuchang and S. Chunyi, "Mining Classification Rules in Multi-strategy Learning Approach," Intelligent Data Analysis, 2: 165-185, 1998.

Vila, M.A., J.C. Cubero, J.M. Medina and O. Pons, "Soft computing: a new perspective for some data mining problems," Vistas in Astronomy, 41: 379-386, 1997.

CHAPTER 3

Data Mining in Scientific Data

Dr.-Ing. Stephan Rudolph
rudolph@isd.uni-stuttgart.de
Institute for Statics and Dynamics of Aerospace Structures
University of Stuttgart, Pfaffenwaldring 27, 70569 Stuttgart

Dipl.-Inform. Peter Hertkorn
hertkorn@isd.uni-stuttgart.de
Institute for Statics and Dynamics of Aerospace Structures
University of Stuttgart, Pfaffenwaldring 27, 70569 Stuttgart

ABSTRACT

Knowledge discovery in scientific data, i.e. the extraction of engineering know-ledge in form of a mathematical model description from experimental data, is currently an important part in the industrial re-engineering effort for an impro-ved knowledge reuse. Despite the fact that large collections of data have been acquired in expensive investigations from numerical simulations and experi-ments in the past, the systematic use of data mining algorithms for the purpose of knowledge extraction from data is still in its infancy.

In contrary to other data sets collected in business and finance, scientific data possess additional properties special to their domain of origin. First, the principle of cause and effect has a strong impact and implies the completeness of the parameter list of the unknown functional model more rigorous than one would assume in other domains, such as in financial credit-worthiness data or client behavior analyses. Secondly, scientific data are usually rich in physical unit information which represents an important piece of structural knowledge in the underlying model formation theory in form of dimensionally homoge-neous functions.

Based on these features of scientific data, a similarity transformation using the measurement unit information of the data can be performed. This similari-ty transformation eliminates the scale-dependency of the numerical data values and creates a set of dimensionless similarity numbers. Together with reaso-ning strategies from artificial intelligence such as case-based reasoning, these

D. Braha (ed.), Data Mining for Design and Manufacturing, 61–85.

similarity numbers may be used to estimate many engineering properties of the technical object or process under consideration. Furthermore, the employed similarity transformation usually reduces the remaining complexity of the resulting unknown similarity function which can be approximated using different techniques.

INTRODUCTION

In industry huge collections of data from expensive experiments, prototypes or numerical simulations have been accumulated over the years due to easy capture and inexpensive storage of digital information.

However, a manual analysis of such large databases is becoming impractical in many domains. Knowledge discovery in scientific data stemming from technical objects or processes represents for these reasons a major challenge and an important part in the industrial re-engineering information processing chain.

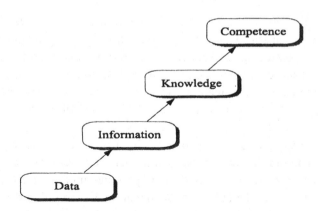

Figure 1: Relation between data, information and knowledge

As shown in figure 1, the raw data coming from sensor information is appropriately condensed into information. The further abstraction of this information can lead to knowledge. The application of this knowledge for solving problems in a given context is the competence of the engineer. All these different data representations are distinct and it is the purpose of this work to motivate starting from figure 1 over figure 2 onto figure 3 the different levels of abstraction along with the development of a formal model for their mathematical representation.

In the natural sciences a common method for helping to understand relations in complex problem domains is to identify characteristic values. These characteristic values describe special phenomena within a problem domain. The relation between the characteristic values is the basis for decisions by deriving a systematic procedure to solve problems from the knowledge of sensitivities, as shown in figure 2. In the following it is therefore described how the various levels of abstraction can be supported in the natural sciences by a functional model.

The extraction of the hidden engineering knowledge in form of a mathematical model of the experimental data is not only an important issue due to the financial savings expected from such re-engineering efforts, but is also a difficult task due to the number of supposedly influencing parameters and the expected nonlinear nature of the governing physical phenomena. A necessary condition to such a knowledge extraction procedure in the natural sciences is its consistency to physics and the condition of dimensional homogeneity.

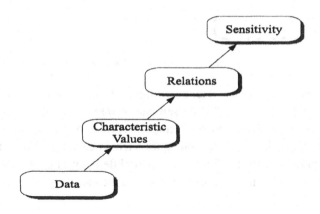

Figure 2: Systematic way to describe complex problem domains

In the framework of the functional model building process in the natural sciences, the data are represented as attribute-value pairs $(x_1, \ldots, x_n)$ with corresponding measurement units, see figure 3. Due to the condition of dimensional homogeneity as explained in the following section, any functional relationship containing physical dimensions can be (re-)written in terms of dimensionless groups $(\pi_1, \ldots, \pi_m)$. This is guaranteed by the Pi-Theorem of Buckingham. It can be shown, that the number of dimensionless groups m is reduced by the rank r of the dimensional matrix compared to the number of variables n in the relevance list and thus leading to a dimensionality reduction with $m = n - r$.

The dimensionless groups $(\pi_1, \ldots, \pi_m)$ are determined by selecting the relevant problem variables $(x_1, \ldots, x_n)$ and applying the Pi-Theorem. In some cases the Pi-Theorem helps to check whether the relevance list is complete and whether there are variables contained in the list which are not relevant for the problem. Once the dimensionless groups are determined it is possible to transform the original dimensional data into a dimensionless space.

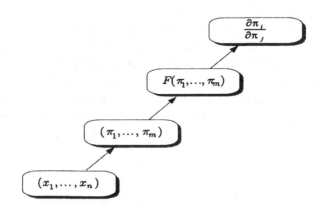

Figure 3: Formal relation between data, information and knowledge

The last two steps of abstraction in figure 3 are as usual: The function approximation $F(\pi_1, \ldots, \pi_m) = 0$ itself can be done by classical approximation techniques using least squares methods, regression analysis etc., but also the use of artificial intelligence techniques such as case-based reasoning and neural networks is possible. Finally, the partial derivatives express the model sensitivity of the influencing dimensionless parameters.

DIMENSIONAL ANALYSIS

In physics and engineering often functions of the type $x_n = f(x_1, \ldots, x_{n-1})$ serve as quantitative models. An explicit function $x_n = f(x_1, \ldots, x_{n-1})$ can hereby always be written in the implicit form $f(x_1, \ldots, x_n) = 0$ (to ease the mathematical notations, the symbol f is used for both the explicit and the implicit form). Since the formal correctness of certain sequences of algebraic operations on an equation can be falsified using the so-called *dimensions check*, it is clear that all possibly correct functions f in physics have to belong to the so-called class of *dimensionally homogeneous functions* (Buckingham, 1914; Goertler, 1975). Due to the property of dimensional homogeneity of all possibly correct functions in their general implicit form $f(x_1, \ldots, x_n) = 0$, the

Pi-Theorem of Buckingham (Buckingham, 1914; Goertler, 1975) holds in all physics and is stated in the following.

Pi-Theorem. *From the existence of a complete and dimensionally homogeneous function f of n physical quantities $x_i \in \mathbb{R}^+$ follows the existence of a dimensionless function F of only $m \leq n$ dimensionless quantities $\pi_j \in \mathbb{R}^+$*

$$f(x_1, ..., x_n) \;=\; 0 \tag{1}$$
$$F(\pi_1, ..., \pi_m) \;=\; 0 \tag{2}$$

where $m = n - r$ is reduced by the rank r of the dimensional matrix formed by the n dimensional quantities. The dimensionless quantities (also dimensionless products or dimensionless groups) have the form

$$\pi_j \;=\; x_j \prod_{i=1}^{r} x_i^{-\alpha_{ji}} \tag{3}$$

for $j = 1, ..., m \in \mathbb{N}^+$ and with the $\alpha_{ji} \in \mathbb{R}$ as constants.

The restriction to positive values of the dimensional parameters $x_i \in \mathbb{R}^+$ can be satisfied by coordinate transformations and is common in physics. Additionally it can be shown that modern proofs of the Pi-Theorem impose no restriction on the specific kind of the operator f and are thus valid for physical equations without exceptions (Bluman and Kumei, 1989).

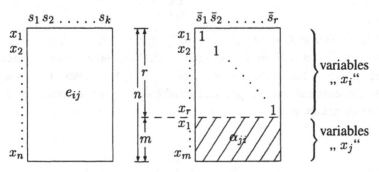

Figure 4: Definition of dimensional matrix

The establishment of the so-called *dimensional matrix* associated with the *relevance list of variables* $x_1, ..., x_n$ is shown in the left hand side of figure 4. This dimensional matrix has n rows for the variables x_i and up to k columns for the representation of the dimensional exponents e_{ij} of the variables x_i in the k base dimensions s_k of the employed unit system. In the current known SI-unit system seven physical dimensions (mass, length, time, temperature, current, amount of substance and intensity of light) are distinguished, thus $k \leq 7$.

To calculate the dimensionless products π_j in equation (3), the dimensional matrix of the relevance list of variables $x_1, \dots, x_n$ as shown in the left hand side of figure 4 needs to be created. By rank preserving operations the upper diagonal form of the dimensional matrix as shown in the right hand side of figure 4 is obtained. This means that either multiples of matrix columns may be added to each other or that matrix rows can be interchanged. The unknown exponents $-\alpha_{ji}$ of the dimensionless products in equation (3) are then automatically determined by negation of the values of the resulting matrix elements α_{ji} in the hatched part of the matrix on the lower right hand side of figure 4.

Similarity

The similarity transformation $\pi : X \mapsto \Pi$ of the space X (of dimensional variables) into the dimensionless space Π (of dimensionless variables) is defined according to equation (3). It represents a surjective mapping, since the similarity transformation π of a space $\mathbb{R}^n$ into a space $\mathbb{R}^m$ with $m = n - r$ represents a dimensionality reduction.

$$\pi_j \;=\; x_j \prod_{i=1}^{r} x_i^{-\alpha_{ji}} \tag{4}$$

This dimensionality reduction has the property that different points in X may be mapped onto the very same point in Π. To complete the mappings, the inverse similarity transformation $\pi^{-1} : \Pi \mapsto X$ of the dimensionless space Π into the dimensional space of physical variables X is given by equation (5) and is therefore not unique.

$$x_j \;=\; \pi_j \prod_{i=1}^{r} x_i^{\alpha_{ji}} \tag{5}$$

The similarity transformation maps each single point p with $(x_1, \dots, x_n)_p$ onto its corresponding point $(\pi_1, \dots, \pi_m)_p$ in dimensionless space as shown in figure 5. The inverse transformation π^{-1} however leads to the whole set of all completely similar points $(x_1, \dots, x_n)_{q=1,\dots,\infty}$ in X defined by the similarity conditions $(\pi_1, \dots, \pi_m)_p = const$.

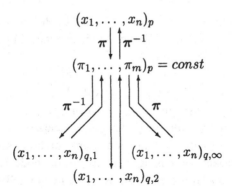

Figure 5: Dimensionality expansion π^{-1} of experiment p onto all completely similar experiments $q = 1, \ldots, \infty$

Similarity Functions

A great advantage in using the dimensionless groups of equation (3) and (4) is that they can be found without explicit knowledge of the physical model by solving a simple homogeneous linear equation system (i.e. the scheme of the dimensional matrix in figure 4) (Goertler, 1975; Rudolph, 1995). As a consequence of the existence of dimensionless groups, the problem of building a correct model $f(x_1, ..., x_n) = 0$ can instead be reduced without loss of generality to the problem of building a correct dimensionless model $F(\pi_1, \ldots, \pi_m) = 0$. Assuming that both models can be written explicitly, one can derive (Hertkorn and Rudolph, 1998a):

$$x_n = f(x_1, ..., x_{n-1}) \qquad (6)$$

$$x_n \prod_{i=1}^{r} x_i^{-\alpha_{mi}} = \prod_{i=1}^{r} x_i^{-\alpha_{mi}} f(x_1, ..., x_{n-1})$$

$$\pi_m = F(\pi_1, \ldots, \pi_{m-1})$$

$$x_n \prod_{i=1}^{r} x_i^{-\alpha_{mi}} = F(\pi_1, \ldots, \pi_{m-1})$$

$$x_n = \prod_{i=1}^{r} x_i^{\alpha_{mi}} \, F\left(x_1 \prod_{i=1}^{r} x_i^{-\alpha_{1i}}, \ldots, x_{m-1} \prod_{i=1}^{r} x_i^{-\alpha_{m-1,i}}\right) \quad (7)$$

Equation (7) is the explicit form of F in equation (2). In physics and engineering, F is known as *similarity function*. Its existence is guaranteed for every dimensionally homogeneous function f as stated by Pi-Theorem (Buckingham, 1914). Both forms of equations (6) and (7) are connected by the forward- and back-transformations in equations (4) and (5).

Figure 6 shows the "data compression" into $m = n - r$ dimensionless parameters in a function approximation framework. The original data $(x_1, \ldots, x_n)$ of the outer coordinate system $(x_1, \ldots, x_n)$ are projected by means of the forward-transformation π onto the dimensionless parameters $(\pi_1, \ldots, \pi_m)$ in the inner coordinate system. The function approximation itself can be done as usual: classical approximation using least squares methods, interpolation techniques and regression analysis, but also using artificial intelligence techniques such as case-based reasoning and neural networks.

In general the use of the dimensionless parameters has been observed to lead to superior approximation results (Hertkorn and Rudolph, 2000). These techniques are shortly described in the application section.

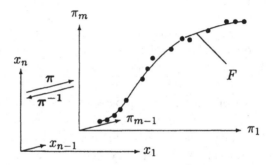

Figure 6: Approximation of f via similarity function F

Analytical models of F (and of f by using the inverse similarity transformation) are often sought after from experimental data using the following polynomial Ansatz from classical function approximation in equation (8):

$$\pi_m \quad = \quad \sum_{i=1}^{\infty} \left(C_i \prod_{j=1}^{m-1} \pi_j^{\gamma_{ij}} \right) \tag{8}$$

$$\xrightarrow{(i \to 1)} \quad C_1 \prod_{j=1}^{m-1} \pi_j^{\gamma_j} \tag{9}$$

with constants $\gamma_j, C_i \in \mathbb{R}$. Very often, also a simplified version of equation (8) using only one single monomial with $(i \to 1)$ as shown in equation (9) is used for the approximation of experimental data.

If used properly, the Pi-Theorem of Buckingham is a powerful tool for the creation, discussion and verification of functional models in the natural sciences. Three typical problems in the creation of functional models are the *remainder* in the establishment of the *parameter list* and the often only approximately known *operator*, which represents the couplings among the in-

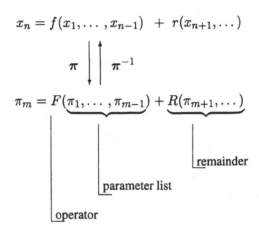

$$x_n = f(x_1, \ldots, x_{n-1}) + r(x_{n+1}, \ldots)$$

$$\pi \left|\right| \pi^{-1}$$

$$\pi_m = F(\underbrace{\pi_1, \ldots, \pi_{m-1}}) + \underbrace{R(\pi_{m+1}, \ldots)}$$

remainder

parameter list

operator

Figure 7: Three typical model approximation errors

dividual model parameters. By means of similarity considerations it can be checked whether the consistency of the model chosen lies inbetween acceptable limits. Since all three error sources are superimposed and nonlinearly coupled, figure 7 emphasizes the generic difficulties to distinguish between these three error sources in a specific application case, if numerical accuracy of the data approximation is the only additionally available source of information.

DIMENSIONAL ANALYSIS IN DATA MINING

The term data mining refers to the most important step in the overall process of discovering useful knowledge from data, called knowledge discovery in databases as shown in figure 8. In scientific data analysis the term data reduction refers to the extraction of essential variables of interest from raw observations (Fayyad et al., 1996). Finding useful features in the data and reducing the dimensionality appears to be the main problem.

In the knowledge discovery in databases process, data reduction is therefore very important for data mining. The advantages of having dimensionally reduced data are for example that it narrows down the search space determined by attributes, allows faster learning for a classifier, helps a classifier produce simpler learned concepts and improves prediction accuracy (Liu and Setiono, 1996). Data in physics and engineering are represented as attribute-value pairs due to the underlying functional model. The knowledge discovery in databases process in figure 8 can be formulated using this data representation (Hertkorn and Rudolph, 2000) as shown in figure 9.

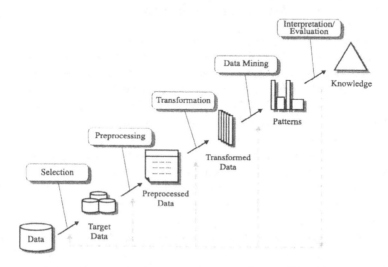

Figure 8: Knowledge discovery in databases (Fayyad et al., 1996)

In the traditional knowledge discovery in databases process, the target dataset is selected from the given dataset. After preprocessing this dataset, the relevant features are selected in the transformation step. This can lead to a data reduction. At this point, the method of dimensional analysis is integrated into the knowledge discovery in databases process (figure 9).

The given variables $(x_1, \ldots, x_n)$ are used to set up the relevance list for the dimensional matrix. By inspecting the dimensional matrix the completeness of the relevance list is checked (Szirtes, 1998). Simultaneously the relevancy of the variables for the problem can be examined. The dimensional irrelevancy of a variable may be detected inspecting the dimensional matrix. However, physical irrelevancy is difficult to detect and can often only be identified by a rigorous analysis of the given problem (Szirtes, 1998).

Having a valid dimensional matrix based on a relevance list $(x_1, \ldots, x_n)$, the dimensionless products π_j are calculated using equation (3). This leads to $m = n - r$ dimensionless products. Hence, the dimensionality of the problem is reduced from n to m. This dimensionality reduction has the effect, that physically completely similar points are mapped onto the same point after applying the similarity transformation. This implies that the number of data is reduced by the number of physically completely similar points and therefore leads to a reduction in data size.

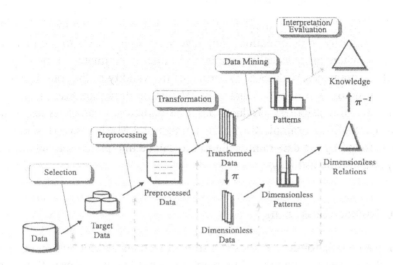

Figure 9: Knowledge discovery in databases using dimensional analysis

The data reduced in this way form the input for the chosen data mining algorithm. Taking case-based reasoning as a data mining algorithm, previous works (Hertkorn and Rudolph, 1998a; Hertkorn and Rudolph, 1998b) examined the advantages when using the method of dimensional analysis for data mining problems in physics and engineering. The most important are:

- *Data representation:* Data is represented using dimensionless groups. This means that the search space is smaller, resulting in a more efficient data retrieval from the database. The dimensionless data representation can also be used as an efficient index for the database.
- *Similarity measure:* The determination of similar data to a given problem is performed with the similarity transformation. Compared to a traditional distance measure, all physically completely similar data have the same dimensionless representation and can therefore be identified, though their values in the dimensional representation may be quite different. The dimensionless data also form the basis for later classification and learning.
- *Function approximation:* Interpolation and approximation techniques are used in the data mining step. In case-based reasoning, case interpo-

lation and case approximation can be used for modifying the solution of similar cases according to the new situation. The results for both techniques in dimensionless space have been observed to be significantly better than in dimensional space (Hertkorn and Rudolph, 1998b).

The patterns found by the data mining algorithm can be verified using the inverse similarity transformation. The inverse mapping leads to a dimension expansion where all physically completely similar experiments in the original dataset are obtained. With these experiments the validity of the found patterns can be verified. A further advantage of calculating dimensionless products is that special dimensionless products describe particular characteristics of the physical world (for example Reynolds' number and Mach number). Thus the patterns found by the data mining algorithm can be interpreted semantically in a very straightforward way.

Case-Based Reasoning

As simple models for the complex and still not well understood reasoning techniques for problem solving used by humans, the techniques of case-based and rule-based reasoning have emerged within the broader field of artificial intelligence (Shapiro, 1987). Especially case-based reasoning has most recently attracted much scientific interest (Kolodner, 1993; Wess, 1995), since the requirements for rule-based systems in terms of knowledge acquisition and knowledge structuring have been repeatedly reported in the past to be difficult.

The current scientific understanding of case-based reasoning (Kolodner, 1993; Maher et al., 1995) establishes four consecutive steps for the successful reuse and applicability of former case knowledge to new unknown cases by means of the so-called *process model* (Aamodt and Plaza, 1994; Kolodner, 1993), e.g. to

☐ *retrieve*,

☐ *reuse*,

☐ *revise and*

☐ *retain*

former case knowledge. A flowchart of the steps required for the intended case-based reasoning procedure is shown in figure 10. According to the description given in (Slade, 1991), the necessary steps can be characterized as follows (see figure 10):

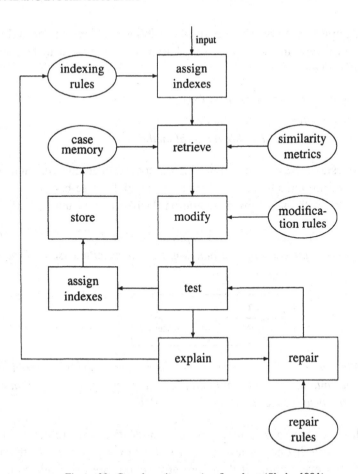

Figure 10: Case-based reasoning flowchart (Slade, 1991)

☐ *Assign Indexes*. Assignment of features characterizing the available case description.

☐ *Retrieve*. The indexes are used to retrieve a similar old case from the case base.

☐ *Modify*. The old case solution is modified appropriately to the new problem case.

☐ *Test*. The proposed solution is tried out and tested.

☐ *Assign and Store*. New solutions are stored in the case database in appropriate form.

☐ *Explain, Repair and Test*. If the suggested solution fails, the cycle of modification, test and explanation is repeated.

For an implementation of these steps the following three major and still unresolved problems in the area of case-based reasoning need to be solved (Wess, 1995). They are:

☐ *finding the similarity measures for cases*
☐ *creation of knowledge representation schemes*
☐ *usage of efficient and stable retrieval methods*

These questions are now investigated in the context of engineering and physics. The foundation of a functional model framework for case-based reasoning in physics starts with the following definition (Hertkorn and Rudolph, 2000):

Definition *From the existence of an implicit function $f(x_1, \ldots , x_n) = 0$ it follows for an instantiation of that function, i.e. one specific case p of the parameter set $(x_1, \ldots , x_n)_p$, that*

$$\underbrace{x_1, \ldots , x_{n-1}}_{\text{premise}} \Rightarrow \underbrace{x_n}_{\text{conclusion}}$$

$(\text{case } p)$

is valid. This is used to define the term case-based reasoning. In the terminology of artificial intelligence it is said that the conclusion (effect) x_n will always occur if the premise (cause) $x_1, \ldots , x_{n-1}$ is fulfilled.

Two distinct cases $p \neq q$ may possess non-identical case representations $(x_1, \ldots , x_n)$ with distinct parameter values x_i and $i = 1, \ldots , n$. Both cases p and q are however similar if their case abstractions are identical. Since in physics the similarity conditions are defined in terms of dimensionless groups, the following main result can be postulated for all case descriptions for which the Pi-Theorem applies (Hertkorn and Rudolph, 2000):

Corollary *If in a case p with the premise $(x_1, \ldots , x_{n-1})_p$ the conclusion x_n holds, then the similarity transformation of this conclusion in the form*

$$\underbrace{\pi_1, \ldots , \pi_{m-1}}_{\text{premise}} \Rightarrow \underbrace{\pi_m}_{\text{conclusion}}$$

$(\text{case } p \text{ completely similar } q)$

does not only hold for one case p, but also holds for all cases q which are completely similar to p and possess therefore the identical premise $\pi_1, \ldots, \pi_{m-1}$. This represents the constructive proof for the correctness of case-based reasoning.

From a scientific point of view this corollary represents the main result. It shows the construction of similarity conditions for the purpose of case-based reasoning in physics, since the Pi-Theorem provides a proof for its validity based on the underlying functional model framework shown in equation (7).

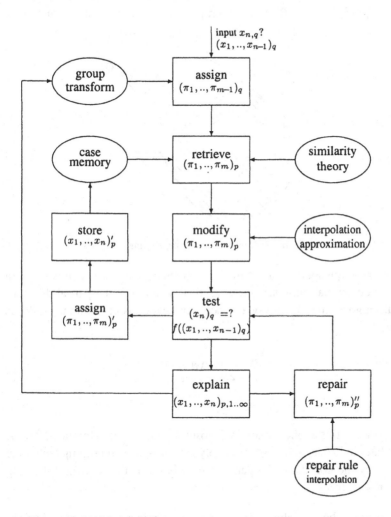

Figure 11: Modified case-based reasoning flowchart (in comparison to figure 10)

In respect to the detailed flow-chart of case-based reasoning in figure 10, the updated figure 11 is now obtained. It contains the respective formulas of the new case-based reasoning method and is applied in the following section.

Heat Transfer Example

Figure 12 shows a model of the heat loss $\dot{q} = \dot{Q}/L$ (measured in $[W/m]$) of a pipe of length L (measured in $[m]$) with an inner radius r_i and an outer radius r_a (both measured in $[m]$). The material has a heat-transfer coefficient λ (measured in $[W/mK]$). The difference in temperature between the inner and outer

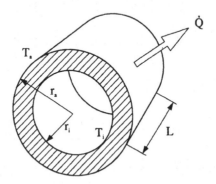

Figure 12: Heat transfer through a pipe

side of the pipe transporting a hot media is given by $\vartheta = T_a - T_i$ (measured in $[K]$). For educational purposes, the analytical solution of the partial differential equation governing the physical process of heat transfer (Holman, 1986) is stated here:

$$\dot{q} = 2\pi \frac{\lambda \vartheta}{\ln\left(\frac{r_a}{r_i}\right)} \tag{10}$$

By no means this explicit analytical closed form solution (Holman, 1986) is assumed later to be known in order to apply the case-based reasoning technique as defined by the corollary. To perform case-based reasoning, the following steps are necessary:

Input. Based on the available case description of relevant parameters in figure 12, the variables $\lambda, \vartheta, r_a, r_i$ are the premise, whereas the variable $\dot{q}$ is the unknown conclusion x_n, see figure 11.

Assign. Using the physical dimensions of the variables $\lambda, \vartheta, r_a, r_i, \dot{q}$ in the relevance list of variables in figure 12, the dimensional matrix in figure 13 left is established.

variable	SI-units	$[M]$	$[L]$	$[T]$	$[\Theta]$		$[S_1]$	$[S_2]$	$[S_3]$
λ	[W/m K]	1	1	-3	-1		1		
r_i	[m]		1					1	
ϑ	[K]				1	$\Rightarrow$			1
r_a	[m]		1					1	
$\dot{q}$	[W/m]	1	1	-3			1		1

Figure 13: Dimensional matrix computations

The rank of the dimensional matrix is $r = 3$ and with $n = 5$ dimensional parameters, $m = n - r = 2$ dimensionless products can be constructed by transforming the dimensional matrix into an upper diagonal form in table 13 right:

$$\pi_1 = \frac{r_a}{r_i} \tag{11}$$

$$\pi_2 = \frac{\dot{q}}{\lambda \vartheta} \tag{12}$$

Retrieve. The dimensionless variable π_1 has been computed in equation (11). It forms the premise for the unknown conclusion π_2 in equation (12). A similar case can now be retrieved from the case memory with this premise: the case base is looked up for the same value of π_1. If such a case is retrieved, the conclusion π_2 is found.

The found completely similar case is then mapped back from the dimensionless space Π onto the space X. The inverse similarity transformation in equation (5) yields in this example

$$r_a = \pi_1 r_i \tag{13}$$
$$\dot{q} = \pi_2 \lambda \vartheta \tag{14}$$

As a case base, the following five (numerical) cases in table 1 are given:

case #	ϑ	λ	r_a	r_i	$\dot{q}$	π_1	π_2
1	20.991937	66	0.006605	0.003	11028	2.2018	7.9604
2	20.405431	66	0.007371	0.003	9412	2.4570	6.9892
3	21.909211	66	0.008733	0.003	8502	2.9112	5.8799
4	18.224435	66	0.008614	0.003	7164	2.8716	5.9563
5	22.230139	66	0.008727	0.003	8632	2.9093	5.8836

Table 1: Case base with 5 known 'old' cases

The case-based reasoning technique is now used to compute the unknown heat loss $\dot{q}$ (i.e. the conclusion x_5) for a new case q with values $\lambda, \vartheta, r_a, r_i$ (i.e. the premise $(x_1, \ldots, x_4)$) not contained in table 1:

case #	ϑ	λ	r_a	r_i	$\dot{q}$	π_1	π_2
query	23.580256	66	-?-	0.003	12388.85251	-?-	7.9604

Table 2: Unknown 'new' design case query

By inspection of table 2 it is evident that the unknown query q lies even outside the known data of the unknown function $f(\vartheta, \lambda, r_a, r_i, \dot{q}) = 0$. With the new case-based reasoning technique however, it appears that the unknown case q in table 2 has the identical transformed premise (π_2) as case 1 in table 1.

According to the corollary, the sought conclusion (π_1) can therefore be taken from case #1 data in table 1. With the inverse similarity transformation of equation (5) or equation (13) follows

$$r_a = \pi_1 r_i = 2.2018 \cdot 0.003 = 0.006605 \, [m] \qquad (15)$$

The result in equation (15) relies on the fact that the unknown engineering

☐ *case domain is real-valued and has physical dimensions*, and that a
☐ *case with identical similarity conditions was contained in the case base.*

It cannot always be expected that such a *completely similar case* with identical premises will be contained in the case base. This issue will therefore be dealt with and investigated in the following. In case-based reasoning systems the case adaptation is one of the key aspects, because the given problem rarely matches old cases. Therefore the retrieved similar cases have to be appropriately modified in order to fit to the new situation by applying a suitable adaptation technique (Kolodner, 1993).

The method of function approximation is a well-known technique in domains of numerical data. The application of a function interpolation technique has already been proposed in case-based reasoning (Chatterjee and Campbell, 1997). In the following a real-valued function approximation is used as well, however the function approximation is done in dimensionless space $\pi_1, \ldots, \pi_m$ rather than in dimensional space $x_1, \ldots, x_n$. Both results are then compared to each other.

As an approximation technique, the least squares method (Bronstein and Semendjajew, 1981) is applied on the case base both in X-space and in Π-space. The two-dimensional form of the least squares method is shown in equation (16) (Bronstein and Semendjajew, 1981).

$$a_0(n+1) + a_1 \sum_{i=0}^{n} x_i + \cdots + a_m \sum_{i=0}^{n} x_i^m = \sum_{i=0}^{n} y_i$$

$$a_0 \sum_{i=0}^{n} x_i + a_1 \sum_{i=0}^{n} x_i^2 + \cdots + a_m \sum_{i=0}^{n} x_i^{m+1} = \sum_{i=0}^{n} x_i y_i$$

$$\cdots\cdots\cdots\cdots\cdots\cdots\cdots\cdots\cdots\cdots\cdots\cdots\cdots\cdots \tag{16}$$

$$a_0 \sum_{i=0}^{n} x_i^m + a_1 \sum_{i=0}^{n} x_i^{m+1} + \cdots + a_m \sum_{i=0}^{n} x_i^{2m} = \sum_{i=0}^{n} x_i^m y_i$$

The case base which contains no completely similar cases is the one shown in table 1. In the current heat transfer example the variables for the inner radius $r_i = 0.3\ [m]$ and the heat-transfer coefficient $\lambda = 66\ [W/mK]$ for pure tin at a temperature of $T = 20\ [°C]$ are now held constant to allow a 3-dimensional plot of the functions used.

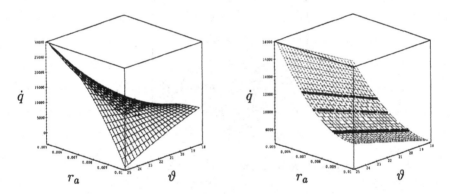

Figure 14: Approximation functions f (left) and F (right) in X-space

The approximation computed in Π-space is then mapped back with the inverse similarity transformation in X-space. In the example, an approximation polynomial of degree 2 is used for the given five cases. Figure 14 shows the approximation function in X on the left hand side, as well as the cases from the case base and the back transformed function on the right hand side, together with the completely similar cases that are used implicitly when approximating in Π. The numerical results of both function approximation methods are shown in table 3. According to the relative error, the approximation in Π yields a numerically better approximation.

ϑ	λ	r_a	r_i	$\dot{q}$	approx. in X	approx. in Π	rel.error [%] in X	rel.error [%] in Π
19.56	66	0.007	0.003	9745	979	9764	89.9	0.2
18.17	66	0.009	0.003	7191	6764	7177	5.9	0.2
19.31	66	0.007	0.003	8982	2573	8988	71.4	0.1
21.80	66	0.008	0.003	9613	17135	9594	78.2	0.2
19.60	66	0.009	0.003	7721	8155	7709	5.6	0.2

Table 3: Relative error in X-space and in Π-space (back transformed)

In figure 15, the relative error in close proximity of the known cases from the case base is shown as well as all completely similar cases of the cases in X which are implicitly used for the approximation in Π-space. The approximation in Π-space is observed to yield numerically to much better balanced results than the approximation in X-space.

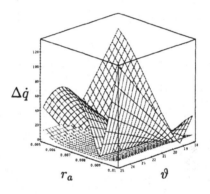

Figure 15: Relative errors of f and F in X-space

As shown in this heat transfer example, the application of the Pi-Theorem seems straightforward if the "dimensions" of the problem are well known. The following example deals with the questions what to do if the data are considered to be "dimensionless" or if their dimensions seem just not so obvious.

Noise Classification Example

There is another lesson to learn from physics: dimensional and dimensionless formulations can be transformed into each other. As an example, statistical thermodynamics has as a basic assumption the dimensionless distribution function $f(x, y)$ of Maxwell, which is is condensed into very few dimensional quantities m_{pq} by means of an integral transformation in equation (17). The integral transformation in form of the higher moments m_{pq} of order $p + q$ is defined as follows (Till and Rudolph, 1999):

$$m_{pq} = \int\limits_{-\infty}^{\infty} \int\limits_{-\infty}^{\infty} x^p y^q f(x, y) \, dx \, dy. \qquad (17)$$

$$= \sum_i \sum_j a_{ij} \int\limits_{\Omega_{ij}} x^p y^q \, dx \, dy. \qquad (18)$$

which can be rewritten by exchanging summation and integration and denoting the picture pixels values with a_{ij} and the integration domain with Ω_{ij} in equation (18) as a sum of monomial powers. If the dimensions $L_x = [x]$ for the x-coordinate and $L_y = [y]$ for the y-coordinate are conceptually introduced (Till and Rudolph, 1999), the dimension of equation (18) is

$$[m_{pq}] = [L_x]^{p+1} [L_y]^{q+1}, \qquad (19)$$

where the p, q are dimensionless constants. It is important to realize that now a dimensional representation is available for the properties of a dimensionless distribution function (i.e. the image function). Due to this reason, the Pi-Theorem of Buckingham can now be applied to the moments in equation (17) in a straightforward way.

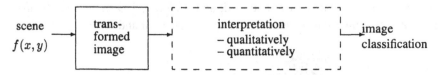

Figure 16: Classical image classification problem setting

The construction of an appropriate classifier for a certain class of objects contained in an image is one of the central questions of pattern recognition and used as a core element in the overall noise classification algorithm. The inner classification procedure of pictures is typically achieved using the following steps as shown in figure 16 and consists of the generation of so-called image features and the construction of a classifier using the features as inputs (Till and Rudolph, 1999).

It is the goal in pattern recognition to find a theoretical procedure for the determination of image features, such as image invariants on the basis of higher moments which may be used for classification in a later stage. An interesting topic of invariants concerns the class definition of images, under which the invariants will remain constant.

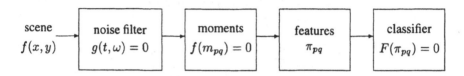

Figure 17: Dimensionally homogeneous image classification (Till and Rudolph, 1999)

By means of similarity theory, a theoretical framework for the construction of invariants can be used for these investigations and is shown in figure 17. Using this completely analytical model, the influence of every parameter and of each procedural step towards the desired classification can be investigated, studied and interpreted.

$$\pi_{pq} = \frac{m_{pq}}{m_{02}^{\frac{-p+3q+2}{8}} \; m_{20}^{\frac{3p-q+2}{8}}} \cdot \tag{20}$$

To compute dimensionless moments π_{pq} in equation (20) from the dimensional moments m_{pq}, their dimensional representation is taken from equation (19). The resulting dimensionless moments π_{pq} from the dimensional matrix procedure are similarity invariants (Till and Rudolph, 1999) and may be used in a later classification stage of the image objects (Till and Rudolph, 1999).

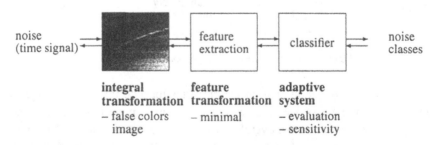

Figure 18: Industrial noise classification (Till and Rudolph, 1999)

In this way, a system for the classification of noise using a neural network classifier can be developed and validated due to the similarity properties of the employed dimensionless features. According to the signal processing sequence shown in figure 18, the noise is transformed into an image using a time-frequency integral transformation. By the generation and use of a minimal set of similarity numbers as image features the noise is classified via a neural network classifier trained with selected experimental sample data.

Through the availability of a complete analytical model for the classification procedure according to the theory of similarity, a sensitivity analysis of the neural network parameters for correct classification purposes can be performed and analyzed. Classification results can even be back-transformed into the original noise space using the bijectivity property of the integral transformation (Till and Rudolph, 1999).

Example Discussion

The procedure of knowledge abstraction starting with the original data as represented in figures 3 and 9 relies essentially on the conceptual notion of physical dimensions. As stated in the previous example of noise classification, it may hereby not always be clear what the true physical dimensions of a problem really are. Philosophically, the true dimension of a problem may even be a disputable question if not a principally open and unresolvable problem. However, if once by a certain consideration such physical dimensions can be attributed to the data representation, the application of the Pi-Theorem becomes mandatory.

The mandatory character of the Pi-Theorem stems from the fact that the similarity postulate, i.e. the definition of the property which distinct objects or processes are similar to each other in a certain situation, is now no longer a question of human user-defined criteria, but a result of a formal invariant analysis. A property of the dimensionless group invariants (i.e. the similarity numbers) is that all the information needed for their computation is only the dimensional representation in form of the SI-units. This dimensional information is considered as the structural knowledge about the technical/scientific domain and is often completely overlooked in artificial intelligence.

This lacking freedom in the definition of similarity in the context of dimensional representations is heavily underestimated or even considered contraintuitive in its formal consequences in many applications of artificial intelligence techniques. But once again it is felt that this technique is just another example for the common experience that a careful data analysis lies at the heart of any sincere and meaningful scientific investigation. Dimensionless methods represent therefore no new dogma (nor an old dogma in just a new suit), but may be conveniently and advantageously combined together with other datamining and processing techniques.

SUMMARY

Huge collections of data from expensive experiments, prototypes or numerical simulations have been stored in databases over the years. Data mining in scientific data from technical processes, i. e. the extraction of the hidden engineering knowledge in form of a mathematical model description of the experimental data is therefore a major challenge and an important part in the industrial re-engineering information processing chain for an improved future knowledge reuse. Scientific data possess special properties because of their domain of origin. In physics and engineering, data is represented as attribute-value pairs with corresponding measurement units.

Due to the model building process in the natural sciences, any functional relationship can be expressed in dimensionless similarity function. The underlying similarity transformations represent object or process invariants and eliminate the scale-dependency of the numerical data values. Together with reasoning strategies from artificial intelligence, such as case-based reasoning, these similarity numbers may be used to estimate the properties of the technical object or process under consideration.

REFERENCES

Aamodt, A., and E. Plaza, "Case-based reasoning: Foundational issues, methodological variations, and system approaches", in AI Communications, 7(1): 39–59, 1994.

Bluman, G. W., and S. Kumei, Symmetries and Differential Equations. New York: Springer, 1989.

Bronstein, I. N., and K. A. Semendjajew, Taschenbuch der Mathematik, 19 edition, Thun: Harri Deutsch, 1981.

Buckingham, E., "On physically similar systems: Illustration of the use of dimensional equations", Physical Review, 4: 345–376, 1914.

Chatterjee, N., and J. A. Campbell, "Interpolation as a means of fast adaptation in case-based problem solving", in Proceedings Fifth German Workshop on Case-Based Reasoning, pp. 65–74, 1997.

Fayyad, U. M., D. Hausler, and P. Stolorz, "Mining scientific data", Communications of the ACM, 39(11): 51–57, 1996.

Fayyad, U. M., G. Piatetsky-Shapiro, and P. Smyth, "From data mining to knowledge discovery: An overview", in Advances in Knowledge Discovery and Data Mining, pp. 1–34, Menlo Park: AAAI/MIT Press, 1996.

Görtler, H., Dimensionsanalyse. Theorie der physikalischen Dimensionen mit Anwendungen. Berlin: Springer, 1975.

Hertkorn, P., and S. Rudolph, "Dimensional analysis in case-based reasoning", in Proceedings International Workshop on Similarity Methods, pp. 163–178, Stuttgart: Insitut für Statik und Dynamik der Luft- und Raumfahrtkonstruktionen, 1998.

Hertkorn, P., and S. Rudolph, "Exploiting similarity theory for case-based reasoning in real-valued engineering design problems", in Proceedings Artificial Intelligence in Design '98, pp. 345–362, Dordrecht: Kluwer, 1998.

Hertkorn, P., and S. Rudolph, "A systematic method to identify patterns in engineering data", in Data Mining and Knowledge Discovery: Theory, Tools, and Technology II, pp. 273–280, 2000.

Holman, J., Heat Transfer. New York: McGraw-Hill, 1986.

Kolodner, J. L, Case-Based Reasoning. San Mateo: Morgan Kaufmann, 1993.

Liu, H., and R. Setiono, "Dimensionality reduction via discretization", Knowledge Based Systems, 9(1): 71–77, 1996.

Maher, M. L, M. B. Balachandran, and D. M. Zhang, Case-Based Reasoning in Design. Mahwah: Lawrence Erlbaum, , 1995.

Rudolph, S., "Eine Methodik zur systematischen Bewertung von Konstruktionen", Düsseldorf: VDI-Verlag, 1995.

Shapiro, S., Encyclopedia of Artificial Intelligence. New York, Wiley, 1987.

Slade, S., "Case-based reasoning", AI Magazine, 91(1): 42–55, 1991.

Szirtes, T., Applied dimensional analysis and modeling. New York: Mc Graw-Hill, 1998.

Till, M., and S. Rudolph, "A discussion of similarity concepts for acoustics based upon dimensional analysis", in Proceedings 2nd International Workshop on Similarity Methods, pp. 181–195, 1999.

Weß, S., Fallbasiertes Problemlösen in wissensbasierten Systemen zur Entscheidungsunterstützung und Diagnostik. Grundlagen, Systeme und Anwendungen. Kaiserslautern: Universität Kaiserslautern, 1995.

CHAPTER 4

Learning to Set Up Numerical Optimizations of Engineering Designs

Mark Schwabacher, Thomas Ellman, and Haym Hirsh
{schwabac, ellman, hirsh}@cs.rutgers.edu
Computer Science Department
Rutgers, The State University of New Jersey
New Brunswick, NJ 08903

ABSTRACT

Gradient-based numerical optimization of complex engineering designs offers the promise of rapidly producing better designs. However, such methods generally assume that the objective function and constraint functions are continuous, smooth, and defined everywhere. Unfortunately, realistic simulators tend to violate these assumptions, making optimization unreliable. Several decisions that need to be made in setting up an optimization, such as the choice of a starting prototype, and the choice of a formulation of the search space, can make a difference in how reliable the optimization is. Machine learning can help by making these choices based on the results of previous optimizations. We demonstrate this idea by using machine learning for four parts of the optimization setup problem: selecting a starting prototype from a database of prototypes, synthesizing a new starting prototype, predicting which design goals are achievable, and selecting a formulation of the search space. We use standard tree-induction algorithms (C4.5 and CART). We present results in two realistic engineering domains: racing yachts, and supersonic aircraft. Our experimental results show that using inductive learning to make setup decisions improves both the speed and the reliability of design optimization.

D. Braha (ed.), Data Mining for Design and Manufacturing, 87–125.

INTRODUCTION

Automated search of a space of candidate designs seems an attractive way to improve the traditional engineering design process. Each step of such automated search requires evaluating the quality of candidate designs, and for complex artifacts such as aircraft, this evaluation must be done by computational simulation.

Gradient-based optimization methods, such as sequential quadratic programming, are reasonably fast and reliable when applied to search spaces that satisfy their assumptions. They generally assume that the objective function and constraint functions are continuous, smooth, and defined everywhere. Unfortunately, realistic simulators tend to violate these assumptions. We call these assumption violations *pathologies*. Non-gradient-based optimization methods, such as simulated annealing and genetic algorithms, are better able to deal with search spaces that have pathologies, but they tend to require many more runs of the simulator than do the gradient based methods. We therefore would like to find a way to reliably use gradient-based methods in the presence of pathologies.

The performance of gradient-based methods depends to a large extent on choices that are made when the optimizations are set up, especially in cases where the search space has pathologies. For example, if a starting prototype is chosen in a less pathological region of the search space, the chance of reaching the optimum is increased. Machine learning can help by learning rules from the results of previous optimizations that map the design goal into these optimization setup choices. We demonstrate this idea by using machine learning for four parts of the optimization setup problem.

When designing a new artifact, it would be desirable to make use of information gleaned from past design sessions. Ideally one would like to learn a function that solves the whole design problem. The training data would consist of design goals and designs that satisfy those goals, and the learning algorithm would learn a function that maps a design goal into a design. We believe this function is too hard to learn. We therefore focused on improving optimization performance by using machine learning to make some of the choices that are involved in setting up an optimization. In the course of our work, we found parts of the optimization setup problem for which machine learning can help: selecting starting prototypes, predicting whether goals are achievable, and selecting formulations of the search space.

Our first effort was in the domain of the design of racing yachts of the type used in the America's Cup race. In this domain, we had success using a technique that we call *prototype selection* which maps the design goal into a selection of a prototype from a database of existing prototypes. We used C4.5, the standard tree-induction algorithm, in this work.

Our second effort was in the domain of the design of supersonic transport aircraft. We tried prototype selection in this domain, and found that it did not perform well, so we decided to try a new idea which we call *prototype synthesis*. Prototype synthesis synthesizes a new prototype by mapping the design goal into the design parameters that define a prototype. It requires continuous-class induction, which is not available in C4.5; hence we used CART. We then realized that we could use the training data that we had collected for prototype synthesis to further enhance optimization performance using a new idea that we call *achievable goal prediction*. Achievable goal prediction uses inductive learning to predict whether a given design goal is achievable, before attempting to synthesize a starting prototype for the goal. Since this decision is discrete, rather than continuous, we used C4.5.

We then had the idea of recognizing when designs are at constraint boundaries, learning to predict this accurately, and using these predictions to reformulate the search space. We call this idea *formulation selection*. This prediction is discrete, so we used C4.5 to make it. We tested this idea in both the yacht and aircraft domains, and found it to be successful in both domains.

This chapter includes sections describing these four techniques for using machine learning to set up optimizations: prototype selection, prototype synthesis, achievable goal prediction, and formulation selection. Each section includes experimental results demonstrating that using the machine learning techniques improves the speed of optimization and/or the quality of the resulting designs.

INDUCTIVE LEARNING

The problem addressed by an inductive-learning system is to take a collection of labeled "training" data and form rules that make accurate predictions on future data. Inductive learning is particularly suitable in the context of an automated design system because training data can be generated in an automated fashion. For example, one can choose a set of training goals and perform an optimization for all combinations of training goals and library prototypes. One can then construct a table that records which prototype was best for each training goal.[1] This table can be used by the inductive-learning algorithm to generate rules mapping the space of all possible goals into the set of prototypes in the library. If learning is successful this mapping extrapolates from the training data and can be used successfully in future design sessions to map a new goal into an appropriate initial prototype in the design library.

The specific inductive-learning systems used in this work are C4.5 (Quinlan, 1993) (release 3.0, with windowing turned off) for problems requiring discrete-class induction, and CART[2] (Breiman, 1984) for problems requiring continuous-class induction. Both of these systems represent the learned knowl-

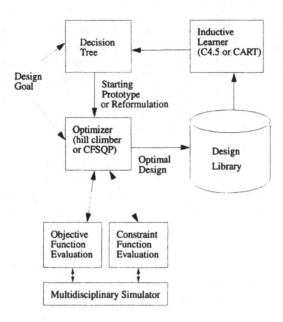

Figure 1. Design Associate block diagram

edge in the form of decision trees. The approach taken by these systems is to find a small decision tree that correctly classifies the training data, then remove lower portions of the tree that appear to fit noise in the data. The resulting tree is then used as a decision procedure for assigning labels to future, unlabeled data.

THE DESIGN ASSOCIATE

Our prototype-selection and formulation-selection techniques have been developed as part of the "Design Associate," a system for assisting human experts in the design of complex physical engineering structures (Ellman et al., 1992). Figure 1 shows a block diagram of the system's software architecture. The inductive learner learns from the design library a decision tree. Given a new design goal, the decision tree is used to map this design goal into a choice of starting prototype from the design library, or a choice of formulation of the search space. The optimizer optimizes this prototype for the new design goal, using the selected formulation. At each iteration of this optimization, the optimizer uses a multidisciplinary[3] simulator to evaluate the objective and constraint functions. At the end of the optimization, the new optimal design is added to the design library.

PROTOTYPE SELECTION

Many automated design systems begin by retrieving an initial prototype from a library of previous designs, using the given design goal as an index to guide the retrieval process (Sycara and Navinchandra, 1992). The retrieved prototype is then modified by a set of design modification operators to tailor the selected design to the given goals. In many cases the quality of competing designs can be assessed using domain-specific evaluation functions, and in such cases the design-modification process is often accomplished by an optimization method such as hill-climbing search (Ramachandran et al., 1992; Ellman et al., 1992). Such a design system can be seen as a *case-based reasoning* system (Kolodner, 1993), in which the prototype-selection method is the *indexing* process, and the optimization method is the *adaptation* process.

In the context of such case-based design systems, the choice of an initial prototype can affect both the quality of the final design and the computational cost of obtaining that design, for three reasons. First, prototype selection may impact quality when the design process is guided by a nonlinear evaluation function with unknown global properties. Since there is no known method that is guaranteed to find the global optimum of an arbitrary nonlinear function (Schwabacher, 1996), most design systems rely on iterative local search methods whose results are sensitive to the initial starting point. Second, prototype selection may impact quality when the prototypes lie in disjoint search spaces. In particular, if the system's design modification operators cannot convert any prototype into any other prototype, the choice of initial prototype will restrict the set of possible designs that can be obtained by *any* search process. A poor choice of initial prototype may therefore lead to a suboptimal final design. Finally, the choice of prototype may have an impact on the time needed to carry out the design modification process — two different starting points may yield the same final design but take very different amounts of time to get there. In design problems where evaluating even just a single design can take tremendous amounts of time, we believe that selecting an appropriate initial prototype can be the determining factor in the success or failure of the design process.

To use inductive learning to form prototype-selection rules, we take as training data a collection of design goals, each labeled with which prototype in the library is best for that goal. "Best" can be defined to mean the prototype that best satisfies the design objectives, the prototype that results in the shortest design time, or the prototype that optimizes some combination of design quality and design time.

The Yacht Domain

We developed and tested our prototype selection methods in the domain of 12-meter racing yachts, which until recently was the class of yachts raced in

Figure 2. Stars & Stripes '87, winner of the 1987 America's Cup competition

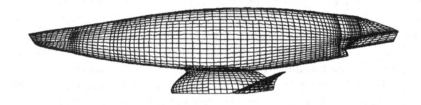

Figure 3. The hull and keel of Stars & Stripes '87.

America's Cup competitions.[4] An example of a 12-meter yacht is the *Stars & Stripes '87*, which is shown in Figure 2; a close-up of its hull and keel is shown in Figure 3.[5]

In the yacht domain, a design is represented by eight design parameters which specify the magnitude with which a set of geometric operators are applied to the B-spline surfaces (Rogers and Adams, 1990) of the starting prototype. The goal is to design the yacht which has the smallest course time for a particular wind speed and race course. Course time is evaluated using a "Velocity-Prediction Program" called "AHVPP" from AeroHydro, Inc., which is a marketed product used in yacht design (Letcher, 1991).

A search space is specified by providing an initial prototype geometry and a set of operators for modifying that prototype. Our current set of shape-modification operators was obtained by asking our yacht-design collaborators for an exhaustive list of all features of a yacht's shape that might be relevant to the racing performance of a yacht. These operators include

- Global-Scaling Operators: *Scale-X*, *Scale-Y* and *Scale-Z* change the over-all dimensions of a racing yacht, by uniformly scaling all surfaces.
- Prismatic-Coefficient Operators: *Prism-X*, *Prism-Y* and *Prism-Z* make a yacht's canoe-body more or less streamlined, when viewed along the X, Y and Z axes respectively.
- Keel Operators: *Scale-Keel* and *Invert-Keel* change the depth and taper ratio of the keel respectively.

These eight operators represent a subset of the full set that were actually developed, focusing on a smaller set suitable for testing our prototype-selection methods.

Prototype Selection Results

We conducted several sets of experiments. In each case we compare our approach with each of four other methods:

1. **Closest goal.** This method requires a measure of the distance between two goals, and knowledge of the goal for which each prototype in the design library was originally optimized. It chooses the prototype whose original goal has minimum distance from the new goal. Intuitively, in our yacht-design problem this method chooses a yacht designed for a course and wind speed most similar to the new course and wind speed.
2. **Best initial evaluation.** This method requires running the evaluation function on each prototype in the database. It chooses the prototype that, according to the evaluation function, is best for the new goal (before any operators have been applied to the prototype). In the case of our yacht-design problem this corresponds to starting the design process from whichever yacht in the library is fastest for the new course and wind speed.
3. **Most frequent class.** This is actually a very simple inductive method that always chooses a fixed prototype, namely the one that is most frequently the best prototype for the training data.
4. **Random.** This method involves simply selecting a random element from the design library, using a uniform distribution over the designs.

We compare these methods using two different evaluation criteria:

1. **Error rate.** How often is the wrong prototype selected?
2. **Course-time increase.** How much worse is the resulting average course time than it would be using the optimal choice that an omniscient selection would make?

In our experiments we focus primarily on the question of how well our inductive-learning prototype-selection method handles problems where the prototypes lie in disjoint search spaces. Our experiments therefore explore how prototype selection affects the quality of the final design.

TABLE 1. *A portion of the input to C4.5 for prototype selection in the yacht domain.*

Long-Leg	Short-Leg	Wind speed	Initial-Design
180	0	8	Design 1
180	0	10	Design 2
180	0	12	Design 2
180	0	14	Design 2
180	0	16	Design 2
180	90	8	Design 1
180	90	10	Design 4
180	90	12	Design 4
180	90	14	Design 4
180	90	16	Design 1

For the prototype selection experiments in the yacht domain, we used the Rutgers Hill-climber as our optimizer. It is an implementation of steepest-descent hill-climbing, that has been augmented so as to allow it to "climb over" bumps in the surface defined by the objective function that have less than a certain width or a certain height.

For our first set of experiments we created a database of four designs that would serve as our sample prototype library (and thus also serve as the class labels for the training data given to our inductive learner). To simulate the effect of having each prototype define a different space, the design library was created by starting from a single prototype (the Stars and Stripes '87) and optimizing for four different goals using all eight of our design-modification operators. All subsequent design episodes used only four of the eight operators, so that each yacht would define a separate space.[6]

We defined a space of goals to use in testing the learned prototype-selection rules. Each goal consists of a wind speed and a race course, where the wind speed is constrained to be 8, 10, 12, 14, or 16 knots, the race course is constrained to be 80% in one direction, and 20% in a second direction, and each direction is constrained to be an integer between 0 and 180 degrees. This space contains 162,900 goals.

To generate training data we defined a set of "training goals" that spans the goal space. This smaller set of goals was defined in the same fashion as for the testing set of goals except that the directions in the race course are restricted to be only 0, 90, or 180 degrees, yielding a smaller space of 30 goals. To label the training data we attempted to find designs for each of the 30 goals starting from each of the four prototypes using the restricted set of operators, and determined which starting point was best.

```
long-leg <= 90 :
|    windspeed > 10 : Design-1
|    windspeed <= 10 :
|    |    short-leg <= 90 : Design-1
|    |    short-leg > 90 : Design-2
long-leg > 90 :
|    windspeed > 14 : Design-2
|    windspeed <= 14 :
|    |    windspeed <= 10 : Design-4
|    |    windspeed > 10 : Design-4
```

Figure 4.   Example of a prototype-selection decision tree generated by C4.5.

TABLE 2.   *Comparison of prototype-selection methods when trained on a set of goals that spans the goal space, using AHVPP.*

Method	Error Rate	Course-Time Increase (sec)
Inductive Learning	30%	24
Most Frequent Class	70%	47
Random Guessing	75%	62
Best Init Eval	70%	64
Closest Goal	70%	78

To generate test data we randomly selected ten "testing goals" from the goal space. We then generated designs starting from each of the four prototypes in the database for each of these testing goals to determine which prototype was best, as well as to determine how much of a loss in course time each incorrect selection would impose. Table 1 shows a portion of the input to C4.5, and Figure 4 gives an example of a decision tree output by C4.5. Table 2 compares the results using C4.5 with the other prototype-selection methods. (Since there are four prototypes, one would expect random guessing to get 75% of the test examples wrong.)

In this experiment, the inductive method (C4.5) performed better than the other methods on both measures of performance. Moreover, we were particularly surprised by how poorly the non-inductive prototype-selection methods (closest goal and smallest initial evaluation) performed — our expectation was that the prototypes chosen by these methods would be close in "design space" to the optimal final design, thus yielding better final designs than starting from the other prototypes.

After studying these results we generated two new hypotheses for why these two prototype-selection methods did not work well. The first is that the shape of the design space may be such that there is little relationship between the distance between two designs and the ability of the hill-climber to climb from one design to the other. If the space contains "bumps" or "ridges" over which the hill-climber cannot climb, then it might be more important for the initial prototype to be on the "right side" of a bump or a ridge than for it to be close to the optimal point. Our second new hypothesis was that some of the prototypes in the database may be "bad" prototypes. This could be the case if the hill-climber got stuck at a local (non-global) optimum during the run that produced the prototype. This latter hypothesis was supported by the fact that one of the four prototypes was never found to be a good starting point for any of the 30 goals in the training data (not even the goal for which it was supposedly optimal, since it wound up being a local optimum and starting from another prototype yielded a superior result). In a realistic design scenario, when there is no control over the source of a design library, there could easily be "bad" prototypes included. Unlike the non-inductive prototype-selection methods, the inductive methods learn to avoid the bad prototypes.

We performed some experiments to test our first new hypothesis that the closest-goal and smallest-initial-evaluation methods performed poorly because of the "bumps" in the evaluation function. We repeated the earlier experiments using a simplified, "smooth" velocity prediction program, called "RUVPP," that we developed at Rutgers. RUVPP differs from the more complex AHVPP in several respects. To begin with, RUVPP represents a yacht as a list of major geometric dimensions such as length, depth, and beam, rather than B-spline surfaces. Furthermore, RUVPP embodies a number of simplifying assumptions about the physics of sailing that are not made in AHVPP. Nevertheless, the simple version, RUVPP, is useful for two reasons: RUVPP is much faster to execute than AHVPP, and RUVPP has fewer of the bumps and ridges that appear in AHVPP. We therefore expect that a hill-climbing search algorithm is less likely to get stuck on the wrong side of a bump or ridge when the simple version, RUVPP, is used as an evaluation function. Table 3 presents the results of experiments comparing the performance of inductively learned prototype-selection rules to the other prototype-selection methods, repeating our earlier experiments, but using RUVPP as the evaluation function, and using forty random test cases instead of just ten.

Because RUVPP is much faster than AHVPP, we conducted additional supporting experiments to test our first new hypothesis, to see if using a spanning set of goals as training data was significant for our results. In particular, rather than using inductive learning on a set of goals that span the space of possible goals, we also performed experiments where C4.5 was trained on a random sample of goals selected from the same space as the testing data. This

TABLE 3. *Comparison of prototype-selection methods when trained on a set of training examples that spans the goal space, using the simplified VPP.*

Method	Error Rate	Course-Time Increase (sec)
Best Init Eval	12%	26
Inductive Learning	37%	57
Closest Goal	40%	76
Most Frequent Class	45%	175
Random Guessing	75%	257

TABLE 4. *Comparison of prototype-selection methods when trained and tested on random goals, using cross-validation and the simplified VPP.*

Method	Error Rate	Course-Time Increase (sec)
Best Init Eval	12%	26
Inductive Learning	30%	35
Closest Goal	40%	76
Most Frequent Class	45%	175
Random Guessing	75%	257

was done using ten trials of four-fold cross-validation on a set of forty random goals. Each such trial involves randomly dividing the data into four sets of size ten, using three of the sets for training data and the remaining one as testing. This is repeated four times, using each ten-element set once for testing, and this process was repeated ten times with different random partitionings of the data. Table 4 reports the results of these experiments.

Consistent with our first new hypothesis, the closest-goal and best-initial-evaluation methods both did much better in both cases with the simplified VPP than they did with AHVPP, while C4.5 did about the same as it had done before. We believe that because the simplified VPP is much smoother than AHVPP, the hill-climber is much less likely to get stuck, so that the distance in goal space or the difference in initial evaluation becomes much more relevant when choosing a prototype. In fact, the improvement in the best-initial-evaluation method was so great that it significantly outperformed the inductive method.

We performed another set of experiments to test our second new hypothesis of why the closest-goal and smallest-initial-evaluation method performed so poorly using AHVPP, namely that they were unable to avoid the "bad" proto-

TABLE 5.   *Comparison of prototype-selection methods when trained on a set of goals that span the space, using the simplified VPP, and a "bad" prototype in the database.*

Method	Error Rate	Course-Time Increase (sec)
Best Init Eval	10%	80
Inductive Learning	30%	82
Closest Goal	32%	89
Most Frequent Class	45%	171
Random Guessing	75%	348

TABLE 6.   *Comparison of prototype-selection methods when trained and tested on a set of random goals, using cross-validation, the simplified VPP, and a "bad" prototype in the database.*

Method	Error Rate	Course-Time Increase (sec)
Inductive Learning	19%	38
Best Init Eval	10%	80
Closest Goal	32%	89
Most Frequent Class	45%	171
Random Guessing	75%	348

type in the database. We repeated our preceding experiments using the simplified VPP, except that we intentionally put a "bad" prototype into the database. To generate a bad prototype, we started with the Stars and Stripes '87, and added a random number between -0.2 and +0.2 to each of the operator parameters. We then randomly chose one of the four prototypes in the database to replace with the bad prototype (but we left the class label the same). The results of repeating the experiments with the bad prototype in the database are presented in Table 5 for training on goals that span the space, and Table 6 for training on random goals.

Consistent with our second new hypothesis, C4.5's ability to avoid the "bad" prototype improved its performance relative to the other methods. When trained on the spanning goals, C4.5 performed only slightly worse than the smallest-initial-evaluation method. When trained on the random goals, C4.5 performed markedly better than any other method as measured by average course-time increase, although the smallest-initial-evaluation method had a lower error rate. This apparent anomaly can be explained as follows: The "bad" prototype was very bad, so that choosing it even a few times resulted in large increases in average course time. C4.5 never chose the bad prototype. The best-initial-evaluation method occasionally chose the bad prototype, so that even though it chose the best prototype more frequently than C4.5, the few times when it chose the bad prototype worsened its average course-time increase.

The Cost of Learning

One important question to answer is whether the inductive prototype-selection method is worth the considerable "off-line" expense of collecting training data — every training example requires one design run for each design in the prototype library. An alternative, possibly cheaper method would be to take an "on-line" approach: for each new design problem optimize starting from every prototype in the database, and then use whichever of the resulting designs is the best.

If the quality of the final design is extremely important and there is ample CPU time available, this "exhaustive" method is the one to use (out of the methods listed in Table 2). On the other hand, if limiting CPU time is important, our inductive learning method becomes cost effective when the computational expense of learning can be amortized over a sufficiently large number of new design goals. More specifically, the inductive prototype-selection method is less expensive than the exhaustive method whenever the number of hill-climbing runs taken by the inductive approach is less than the number of runs taken by the exhaustive approach, i.e., $TP + G < PG$ or

$$G > \frac{T}{1 - \frac{1}{P}}$$

where T is the number of training examples, P is the number of prototypes in the database, and G is the number of new goals for which prototypes need to be selected. (When using the inductive prototype-selection method, TP is the cost of generating the training data, and G is the cost of performing optimizations for the new goals. When using the exhaustive method, each prototype in the database must be optimized for each new goal, at a cost of PG.) In all of the experiments that we performed, there were four prototypes and 30 training examples, so our inductive approach will be less expensive than the exhaustive approach as long as at least 40 out of the more than 150,000 remaining design goals must be attempted.

When doing prototype synthesis rather than prototype selection, it is not necessary to collect training data in which each prototype in a database is used as a starting point of an optimization for each of a collection of goals. (Prototype synthesis takes as training data the optimal design parameters for each goal, rather than the selection of the best prototype from a database for each goal.) Instead, any optimizations that have been previously done (within the

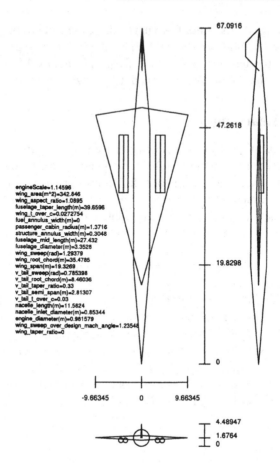

Figure 5. Supersonic transport aircraft designed by our system (dimensions in meters)

same goal space) can be used as training data. Hopefully, such data will already exist in a design library, so additional optimizations will not be needed to generate training data. Prototype synthesis is further described in the next section.

PROTOTYPE SYNTHESIS AND ACHIEVABLE GOAL PREDICTION

Prototype synthesis uses continuous-class induction (also known as regression) to map the design goal directly into the design parameters that define a new prototype, instead of selecting an existing prototype from a database. What is learned is not a set of rules for selecting a prototype, but rather a set of functions that map the design goal into the design parameters. We performed some

Phase	Mach	Altitude		Duration (min)	comment
		m	ft		
1	0.227	0	0	5	"takeoff"
2	0.85	12 192	40 000	50	subsonic cruise (over land)
3	2.0	18 288	60 000	225	supersonic cruise (over ocean)

capacity: 70 passengers.

TABLE 7. *Mission specification for aircraft in Figure 5*

experiments to test prototype synthesis in the domain of supersonic transport aircraft design.

The Aircraft Domain

Figure 5 shows a diagram of a typical airplane automatically designed by our software system to fly the mission shown in Table 7. The optimizer attempts to find a good aircraft conceptual design for a particular mission by varying major aircraft parameters such as wing area, aspect ratio, engine size, etc; using a numerical optimization algorithm. The optimizer evaluates candidate designs using a multidisciplinary simulator. In our current implementation, the optimizer's goal is to minimize the takeoff mass of the aircraft, a measure of merit commonly used in the aircraft industry at the conceptual design stage. Takeoff mass is the sum of fuel mass, which provides a rough approximation of the operating cost of the aircraft, and "dry" mass, which provides a rough approximation of the cost of building the aircraft. The simulator computes the takeoff mass of a particular aircraft design for a particular mission as follows:

1. Compute "dry" mass using historical data to estimate the weight of the aircraft as a function of the design parameters and passenger capacity required for the mission.
2. Compute the landing mass $m(t_{final})$ which is the sum of the fuel reserve plus the "dry" mass.
3. Compute the takeoff mass by numerically solving the ordinary differential equation

$$\frac{dm}{dt} = f(m, t)$$

which indicates that the rate at which the mass of the aircraft changes is equal to the rate of fuel consumption, which in turn is a function of the current mass of the aircraft and the current time in the mission. At each time step, the simulator's aerodynamic model is used to compute the current drag, and the simulator's propulsion model is used to compute the fuel consumption required to generate the thrust which will compensate for the current drag.

A complete mission simulation requires about 1/4 second of CPU time on a DEC Alpha 250 4/266 desktop workstation.

The numerical optimizer used in the prototype synthesis experiments is CF-SQP (Lawrence et al., 1995)[7], a state-of-the-art implementation of the Sequential Quadratic Programming (SQP) method. SQP is a quasi-Newton method that solves a nonlinear constrained optimization problem by fitting a sequence of quadratic programming problems[8] to it, and then solving each of these problems using a quadratic programming method. We have supplemented CFSQP with *rule-based gradients* (Schwabacher and Gelsey, 1997) and *model constraints* (Gelsey et al., 1998).

Because the search space has many local optima, we use a technique that we call "random multistart" to attempt to find the global optimum. In an n-point random multistart, the system randomly generates starting points within a particular box until it finds n evaluable points[9], and then performs an SQP optimization from each of these points. The best design found in these n optimizations is taken to be the global optimum.

In the airframe domain, the design goal is to minimize take-off mass (a rough estimate of life-cycle cost) for a specified mission. We defined the following space of missions:

```
distance between 1609 km (1000 miles)
  and 16 090 km (10 000 miles)
percentage over land between 0 and 100%
mach number over land of 0.85
altitude over land 12 192 m (40 000 ft)
mach number over water between 1.5 and 2.2
altitude over water 18 288 m (60 000 ft)
optional takeoff phase, no climb phase
```

A mission within this space can be represented using three real numbers (distance, percentage over land, and mach number) and one Boolean value (whether the takeoff phase is included). We generated 100 random missions as follows: The distance and mach number were uniformly distributed over their possible ranges. There was a 1/3 probability of having the mission entirely over land, a 1/3 probability of having it entirely over water, and a 1/3 probability of having the percentage over land uniformly distributed between 0 and 100%. There was a 1/2 probability of including the takeoff phase.

```
distance > 14 456 km (8982.46 miles): infeasible
distance <= 14 456 km (8982.46 miles):
|    distance <= 10 276 km (6384.94 miles): feasible
|    distance > 10 276 km (6384.94 miles):
|    |    overland <= 23.6023% : feasible
|    |    overland > 23.6023% : infeasible
```

Figure 6. Learned decision tree for deciding if a mission is feasible.

Achievable Goal Prediction

In order to generate training data to test our techniques in the airframe domain, we performed a 10-point random multistart CFSQP optimization for each of the 100 random missions. We found that for many of these missions, CFSQP was unable to find a feasible design in any of the ten runs — that is, it was unable to design a plane that could fly the mission. It occurred to us that it would be valuable if we could predict in advance whether a given mission was achievable, so that we could avoid attempting to synthesize prototypes for infeasible missions. We hypothesized that C4.5 would be able to make this prediction, and that it would be able to do so with greater accuracy than MFC.

To test this new *achievable goal prediction* idea, we trained C4.5 on a set of training examples showing whether each of our 100 missions was feasible. It produced the decision tree in Figure 6. This decision tree shows that missions are infeasible if they are very long, or if they are moderately long and have a significant portion over land. Further analysis revealed that building a plane to fly such a mission would require an engine larger than the largest engine that we allowed. Our upper bound on engine size can be considered to be representative of the largest commercially available engine.

Tenfold cross validation showed that C4.5 has a 4% error rate on this learning task, compared with 50% for random guessing and 24% for most frequent class. The decision tree of figure 6 can be used to predict, without doing any simulation or optimization, whether a new proposed mission is feasible.

Prototype Synthesis

In order to map the new mission into the numerical design parameters that define a prototype, we need to use *continuous-class induction* (which is also known as regression). We used CART (Classification And Regression Trees), which builds a "regression tree" that has a numerical constant at each leaf (Breiman, 1984). We trained CART on the 100 randomly generated training goals as follows: For each design parameter, we gave CART a set of training data, where each item in the training data included the goal and the optimal

TABLE 8. *Accuracy of CART in predicting each design parameter in the airframe domain.*

Design Parameter	Relative RMSE
engine size	0.65
wing area	0.59
wing aspect ratio	0.06
fuselage taper length	0.07
effective structural thickness over chord	0.08
wing sweep over design mach angle	0.08
wing taper ratio	0.21
fuel annulus width	1.02

TABLE 9. *Comparison of prototype-synthesis methods.*

Method	Success	Cost (number of simulations)
CART	13/16	7394
mean	14/16	11963
1 random	8/16	16893
2 random	13/16	33883
3 random	14/16	47395

value of the design parameter. CART thus generated a set of trees to map the design goal into a set of design parameters that we hope will be near the optimal values for that goal. Table 8 shows the root mean squared error in CART's prediction of each design parameter, relative to the error of "constant regression," which always uses the mean of the training data. A value less than one in this table indicates that CART's prediction was more accurate than that of constant regression. Our expectation that these relative errors would be low was confirmed for all of the parameters except fuel annulus width.

We performed a set of experiments to test whether using these trees to do prototype synthesis would produce better optimization performance than using the mean prototype or a random prototype. We used 25 randomly generated testing goals. Table 9 compares using the prototypes synthesized by CART with using a 1-, 2-, or 3-point random multistart, or always using the prototype which is the mean of all the optimized prototypes in the training data. Of the 25 randomly generated test goals, 16 were feasible. The "success" column shows the number of optimizations that came within 1% of the point that we believe to be the global optimum.[10] Some of the failures occurred because the learning method produced an unevaluable prototype that could not be simulated, and therefore could not be optimized. Other failures occurred because

TABLE 10. *Performance of one random probe, averaged over ten trials.*

Measure	Success	Cost (number of simulations)
Mean	8.8/16	15062
Standard Deviation	1.9/16	3102

the optimizer, when started from the synthesized point, failed to get within 1% of the apparent global optimum. The "cost" column shows the total number of simulations used in the 16 optimizations. Using the mean prototype instead of a single random prototype resulted in much greater success, at 33% lower cost. Using CART produced a success rate about the same as using the mean prototype, with an additional 38% cost reduction. Using a 2-point random multistart produced the same success rate as using CART, but it required more than four times as many simulations.

To test the significance of the result that CART performed better than one random probe, we repeated the one-random-probe test ten times, with ten different seeds to the random number generator. The mean and standard deviation of the success rate and cost are shown in Table 10. CART's success rate was more than two standard deviations higher than that of one random probe, and its cost was more than two standard deviations lower than that of one random probe.

FORMULATION SELECTION

Besides the selection of a starting prototype, another important decision in setting up an optimization is the decision on how to formulate the search space. This decision can substantially affect the performance of the optimizer in two ways. First, using a lower-dimensional formulation of the search space makes optimization faster, since each gradient computation requires fewer runs of the simulator, and the distance in design space from the starting point to the optimum is smaller. Second, different formulations of the search space can result in different degrees of "smoothness" of the search space, which can impact not only the speed of the optimizer, but also the ability of the optimizer to get to the optimum, and therefore the quality of the resulting designs.

We present a method of reformulation called "constraint incorporation," which reduces the dimensionality of the search space and increases its smoothness by incorporating constraints into the search space.

Traditionally, numerical optimization has dealt with explicit, "hard" constraints. The optimizer assumes that these constraints can never be violated. A hard constraint can be expressed as

$$f(x_1, x_2, \ldots, x_n) \leq k$$

(Here $x_1, x_2, \ldots, x_n$ are the *design parameters* that represent the design.)
The constraint is said to be *inactive* if $f(x_1, x_2, \ldots, x_n) < k$, *active* if
$f(x_1, x_2, \ldots, x_n) = k$, and *violated* if $f(x_1, x_2, \ldots, x_n) > k$. Hard constraints
can result from the laws of physics, for example.

Another type of constraint is the "soft" constraint, for which there is some
sort of known penalty for violating the constraint. A soft constraint can be
expressed as

if $f(x_1, x_2, \ldots, x_n) > k$ then apply penalty $P(x_1, x_2, \ldots, x_n)$

These usually arise from human-written laws, such as regulations specifying
a monetary penalty for exceeding a certain noise level. In either case, if it is
known that the constraint will be active at the optimal design point, and the
constraint function f is invertible, then the constraint can be *incorporated* into
the search space by using the inverse of f to eliminate one of the design parameters. This incorporation is done by making the inequality constraint into
an equality constraint, and then solving for one of the design parameters in
terms of the other design parameters. (Papalambros and Wilde, 1988) describe
how monotonicity knowledge can be used to determine that certain constraints
will be active at the optimum. Incorporating these constraints produces a new
search space with lower dimensionality, since the incorporation eliminates a design parameter, and greater smoothness, since the incorporation eliminates the
"ridge" (or non-smoothness) in the search space caused by the "if" statement
in the constraint. If there are n constraints that can be incorporated in this way,
then there are 2^n possible formulations that can be produced by incorporating
different subsets of constraints.

Constraint activity depends on the goal (some constraints are active at the
optimum for only some design goals), for two reasons: First, the constraint
thresholds are part of the design goal. Second, different design goals will result
in different optimal values of the design parameters on which the constraint
functions depend.

Because constraint activity depends on the goal, different formulations of
the search space are appropriate for different design goals. We describe a way
in which inductive learning can be used to map the design goal into the appropriate formulation.

To use inductive learning to form formulation-selection rules, we take as
training data a collection of design goals, each labeled with the set of constraints that are active (within a threshold) at the optimal design point. We run
the inductive learner once for each constraint, producing for each constraint a
set of rules that can be used to predict whether the constraint will be active for
new design goals.

The training data can be generated in an automated fashion. For example, one can choose a set of training goals and perform an optimization for each goal. One can then evaluate each constraint function for each optimal design, and then construct a table that records which constraints were active (within a threshold) for each training goal. This table can be used by the inductive-learning algorithm to generate a set of rules for each constraint, mapping the space of all possible goals into a prediction of whether or not that constraint will be active at the optimal design point for that goal. If learning is successful, these mappings extrapolate from the training data and can be used successfully in future design sessions to map a new goal into an appropriate formulation.

Formulation Selection Results in Yacht Domain

We performed some experiments to test the performance of formulation selection in the yacht domain. In the experiments described in this subsection, we used CFSQP as the optimizer, with *course-time*, computed by RUVPP, as the objective function, and with one explicit, nonlinear, "hard" model constraint. This constraint specifies that the mass of the yacht, before adding any ballast, must be less than or equal to the mass of the water that it displaces. (In other words, the boat must not sink.)

Yachts entered in the 1987 America's Cup race had to satisfy a hard constraint known as the 12-Meter Rule (IYRU, 1985). Instead of using this rule as an explicit constraint, we incorporated it into the search space. (How we incorporated it is described below.) The basic formula in the rule is:

$$\frac{length - freeboard + \sqrt{sailarea}}{2.37} \leq 12m$$

In addition to the basic formula, the rule contains several soft constraints, along with associated penalties for violating these constraints. These soft constraints are:

- draft constraint
- beam constraint
- displacement constraint
- winglet span constraint

For example, the *beam constraint* states

if *beam* < 3.6m, then add four times the difference to *length*

While constructing the simulator, we used a reasoning process similar to that described in (Papalambros and Wilde, 1988) to determine that the constraint described by the basic formula of the 12-Meter Rule, above, will always be active, since the objective function being minimized, *course-time*, is monotonically non-increasing in *sail-area*,[11] and the left-hand-side of the constraint is monotonically increasing in *sail-area*. We therefore *incorporated* this constraint into

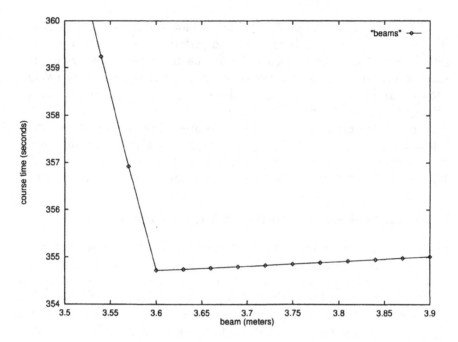

Figure 7. The nonsmoothness in the search space caused by the beam constraint.

the simulator by solving for *sail-area* in terms of the other design parameters. So, for example, when the optimizer makes *length* bigger, *sail-area* is automatically made smaller. In addition, because we also implemented the soft constraints as penalty functions, reducing *beam* beyond $3.6m$ causes the quantity *length* in the formula to increase, which causes *sail-area* to decrease.[12]

Because the beam constraint contains an *if* statement, this incorporation causes a nonsmoothness in *course-time* as a function of *beam*. That is, there is a discontinuity in the first derivative of *course-time* with respect to *beam*. Figure 7 illustrates this nonsmoothness by showing the cross-section of the search space corresponding to the *beam* design parameter.[13] This nonsmoothness can cause a gradient-based optimizer such as CFSQP to get stuck, and to fail to get to the optimum.

For many design goals, the optimal design is right on the constraint boundary. The optimal beam is often 3.6 m. If we expect the optimal beam to be 3.6 m, then we can incorporate the beam constraint into the operators. In the case of the beam constraint, this incorporation is trivial — we simply set *beam* to 3.6 m and leave it there. For other constraints, the incorporation is more complicated. For example, there is a constraint that specifies a penalty if *displacement* does not vary with a certain cubic polynomial in *length*. *Displacement* is not a design parameter; rather, it is a quantity computed from all of the design parameters. In order to incorporate the displacement constraint, we used Maple (Char et al., 1992), a symbolic algebra package, to invert the displacement formula, and created a new set of operators that vary certain parameters while maintaining *displacement* at the minimum displacement allowed by the constraint. For still-more-complicated constraints, it might not be possible to invert the constraint function using Maple; it might therefore be necessary for the operators to contain numerical solvers that find the right values of the incorporated design parameters so as to put the design on the constraint boundary.[14]

We created operators to incorporate all four of the above-listed 12-Meter Rule constraints: the draft constraint, the beam constraint, the displacement constraint, and the winglet constraint. Using these operators, we are able to either incorporate or not incorporate each of these four constraints independently. We thus defined a set of sixteen (2^4) possible formulations of the search space. From our initial experiments with these operators, we determined empirically that incorporating the draft constraint substantially improved the reliability and speed of optimization for any design goal. We therefore decided to always incorporate the draft constraint, leaving us with a space of eight possible formulations that we used in the experiments described below.

Having defined eight formulations of the search space, we used inductive learning to decide, based on the design goal, which formulation to use. As training data, we used 100 previous optimizations. The optimizer failed for one of these goals, so we used the remaining 99 goals as training data in the results that follow. For each previous optimization, we evaluated each 12-Meter Rule constraint function at the optimum, and determined if the constraint was active (within a tolerance). Each of these previous optimizations had as its design goal minimizing course time for a single-leg race course, which can be represented using two numbers: the wind speed, and the heading (the angle between the yacht's direction and the wind direction). The design goal can therefore be represented using these two numbers. We ran the inductive learner once for each of the three constraints. Each time, the inductive learner was provided with a set of triples: the wind speed, the heading, and a ternary value indicating whether the constraint was inactive, active, or violated. One of the constraints was violated at the optimum in 10 of these optimizations. Figure 8 gives an example of a decision tree output by C4.5. This decision tree predicts whether the displacement constraint will be active at the optimum, based on the design goal. By running a new design goal down three decision trees, one for each of the three constraints that can be incorporated, the system can make predictions of whether each constraint will be active at the optimum. These three yes/no predictions directly map into one of the eight (2^3) formulations of the search space.

We used C4.5 to perform tenfold cross-validation, and obtained the error rates shown in Table 11. Here we compare the error rates of C4.5 with and

```
heading <= 109 :
|    windspeed <= 6.3 : active
|    windspeed > 6.3 :
|    |    windspeed > 8.2 : violated
|    |    windspeed <= 8.2 :
|    |    |    heading <= 65 : violated
|    |    |    heading > 65 : active
heading > 109 :
|    windspeed > 11.5 : active
|    windspeed <= 11.5 :
|    |    heading <= 135 : active
|    |    heading > 135 : inactive
```

Figure 8.   Learned decision tree for the displacement constraint.

TABLE 11.   *Cross-validated error rates for selecting whether to incorporate each constraint.*

method	Beam	Displacement	Winglet
C4.5 w/ pruning	11.1%	15.1%	7.0%
C4.5 w/o pruning	11.1%	15.1%	10.0%
C4.5rules	11.1%	15.1%	10.0%
MFC	33.3%	53.5%	13.1%
Random	66.7%	66.7%	66.7%

without pruning, and of C4.5rules, a variant of C4.5 that extracts rules from the trees, with the expected error rate of random guessing (which is two-thirds since there are three classes from which to guess), and the error rate of the Most Frequent Class (MFC) learning method. MFC always chooses the class that occurs most frequently in the training data. In this case, that means that it always chooses the same formulation, namely the one that is most often the best formulation in the training data.

As Table 11 shows, C4.5 with pruning performed slightly better than C4.5 without pruning or C4.5rules (and so in our further experiments reported below we use only C4.5 with pruning), and all three substantially outperformed MFC, which in turn substantially outperformed random guessing.

These results are for error rates, the proportion of cases where learning makes an incorrect guess. A more important question in this domain is how learning affects the overall problem-solving task, namely how it improves the speed and reliability of the design optimization process. Does learning make the design process faster or slower? Are the resulting designs better or worse?

TABLE 12. *Effect of using formulations chosen by learner on optimization performance. A positive quality change indicates an improvement in quality (which is a reduction in course time).*

method	quality change	CPU time change
omniscient	+0.085%	-36%
exhaustive	+0.085%	+384%
C4.5	+0.080%	-35%
MFC	+0.029%	-32%
none	0	0
random	-0.276%	-40%
all	-0.599%	-74%

To measure these effects, we performed optimizations for 25 new randomly generated goals using the formulations suggested by each learning method. Table 12 shows the effect that C4.5 (with pruning) and MFC had on the average course time (the quality of the design), and average number of evaluations (the speed of the optimization), as compared with the "old way" of doing optimization without incorporating any of the three constraints into the operators. The first column in the table shows the percentage difference between the optimized course-time produced with the original formulation, and the optimized course time produced with the specified formulation. The second column shows the percentage difference between the cost of performing the optimization with the original formulation, and the cost of performing it with the specified formulation.

We also include in this table the performance of several other methods. A hypothetical "omniscient" problem solver always magically guesses the best possible choice (the one that results in the best course time).[15] No learning method will enable results superior to those produced by this method. The "exhaustive" optimization method performs eight optimizations for each goal, using all eight possible formulations, and then chooses the best resulting design. Incorporating "all" constraints all the time results in the fastest possible optimization within this set of formulations (at the cost of quality loss).

C4.5 produced a significant speedup in optimization, with no quality loss. In fact, it produced a small quality increase. (This quality increase suggests that with the original formulation, the optimizer gets "stuck" on the "ridges" that the constraints cause the search space to have, and therefore sometimes fails to get the optimum.) MFC produced a slightly smaller speedup and a slightly smaller quality improvement. The difference between C4.5 and MFC in quality change was, however, statistically significant at the 99% confidence level, according to the paired t-test. Both learning methods performed substantially better than random guessing. C4.5 performed almost as well as the hypothetical omniscient learner, which means it performed almost as well as any learner could possibly do.[16]

Incorporating all of the constraints all of the time resulted in a very large speedup, with a modest quality loss. This method may be appropriate if one wants a quick and approximate optimization. It might, for example, be used in the early stages of design when the engineer wants to get a feel for the search space by asking "what-if" questions.

One question that these results raise is how training-data quantity affects performance. If one does not have results from a large number of previous optimizations available, then one can either run some extra optimizations to generate training data (which is expensive), or do the learning with less training data (which is likely to produce higher error rates and lower optimization performance). We ran some experiments to determine how C4.5's performance varies with training-set size, and how its performance compares with that of MFC for various training-set sizes. We applied our learning approach to datasets of varying sizes, with the error rates shown in Figure 9. For each training-set size in the figure, we randomly chose 10 different subsets of our training data of that size, and performed 10-fold cross-validation on each subset. The figure shows the averages. The three symbols at the right side of the figure show MFC's performance on the full training set. C4.5 outperformed MFC for every training-set size, but C4.5's error rate on smaller training sets was significantly larger than C4.5's error rate for larger training sets (with performance reaching an asymptote for training sets of about 60 cases or more).

Formulation Selection Results in Airframe Domain

We believe that our formulation selection technique is applicable to a broad range of design optimization problems. To test the domain-independence of the formulation selection technique, we performed additional experiments in the airframe domain, and compared the impact on optimization performance of C4.5 with that of MFC.

In the airframe domain, there are eight design parameters, each of which can have an upper and lower bound. The optimal design sometimes lies at the bounds of some of these parameters, depending on the mission.

We used CFSQP as the optimizer, and used the same simulator and the same space of missions as in the prototype synthesis experiments. We used the same C4.5 decision tree to predict which missions are feasible. As training data, we used the same 100 10-point random multistart CFSQP optimizations, 76 of which are feasible.

We used the 76 feasible missions to train C4.5 for formulation selection.

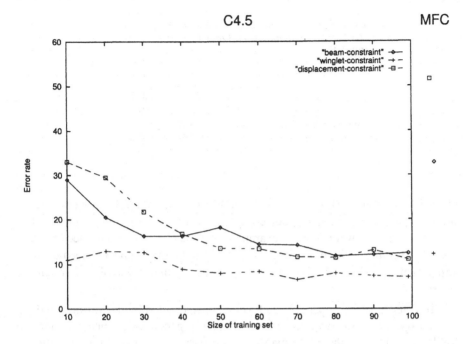

Figure 9.   Effect of training set size on learner performance.

```
overland <= 95.0872% : zero (54.0)
overland > 95.0872% :
|    takeoff = no: zero (12.0/1.0)
|    takeoff = yes: nonzero (10.0)
```

Figure 10.   Learned decision tree for predicting if the taper ratio will be at its lower bound of zero.

Of the eight design parameters, four were never at their upper or lower bounds at the apparent optima for any of the 76 missions. The other four had optima at their lower bounds for some missions. We trained C4.5 to predict whether these four design parameters would be at their lower bounds, depending on the mission. C4.5 produced a separate decision tree for each of these four design variables. For example, Figure 10 shows the decision tree for wing taper ratio. This decision tree says that wing taper ratio will be at its lower bound of zero, unless the mission includes a takeoff phase and is almost entirely over land. The four decision trees can be used to select among 16 (2^4) possible formulations.

Table 13 compares the cross-validated error rates of C4.5 with those of most frequent class and random guessing for each of the four design parameters. For the first three parameters, C4.5 did much better than most frequent

TABLE 13. *Cross-validated error rates for selecting whether to incorporate each lower bound, in the airframe domain.*

design parameter	C4.5	MFC	Random
wing taper ratio	2.7%	14.5%	50.0%
wing sweep	2.5%	27.6%	50.0%
fuselage taper length	3.9%	22.4%	50.0%
fuel annulus width	13.6%	5.3%	50.0%

class. For the fourth parameter, fuel annulus width, C4.5 did much worse than most frequent class, violating our expectations. In this case, only 4 of the 76 training examples were positive examples. We suspect C4.5 would need more training examples to be more accurate. Interestingly, in our prototype synthesis experiments, CART had difficulty predicting the optimal value of fuel annulus width.

To determine the impact of the formulations selected by the various methods on optimization performance, we randomly generated 25 new missions. Table 14 compares the performance of the various methods of formulation selection when doing optimizations for these new missions. For the methods that used C4.5, we used the decision tree of Figure 6 to predict whether each new mission was feasible, and only performed optimizations for those missions that were predicted to be feasible. For the other methods, we performed optimizations for all 25 missions. Each optimization was a 10-point multistart. The "success" column indicates for how many of the missions the specified method came within 1% in takeoff mass of the best design found.[17] The "time change" column shows the change in total number of simulations used in all of the optimizations performed, compared with not incorporating any constraints.

Because cross-validation showed that C4.5 under-performs MFC for predicting whether to incorporate fuel annulus width, we did not use C4.5 to decide whether to incorporate this parameter. We used C4.5 to decide whether to incorporate the other three parameters, and used two different methods to decide whether to incorporate fuel annulus width. The first method used MFC to decide whether to incorporate the fuel annulus width, which resulted in always incorporating it. The results of this method are labeled "C4.5/MFC" in Table 14. For the second method, we decided to play it safe and never incorporate fuel annulus width, since cross validation suggests that we are not able to accurately predict when this parameter will be at its bound. The results of this method are labeled "C4.5/none" in Table 14. We compare these methods with most frequent class, and with the exhaustive method that does optimizations for all 16 (2^4) formulations, and the omniscient method which magically guesses

TABLE 14. *Effect of using formulations chosen by learner on optimization performance, in airframe domain.*

method	success	time change
omniscient	16	-51%
exhaustive	16	+1206%
C4.5/none	15	-36%
none	15	0
C4.5/MFC	13	-57%
MFC	13	-21%
all	3	-55%

the best formulation.

The first interesting thing to note about Table 14 is that there is one mission for which CFSQP failed to reach the optimum without reformulation. The only way to reach the optimum for this mission is to use the "omniscient" method (which does not exist), or the "exhaustive" method (which is extremely expensive). The next thing to note is that using the formulations selected by C4.5 for the first three parameters, while not incorporating fuel annulus width ("C4.5/none"), reduces cost by 36% compared with not incorporating any constraints ("none"), without any loss of quality. Using C4.5 for the first three parameters, and MFC for fuel annulus width (C4.5/MFC), causes CFSQP to fail to find the optimum in two additional cases. Using MFC for all parameters causes the same number of missed optima, at a higher cost. And incorporating all of the parameter bounds all of the time results in CFSQP almost always failing to get to the optimum.

The airframe domain results are surprisingly similar to the yacht domain results. In the yacht domain, using the formulations selected by C4.5 reduced the cost of optimization by 35% (Table 12), while in the airframe domain the speedup was 36%. In the yacht domain, using C4.5 also resulted in a small quality increase, while in the airframe domain, quality remained the same. This may be because the yacht domain reformulations increase the smoothness of the search space (by eliminating the 12m-rule penalties), while the airframe domain reformulations do not. Another interesting thing to note is that while the difference between MFC and C4.5 was small (but statistically significant) in the yacht domain, it was much larger in the airframe domain.

RELATED WORK

Cerbone (Cerbone, 1992) has reported work which applied machine-learning techniques to a problem similar to our prototype-selection problem. His design space, in the domain of truss design, has an exponential number of disconnected search spaces. He uses inductive learning techniques to learn rules for selecting a subset of these search spaces for further exploration. In contrast, our system has a smaller number of prototypes (each of which defines a search space) from which to choose, and it just chooses one of them. Cerbone uses an ad-hoc utility function to combine solution quality and search time when evaluating his learning methods, while we only consider solution quality in this chapter. Cerbone also presents two learners that incorporate background knowledge by incorporating the objective function into the learner.

Research on prototype-retrieval strategies for hill-climbing design optimization is reported by Ramachandran *et al.* (Ramachandran et al., 1992), who investigated a number of library-based methods for finding starting points for the DPMED iterative parameter-design system. These included a nearest-neighbor method, a curve-fitting method, and a hybrid method. The curve-fitting method is similar to our prototype synthesis method. It uses regression to find a function mapping goal parameters to initial design parameters, whereas our approach uses inductive learning to find a regression tree mapping goal parameters to initial design parameters. Ramachandran *et al.* compared their retrieval strategies in terms of the numbers of iterations needed to carry out the hill-climbing design-optimization process. They showed that starting points obtained by curve fitting led to fewer iterations than were required when the nearest-neighbor method was used. In contrast to this work, our work has evaluated retrieval strategies in terms of the quality of the resulting designs, in addition to the number of iterations needed to find them.

Researchers in case-based reasoning have investigated the use of library-retrieval techniques for case-based design (Sycara and Navinchandra, 1992; Kolodner, 1993), but have not used them to initialize an iterative design process. (Bhatta and Goel, 1995) describe a system that learns to retrieve a starting point for the design of a high-acidity sulfuric acid cooler. They evaluate the performance of this indexing system based on its effect on retrieval time, and not based on its impact on optimization performance.

In (Rasheed and Hirsh, 1999), Rasheed and Hirsh describe a *screening module* that they added to a genetic algorithm used for engineering design optimization. The screening module uses a nearest-neighbor approach to learn to avoid evaluating points in the search space that are likely to have poor evaluation functions because they are near other "bad" points. They tested their algorithm in multiple domains, including the supersonic aircraft domain described in this chapter, and found that the screening module makes the optimization significantly faster.

Gelsey *et al.* (Gelsey et al., 1996) describe a Search Space Toolkit which assists in determining properties of the search space that can be used for reformulation. (Choy and Agogino, 1986) describe a system that automates (Papalambros and Wilde, 1988)'s method of using monotonicity analysis to detect constraint activity.

In (Williams and Cagan, 1994), Williams and Cagan present *activity analysis*, a technique inspired by monotonicity analysis. Their technique is similar to the formulation selection technique described in this chapter, except that they use qualitative reasoning instead of machine learning to determine which constraints will be active at the optimum. Their technique has the advantages that it does not require training data, and that the reformulation is guaranteed not to lose the global optimum. It has the disadvantage that it requires that the objective function and constraint functions be symbolically differentiable and composed of simple arithmetic operations; it would therefore not be applicable to the complex simulators used in the experiments described in this chapter.

A number of research efforts have combined AI techniques with numerical optimization. (Ellman et al., 1993) describes a method for switching between a less expensive, less accurate simulator, and a more expensive, more accurate simulator during optimization, based on the magnitude of the gradient. (Bouchard et al., 1988) describes ways in which expert systems could be applied to the parametric design of aeronautical systems. (Hoeltzel and Chieng, 1987) describe a system for digital chip design in which design is done at an abstract level, using machine learning to estimate the performance that would be obtained if the design were carried out at a more detailed level. (Orelup et al., 1988) describes a system called Dominic II that uses an expert system to switch among various strategies during numerical optimization. None of these efforts is focused directly on the problems of prototype selection and formulation selection addressed in this chapter.

Simulated annealing (SA) and genetic algorithms (GA) are able to deal with certain pathologies, such as nonsmoothness, but they tend to be much slower than gradient-based optimization. They tend to require thousands, or even tens of thousands, of simulations, and are thus not practical when each simulation is expensive.

Powell (Powell, 1990; Tong et al., 1992; Powell and Skolnick, 1993) has built a module called Inter-GEN, part of the ENGINEOUS system (Tong, 1988), that seeks to combine the ability of genetic algorithms to handle multiple local optima with the speed of numerical optimization algorithms. It contains a genetic algorithm, and a numerical optimizer, and uses a rule-based expert system to decide when to switch between the two. Powell has tested his system on a realistic jet engine design problem. He does not, however, address the issues of prototype selection or formulation selection.

FUTURE WORK

This chapter presents an initial exploration of the use of inductive learning to set up an optimization, and there are a number of directions for future work. These directions for future work fall into three groups: extending this work to more difficult design tasks, improving results by using other learning methods, and applying inductive learning to other choices that must be made in setting up an optimization.

Other Design Tasks

First, the experiments reported here explore the sensitivity of our prototype-selection method to the nature of the design library, specifically with respect to the quality of the stored designs. It would be helpful to more fully explore the sensitivity of our approach to the design library, for example by studying how our approach scales up as the library size increases.

The yacht domain results presented here apply to a constrained class of yacht-design goals, those comprised of a single leg (for formulation selection) or two legs (for prototype selection). One question is how this approach can be applied to courses comprised of varying numbers of legs. We believe that we could get reasonable optimization performance by using the trees learned from single-leg courses to perform multi-leg formulation selection in the following way: If a constraint should be incorporated for every leg of the race-course, then incorporate it for the full, multi-leg course. We need to test how well optimization performs when handling race-courses in this manner. We could also attempt to learn directly for multi-leg race-courses. Doing so would raise an interesting machine-learning question, since describing a multi-leg race-course requires a variable number of attributes, and thus traditional learners such as C4.5 do not directly apply. Learning methods operating on more expressive representations, such as inductive logic programming systems like FOIL (Quinlan, 1990), may enable going beyond the simple representation of goals used here and handling more complicated goals, including those involving multi-leg race courses or multiple disciplines.

In the results presented here, we assume that the only change between the previous design sessions and the current design session is the design goal (for example, expressed as a (*wind speed, heading*) pair for formulation selection in the yacht domain). An interesting question is what would happen if in addition to changing the goal, we also changed the constraints, or the simulator, or the form of the goal. We would need to find a way to encode as a set of attributes for the learner whatever had changed.

We believe that the formulation selection results presented here will easily generalize to situations in which there are more than sixteen formulations. We used the results from the same set of 100 optimizations to perform three separate learning tasks (for three constraints), and then combined the rules generated by these three learning sessions to select one of the eight formulations. As the number of formulations grows, the number of constraints, and therefore the amount of CPU time needed for the learning, will grow logarithmically with the number of formulations. The CPU time needed for learning is currently insignificant compared with the CPU time needed for the subsequent optimizations. We expect that as the number of formulations grows, the number of training examples needed will remain constant (since the same training examples are used for each constraint), and the amount of CPU time needed for learning will remain insignificant. We plan to test this hypothesis by using other constraints within the yacht design domain, such as the "boat doesn't sink" constraint.

The learning approach could also be used to decide when to reformulate soft constraints as hard constraints. If it were known with a high degree of confidence that a certain soft constraint will not be violated at the optimum for certain goals, then this soft constraint could be converted into a hard constraint for those goals, which would eliminate a ridge from the search space and thereby make optimization more robust (although it would not reduce the dimensionality of the search space). For example, in the training data that we collected, the *beam constraint* was never violated, so it might be replaced safely with a hard constraint.

Other more-difficult problems might involve a less-smooth search space, a higher-dimensional goal space, or a less reliable optimizer. Such problems may arise when we test this method in still other domains.

Other Learning Methods

We found that C4.5 performed nearly as well as a hypothetical "omniscient" learner, when doing formulation selection for the fairly simple design problems that we used in our experiments. When doing prototype selection, however, there was room for improvement. Other learning methods might prove useful in attacking the prototype selection problem, and might also prove useful when doing formulation selection for some of the harder design problems described in the previous subsection. For example, it would be interesting to see how well neural networks, nearest-neighbor methods, or statistical regression would perform. In particular, C4.5, like most decision-tree learners, uses linear, axis-parallel cuts in its decision trees. However, Figure 11 shows how the activity of the beam constraint varies over the goal space in the training data we used — the space is clearly divided into two regions (except for one point which we believe is noise). The border between these regions does not appear to be axis parallel, and appears to be nonlinear. This suggests that better formulation-selection performance might be achieved using an "oblique" deci-

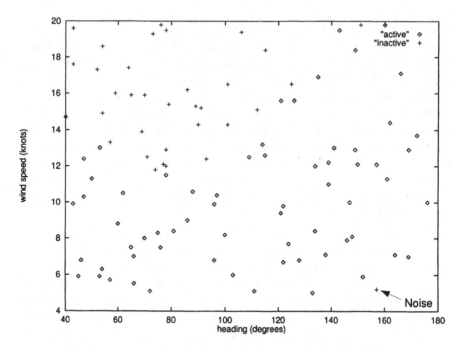

Figure 11. Activity of the beam constraint over the goal space.

sion tree learner, such as OC1 (Murthy et al., 1994), or by attempting to learn nonlinear region boundaries.

As would be expected, even though our yacht-domain formulation-selection results with C4.5 were nearly optimal for 100 examples, results degrade when given less training data. Although it would be interesting to see if other learning methods would have better small-dataset performance, for any learner we would expect performance to be inferior for small enough datasets. One approach for improving results in such small-dataset cases — as well as in other cases where off-the-shelf learners such as C4.5 may not perform well even if given larger datasets — is to integrate background knowledge into the learning process. One form of background knowledge that is often available, such as in the yacht-design domain, is *modality constraints*. This is knowledge that expresses the modality of the learned class with respect to the attributes. For example, we believe that optimal *beam* is monotonically increasing in wind speed, and monotonically decreasing in heading. We also know that the activity of any constraint of the form $f(x_1, x_2, \ldots, x_n) \leq k$ must be monotonic in k, so, for example, the activity of a cost constraint must be monotonic in the cost threshold. One open question is how such knowledge could be integrated into learning. One approach would be to use such modality constraints to remove from the training data points that violate the constraints (on the assumption that these points are noise). A second approach is to modify the tree induction algorithm so that it will never construct a tree that violates the constraints. A similar approach was used to constrain decision lists in (Clark and Matwin, 1993).

Finally, even after our learning approach is applied, every additional future optimization can serve as an additional training point for the learning. Thus learning methods that can work in an incremental fashion might also prove useful for this task. In addition, it may prove useful to develop methods that select suitable data prior to learning. For example, when there are not enough existing optimizations to achieve adequate learning results, additional optimizations can be performed to generate further training data. Rather than performing these new optimizations for random goals or for a set of goals that span the goal space, one could allow the learner to choose the goals to be used in the new training data. Background knowledge — such as modality constraints — could prove particularly useful in selecting such goals.

Other Setup Choices

We have applied inductive learning to several decisions that must be made when setting up an optimization, including choosing a starting protype and a formulation of the search space, and predicting whether a design goal is achievable. There are other parts of the setup process to which inductive learning might be applicable. For example, one might try to use inductive learning to choose an optimization algorithm, or a good value of the optimizer's stopping tolerance, or a good step size to use in gradient computation, or a good box within which to randomly generate starting prototypes, or a good number of random starting prototypes to generate, or the right level of accuracy to use in the simulator. For each of these decisions, it would need to be determined whether the best choice depends on the design goal. Finally, more experiments need to be done to explore the impact on optimization performance of using inductive learning to simulataneously make multiple choices within the optimization setup problem.

CONCLUSION

Gradient-based methods do not perform well when optimizing designs using simulators that have pathologies. We have described and demonstrated the utility of four techniques that improve optimization performance in such situations by using inductive learning to make decisions when setting up the design optimization. Two of these are methods of choosing an initial prototype for optimization. Prototype selection is especially appropriate in domains such as the yacht domain in which there is a database of previous designs available, and the available simulators are noisy. Prototype synthesis is especially appropriate in domains such as the aircraft domain in which finding a feasible design is difficult. The third technique, feasible goal prediction, is similarly useful in such a domain.

We tested the fouth technique, formulation selection, in both the yacht domain and the aircraft domain. We showed that using this technique can make design optimization faster, because the reformulation reduces the dimensionality of the search space, and more reliable, because the reformulation can make the search space smoother.

NOTES

[1] The cost of generating this table is discussed in the "Cost of Learning" subsection.

[2] CART stands for Classification And Regression Trees

[3] We call the simulator *multidisciplinary* because it contains code to evaluate the design using several engineering disciplines. For example, our aircraft simulator includes weights, aerodynamics, and propulsion.

[4] In 1992, the 12-meter class was replaced with the new America's Cup Class.

[5] This is the boat that won the 1987 America's Cup competition, returning the trophy to the United States after an Australian win in 1983 (Letcher et al., 1987).)

[6] The four operators we chose were *Scale-X*, *Scale-Y*, *Prism-Y*, and *Scale-Keel*. We chose these operators because the results of our earlier work on operator-importance analysis suggested that these are the four most important operators (Ellman and Schwabacher, 1993).

[7] CFSQP stands for "C code for Feasible Sequential Quadratic Programming."

[8] A quadratic programming problem consists of a quadratic objective function to be optimized, and a set of linear constraints.

[9] Some randomly generated designs, which we call "unevaluable points," cannot be simulated, either because the designs are meaningless or because of limitations of the simulator.

[10] Because CFSQP failed to find a feasible point in some of these optimizations, it was not possible to compute the average design quality.

[11] The simulator assumes that there is perfect reefing, so additional sail area can never hurt the yacht's performance.

[12] Because we incorporated the 12-Meter rule into the simulator, we did not need to use it as an explicit constraint.

[13] Although this figure shows only a "snapshot" of the search space for specific values of the other design parameters, we believe that the trend shown in the figure is generally applicable.

[14] Operators containing numerical solvers would probably be more computationally expensive than operators containing the algebraic solutions of the constraint functions, so the CPU time

savings from reformulation would probably be smaller.

[15] We simulated the omniscient learner by performing optimizations using all eight formulations for each goal (as in the "exhaustive" method), and then ignoring the cost of the seven optimizations that turned out not to be best.

[16] Interestingly, according to the t-test, the difference between C4.5 and the omniscient method was not statistically significant, but this just illustrates a limitation of the t-test, since we know that the omniscient method really is better, on average, than C4.5.

[17] Because CFSQP failed to find a feasible point in some of these optimizations, it was not possible to compute the average design quality.

ACKNOWLEDGMENTS

This research has benefited from numerous discussions with members of the Rutgers HPCD project. We especially thank Gerard Richter for his contributions to the formulation selection work, Andrew Gelsey for helping with the cross-validation code, John Keane for helping with RUVPP, and Andrew Gelsey, Brian Davison, and Tim Weinrich for comments on previous drafts of this chapter. This research was supported by the Advanced Research Projects Agency of the Department of Defense under ARPA-funded NASA grant NAG 2-645 and under contract ARPA-DABT 63-93-C-0064. Mark Schwabacher was a Postdoctoral Research Associate at the National Institute of Standards and Technology for a portion of the time that he worked on this research; during that time he was supported by a National Research Council Postdoctoral Research Associateship. This chapter was previously published in a slightly different version as a journal article in *AI EDAM* 12:2; we thank Cambridge University Press for giving us permission to republish it here.

CURRENT ADDRESSES

Mark Schwabacher
NASA Ames Research Center
MS 269-1
Moffett Field, CA 94035

Thomas Ellman
Department of Computer Science
Vassar College
Poughkeepsie, New York 12601

REFERENCES

Bhatta, S. and Goel, A., Model-based design indexing and index learning in engineering design, in *Working Notes of the IJCAI Workshop on Machine Learning in Engineering*, 1995.

Bouchard, E. E., Kidwell, G. H., and Rogan, J. E., The Application of Artificial Intelligence Technology to Aeronautical System Design, in *AIAA/AHS/ASEE Aircraft Design Systems and Operations Meeting*, Atlanta, Georgia, 1988, AIAA-88-4426.

Breiman, L., Classification And Regression Trees, Belmont, Calif.: Wadsworth International Group, 1984.

Cerbone, G., Machine learning in engineering: Techniques to speed up numerical optimization, Technical Report 92-30-09, Oregon State University Department of Computer Science, 1992, Ph.D. Thesis.

Char, B., Geddes, K., Gonnet, G., Leong, B., Monagan, M., and Watt, S., First Leaves: A Tutorial Introduction to Maple V, Springer-Verlag and Waterloo Maple Publishing, 1992.

Choy, J. and Agogino, A., SYMON: Automated Symbolic Monotonicity Analysis System for Qualitative Design Optimization, in *Proceedings ASME International Computers in Engineering Conference*, 1986.

Clark, P. and Matwin, S., Using qualitative models to guide inductive learning, in *Proceedings of the tenth international machine learning conference*, 49–56, Morgan Kaufmann, 1993.

Ellman, T., Keane, J., and Schwabacher, M., The Rutgers CAP Project Design Associate, Technical Report CAP-TR-7, Department of Computer Science, Rutgers University, New Brunswick, NJ, 1992, ftp://ftp.cs.rutgers.edu/pub/technical-reports/cap-tr-7.ps.Z.

Ellman, T., Keane, J., and Schwabacher, M., Intelligent Model Selection for Hillclimbing Search in Computer-Aided Design, in *Proceedings of the Eleventh National Conference on Artificial Intelligence*, 594–599, Washington, DC: MIT Press, Cambridge, MA, 1993.

Ellman, T. and Schwabacher, M., Abstraction and Decomposition in Hillclimbing Design Optimization, Technical Report CAP-TR-14, Department of Computer Science, Rutgers University, New Brunswick, NJ, 1993, ftp://ftp.cs.rutgers.edu/pub/technical-reports/cap-tr-14.ps.Z.

Gelsey, A., Schwabacher, M., and Smith, D., Using Modeling Knowledge to Guide Design Space Search, AI Journal, 101(1-2), 35–62, 1998.

Gelsey, A., Smith, D., Schwabacher, M., Rasheed, K., and Miyake, K., A Search Space Toolkit: SST, Decision Support Systems, 18, 341–356, 1996.

Hoeltzel, D. and Chieng, W., Statistical Machine Learning for the Cognitive Selection of Nonlinear Programming Algorithms in Engineering Design Optimization, in *Advances in Design Automation*, Boston, MA, 1987.

IYRU, The Rating Rule and Measurement Instructions of the International Twelve Metre Class, International Yacht Racing Union, 1985.

Kolodner, J., Case-Based Reasoning, San Mateo, CA: Morgan Kaufmann Publishers, 1993.

Lawrence, C., Zhou, J., and Tits, A., User's Guide for CFSQP Version 2.3: A C Code for Solving (Large Scale) Constrained Nonlinear (Minimax) Optimization Problems, Generating Iterates Satisfying All Inequality Constraints, Technical Report TR-94-16r1, Institute for Systems Research, University of Maryland, College Park, MD, 1995.

Letcher, J., The Aero/Hydro VPP Manual, Southwest Harbor, ME: Aero/Hydro, Inc., 1991.

Letcher, J., Marshall, J., Oliver, J., and Salvesen, N., Stars and Stripes, Scientific American, 257(2), 1987.

Murthy, S., Kasif, S., Salzberg, S., and Beigel, R., A System for Induction of Oblique Decision Trees, Journal of Artificial Intelligence Research, 2, 1–32, 1994.

Orelup, M. F., Dixon, J. R., Cohen, P. R., and Simmons, M. K., Dominic II: Meta-Level Control in Iterative Redesign, in *Proceedings of the National Conference on Artificial Intelligence*, 25–30, St. Paul, MN: MIT Press, Cambridge, MA, 1988.

Papalambros, P. and Wilde, J., Principles of Optimal Design, New York, NY: Cambridge University Press, 1988.

Powell, D., Inter-GEN: A Hybrid Approach to Engineering Design Optimization, Ph.D. thesis, Rensselaer Polytechnic Institute Department of Computer Science, Troy, NY, 1990.

Powell, D. and Skolnick, M., Using genetic algorithms in engineering design optimization with non-linear constraints, in *Proceedings of the Fifth International Conference on Genetic Algorithms*, 424–431, Univeristy of Illinois at Urbana-Champaign: Morgan Kaufmann, Los Altos,

CA, 1993.

Quinlan, J. R., Learning logical definitions from relations, Machine Learning, 5, 239–266, 1990.

Quinlan, J. R., C4.5: Programs for Machine Learning, San Mateo, CA: Morgan Kaufmann, 1993.

Ramachandran, N., Langrana, N., Steinberg, L., and Jamalabad, V., Initial Design Strategies for Iterative Design, Research in Engineering Design, 4, 159–169, 1992.

Rasheed, K. and Hirsh, H., Learning to be Selective in Genetic-Algorithm-Based Design Optimization, Artificial Intelligence for Engineering Design, Analysis, and Manufacturing, 13, 1999.

Rogers, D. and Adams, J., Mathematical elements for computer graphics, McGraw-Hill, second edition, 1990.

Schwabacher, M., The Use of Artificial Intelligence to Improve the Numerical Optimization of Complex Engineering Designs, Technical Report HPCD-TR-45, Department of Computer Science, Rutgers University, New Brunswick, NJ, 1996, Ph.D. Thesis. http://www.cs.rutgers.edu/~schwabac/thesis.html.

Schwabacher, M. and Gelsey, A., Intelligent Gradient-Based Search of Incompletely Defined Design Spaces, Artificial Intelligence for Engineering Design, Analysis and Manufacturing, 11(3), 199–210, 1997.

Sycara, K. and Navinchandra, D., Retrieval Strategies in a Case-Based Design System, in C. Tong and D. Sriram, editors, Artificial Intelligence in Engineering Design (Volume II), 145 – 164, New York, NY: Academic Press, 1992.

Tong, S. S., Coupling Symbolic Manipulation and Numerical Simulation for Complex Engineering Designs, in International Association of Mathematics and Computers in Simulation Conference on Expert Systems for Numerical Computing, Purdue University, West Lafayette, IN, 1988.

Tong, S. S., Powell, D., and Goel, S., Integration of Artificial Intelligence and Numerical Optimization Techniques for the Design of Complex Aerospace Systems, in 1992 Aerospace Design Conference, Irvine, CA, 1992, AIAA-92-1189.

Williams, B. and Cagan, J., Activity Analysis: The Qualitative Analysis of Stationary Points for Optimal Reasoning, in Proceedings of the Twelfth National Conference on Artificial Intelligence, 1217–1223, Seattle, WA: MIT Press, 1994.

CHAPTER 5

Automatic Classification and Creation of Classification Systems Using Methodologies of "Knowledge Discovery in Databases (KDD)"

Prof. Dr. -Ing. Dr. h. c. Hans Grabowski
gr@rpk.mach.uni-karlsruhe.de
Institute for Applied Computer Science in Mechanical Engineering (RPK),
University of Fridericiana, TH Karlsruhe, Kaiserstr. 12, D 76128 Karlsruhe

Dr. -Ing. Ralf-Stefan Lossack
lossack@rpk.mach.uni-karlsruhe.de
Institute for Applied Computer Science in Mechanical Engineering (RPK),
University of Fridericiana, TH Karlsruhe, Kaiserstr. 12, D 76128 Karlsruhe

Dipl.-Ing. Jörg Weißkopf
weiskopf@rpk.mach.uni-karlsruhe.de
Institute for Applied Computer Science in Mechanical Engineering (RPK),
University of Fridericiana, TH Karlsruhe, Kaiserstr. 12, D 76128 Karlsruhe

ABSTRACT

Automatic classification was needed in industrial practice for a long time. At present this topic is addressed in the context of the project "Automatic Classification of Products". Objectives of this project are the conception and implementation of a program system for automatic classification of products. A software system was developed to automatically compute classifications and encodings on basis of classification schemes, that can be flexible specified. These classification schemes at present have to be specified manually by using features, feature combinations, and calculation rules.

In this paper concepts and results of the project will be described. Furthermore, it will be discussed how methodologies of KDD (such as Data Mining and Cluster Analysis) can be used to automate the generation of classification schemes. These classification schemes should then be used as input data for the software system for automatic classification of products. The entire process from preparation of the classification schemes up to the classification activity can be automated this way.

D. Braha (ed.), Data Mining for Design and Manufacturing, 127–143.

INTRODUCTION

Competition between enterprises will increase in the future, caused by higher quality and performance requirements of tomorrow's customers. Even today enterprises encounter this development with a stronger differentiation of their products. This fact causes higher development and production cost, the development of new innovative products requires more and more time and therefore, resources must be saved elsewhere.

Within the product development process the direct access of existing data gains strategic importance by increasing information demand and volume. Thus, the classification of products is a suitable possibility to supply the product development process with already available data by falling back over compressed and high order information. Within the cooperate project PAK (*"Produkte Automatisch Klassifizieren"* - Automatic classification of products) a concept and system was developed, which enables the access to experience by automatic classification of different data sources. The system supports rationalizing the technical and organizational areas of an enterprise.

The objective of the automatic classification is to enforce the re-use of design solutions that were created in the past. Reusing existing design solutions can reduce time and cost of the entire product development process by 10 – 50%. It is difficult to quantify the benefit exactly, in general the benefit is attained by [1]:

- Replacement of manual or semi-automatic classification activities reduces personnel cost.
- Re-use of identical or similar design solutions shortens the "time to market" process and reduces production cost.
- The extraction of semantically high-quality information from few product attributes enables a quick location of any documents.
- Automatic recording of conventional and interpreted product data simplifies long-term archiving.
- Flexible specification of product features enables quick search in data archives.

AUTOMATIC CLASSIFICATION

Assumptions

Some prerequisites have to be fulfilled until a system for automatic classification can be used. First of all, attributes relevant for product classification and relations between these attributes have to be determined. In the past several generic classification systems [5], [6] were developed, e.g.

OPITZ [2] and Zimmermann-ZAFO [3]. If such a generic classification system should be used, the attributes relevant for classification are already determined by this system.

The decision whether the system for automatic classification should be used depends on the following factors:

- What's the amount of work that has to be spent on defining the classification system? Perhaps an existing classification system can be used or must only be slightly modified. Or is it necessary to create a complete new classification system? The PAK-System supports the modification and creation of classification systems with easy to use graphical tools.
- Which information is needed by the user? What's the amount of work to achieve this information manually – if this is possible at all? The PAK-System presents to each user exactly the information he needs to make a decision.
- In which way and when should the information be presented to the user? Is it e.g. necessary for a designer to leave his CAD-system to search for relevant information? The PAK-system presents the relevant information to each user in his own working environment as soon as possible.
- Which effort is necessary for classification activities? The PAK-system for example is able to automatically fill tabular layouts of article characters ("Sachmerkmalleisten") within a PDM-system.

Considering these facts, the benefit of the PAK-system can be estimated easily.

Concept for a System for Automatic Classification of Products

Figure 1 illustrates the simplified architecture of the PAK-system. In order to distinguish between the data to be classified and the data that is needed for the classification process the terms user data and meta data were introduced.

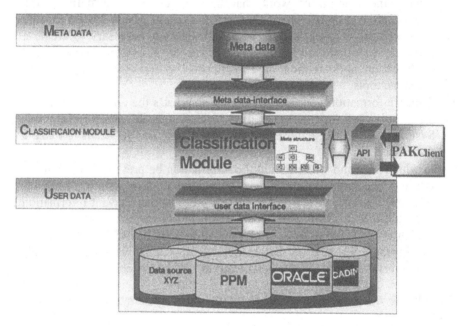

Figure 1: Architecture of the PAK-System

User Data

The term user data describes all user-specific data that can be classified. Depending on how and where the data is stored in information systems, many different user data sources exist. For example an information system for process planning contains a different pack of data than a CAD-system. Currently, the PAK-system supports the following groups of IT-systems:

- CAD-systems: Pro/ENGINEER, ACIS
- EDM/PDM-systems: CADIM/EDB
- ERP-systems: SAP R/3
- Database systems: Oracle, mSQL, MSAccess, Postgres, Centura, mySQL
- Other data sources: MSExcel, text documents (ASCII, MSWord), XML documents, DIN V 4000-100 documents

All data that is stored in one of these IT-systems can be accessed and classified. It is easy to implement interfaces for further IT-systems. The interfaces do not need very "intelligent" algorithms, as the interpretation system is provided by the PAK-system itself.

Meta Data

Meta data provides a view on the information, that describes a product, in a structure and information density that is suitable for classification [4]. The meta data that is needed for classification – such as attributes, classes, class hierarchies and classification schemes – is held in an own database and is accessible to other system modules via an internal interface:

- Product features, that are stored in the user data, are mapped to attributes within the PAK-system. Synonymous attributes of one or more objects can be organized and pooled in attribute structures. Furthermore, it is possible to create new attributes by defining calculation rules using other attributes, constants, variables and many different operator types such as math operators, boolean operators, compare operators and so on. For example the length and width of objects are stored in a user data source, but not the area, that may be needed for classification. So it is possible to create an attribute called area and to define the calculation rule area=length*width for it.
- Basis for the creation of a classification system is the possibility to define classes. The most important characteristic of a class is the so called defining rule. By evaluating this defining rule the classification module decides whether an object stored in a user data source has to be assigned to this class or not. The assignment of an object to a class does not have to be "sharp". The PAK-system allows to handle unsharp so called fuzzy-attributes. So qualitative terms can be used to define classes. The assignment of an object to a class is described by a fuzzy value between zero and one that describes, how exact the object meets the classification criteria specified in the defining rule of the class.
- The specified classes can be hierarchically structured in class hierarchies. Within a hierarchy the characteristics of a class are inherited to the subclasses. This means, that an object has to meet the classification criteria of a class and the criteria of all superclasses to be assigned to that class.

Class hierarchies are flexible classification schemes that can be easily modified or extended, if required. By assigning key numbers to the classes of a hierarchy the creation of the classification system is completed.

Automated Creation of Classification Schemes

In many enterprises a huge amount of data exists, that is even growing daily. All this data has to be analyzed in order to extract *classification relevant features*. Therefore, the creation of an enterprise specific classification system is accompanied by a large effort. Whether a feature is classification relevant or not is often decided subjectively. In fact, classification relevant features are often the result of a combination of other features, for example by a mathematic operation (e.g. area=length*width). Such correlations are very often not obvious and can be easily overlooked when creating a classification system manually.

The combination of classification relevant features to classes is a subjective process, too. Particularly the delimitation of one class against another class is difficult. Also the definition of the granularity of the classes is difficult, this is particularly a problem if many thousands of objects are assigned to one class.

Finally, the classes have to be structured in order to build a complete classification system. Here, decisions have to be made in which way the classes have to be structured (hierarchical, non-hierarchical, meshed and so on). Inheritance structures have to be defined, dependencies between classes have to be considered. This process also necessitates the analysis of potentially huge databases in order to ensure the creation of an effectively useable classification system.

KDD (Knowledge Discovery in Databases) – Methodologies

Today, many methodologies are developed and in use to analyze huge databases, to identify correlations and similarities, and to simplify the access to information in this way. The most important methodologies of automatic identification of features and classification are the methodologies of data mining and cluster analysis.

Data Mining

The term KDD is used as a genus for the term data mining. The basic topic of data mining is the identification of patterns in data, whereas KDD enfolds the entire process, including preliminary works and evaluation of identified patterns.

Data mining is used to identify patterns in existing data. The decision whether the results contain interesting or useful knowledge about correlations in the data is part of the entire process, and often subjective, then manual judgment is needed [7].

Cluster Analysis

Cluster analysis are mathematic-statistical, multivariate methods that are used for computer aided classification of objects. They are suitable for analyses, comparisons and groupings of objects in a multi dimensional vector space.

The focus of cluster analysis is to build classes of objects. Objects within a class should be as similar as possible, while the classes among themselves should differ as much as possible. The specific features of objects are represented by vectors in a multi dimensional space, each object is then represented by a point in this vector space. The distance between the points is the measure for group building. For classification, the relevant features, that have to be the same for all objects, are selected, and data and distance matrices are created:

- Data matrix: As features can appear with different scaling (metric, nominal, ordinal) and as they can have different value ranges, a feature transformation or standardization is necessary in order to make them comparable. After transforming and weighing the features results in a n x m matrix. In this matrix for each element "n" the characteristic "m" attributes are stored.
- Distance matrix: For any object similarity coefficients have to be calculated. They define the degree of similarity between objects.

The result of a cluster analysis can be displayed in a *dendrogram.*

Automated Identification of Classification Relevant Attributes

In classic cluster analysis for each *classification relevant feature* a vector is defined. All vectors span a n-dimensional vector space. As already described, each object is now represented by a point in this vector space. The number and types of relevant features have to be known before starting the cluster analysis. In the approach described here, *all* features, that can be determined, are used to build the vector in a first step. This number of features now has to be reduced to the *classification relevant features*, and it has to be extended by further *classification relevant features*.

Determination of Non Classification Relevant Features

This points that represent objects are projected to the axis of the vector space. The distribution of the projected points allows conclusions about the classification relevance of this feature (see Figure 2).

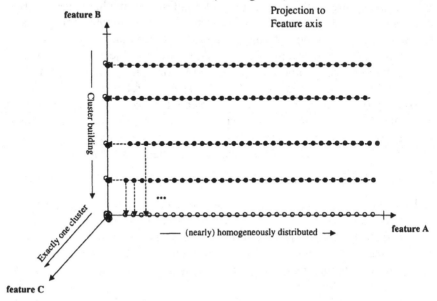

Figure 2: Determination of non classification relevant features

Determination of Further Classification Relevant Features

When only the distribution along the single axis is analyzed while determining the classification relevance of a feature, important information for classification can be lost. Figure 3 illustrates an example, where all objects are distributed homogeneously both concerning the axis of feature A and feature B. But a projection of the points on the plane spanned by feature A and B shows, that the distribution of the points on that plane is not homogeneous.

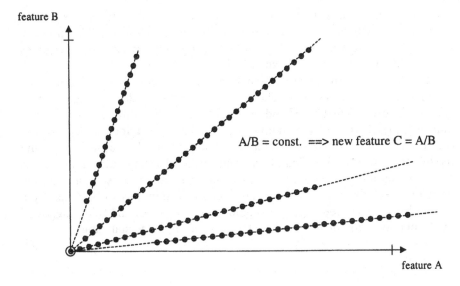

Figure 3: Determination of further classification relevant features

Using suitable methods, e.g. of pattern matching, it can be determined, that the points are distributed linearly on straight lines through the origin. That leads to the conclusion, that the combination of the features A and B results in a feature C that is classification relevant. The type of combination is indicated by the pattern. In this example, the contribution is linear through the origin, so all points can by described by the function $y = a_n * x$ where n is the number of straight lines. The correlation can be achieved by building the inverse function, in this case: A/B nearly constant (for each straight line in the A-B plane). That means, along an axis for a feature C = A/B, that is added to the vector space, n clusters will be found. A new, with high probability classification relevant feature C is identified.

Automated Creation of Classes and Classification Schemes

A conventional cluster analysis results in a number of object groups. The cluster membership of an object represented by a point p in the vector space is determined by the distance of the point to the center of the cluster. There exist several methods to compute this distance and the position of the cluster center. A cluster, and therefore its center, is changed by each point assigned to that cluster. So its possible – at a definite time – to determine exactly the center and geometry of a cluster. Therefore, the first approach to define a class is the geometry of the cluster volume V. For each determined cluster C_n

with the cluster volume V_n a class K_n is defined. Each of these classes has the defining rule, where p is the point representation of an object to classify. The evaluation of this defining rule results in a boolean value, that determines whether an object has to be assigned to that class or not.

The mathematical representation of a cluster volume in a n-dimensional vector space is very complex. Therefore, the defining rules, that are derived from the cluster volumes, - if understandable at all - are difficult to read for a human being. This results in an unwanted blackbox behavior, the decisions of the classification module cannot be comprehended. Furthermore, any modification to the classification system or the defining rule of a class is nearly impossible. Such modifications will be normally necessary, as the final decision about the structure of a classification system has to be taken by a human being. Therefore, it is inevitable, that the classes and rules suggested by the classification module have to be readable and understandable.

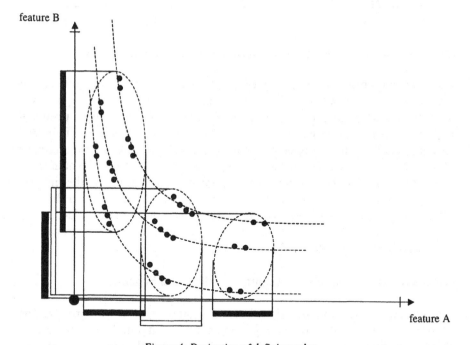

Figure 4: Derivation of defining rules

Figure 4 illustrates an approach to derivate comprehensible defining rules. For each feature and for each cluster an interval of valid values is determined. If a point representation of an object to classify lies inside all intervals of a cluster, it can be assigned to that class. Here, the cluster geometries are approximated by enfolding, cubic vector spaces. This simplification will already lead to satisfying results in many cases.

An optimization of the precision of this method can be achieved by combining several cubic spaces to approximate a cluster geometry (see Figure 5).

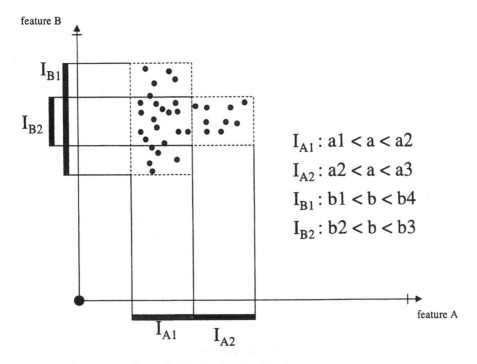

$I_{A1}: a1 < a < a2$

$I_{A2}: a2 < a < a3$

$I_{B1}: b1 < b < b4$

$I_{B2}: b2 < b < b3$

Figure 5: Combination of cubic cluster spaces

The class derived from the cluster shown in Figure 5 in this case has the defining rule:

$$(P \in I_{A1} \text{ AND } P \in I_{B1}) \text{ OR } (P \in I_{A2} \text{ AND } P \in I_{B2}).$$

In this manner it is possible to describe cluster geometries by comprehensive rules as exact as wanted.

When deriving defining rules from cluster geometries, the sharpness of the cluster borders is a problem. The cluster border is determined by the envelopment of a group of points that are nearby to each other, but finally the shape of the bounding geometry is chosen arbitrarily. Therefore, a strict exclusion of a point that lies immediately outside the cluster geometry is not tolerable. The solution is to build "unsharp" cluster geometries, so called *fuzzy-clusters*.

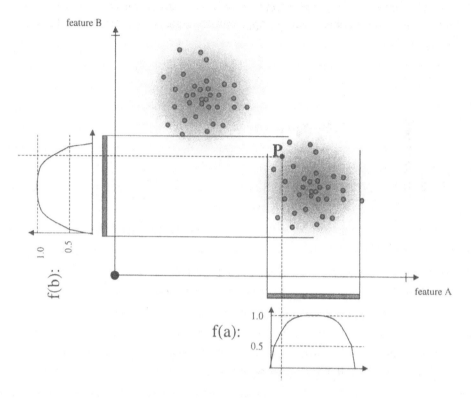

Figure 6: Definition of fuzzy-clusters

Figure 6 shows an example for the unsharp definition of a cluster border. For each feature a *fuzzy function* is defined over the interval of valid feature values. The membership of a point p to a cluster is computed by:

$$f(a) \;*\; f(b) \;=\; 0.7 \;*\; 0.8 \;=\; 0.56.$$

This means, that the point p can be assigned to the class derived from this cluster with a probability of 56%.

Hierarchical Structuring of Classes

In order to create a complete classification system, it is necessary to correlate the classes to each other. There exist several possibilities to structure classification systems: hierarchical, non-hierarchical, hybrid (combining hierarchical and non-hierarchical parts), hierarchical with multiple inheritance or meshed. The industrial practice has shown that hierarchical classification systems are accepted best, because they provide sufficient possibilities of structuring, and the handling of even numerous classes is ensured. Therefore, an approach will be described, how classes found on basis of cluster analysis can be structured hierarchically.

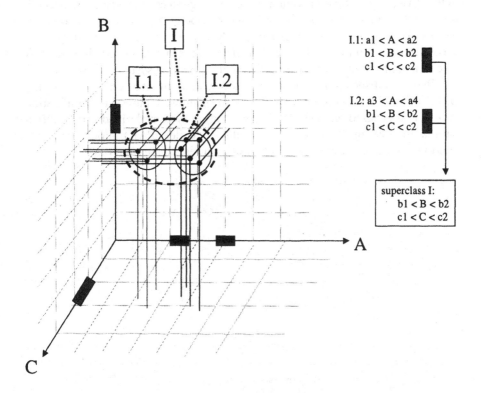

Figure 7: Hierarchical structuring using cluster projection

Figure 7 shows an example where two classes *I.1* and *I.2* as well as their defining rules were determined. It is possible to determine, that both underlying clusters fuse to one cluster in a vector space reduced by feature vector *A*. This is achieved y an iterative reduction of the dimensionality of the vector space. Deriving a class from this cluster leads to a class *I* that is a superclass of *I.1* and *I.2*.

Example for Automatic Classification of Geometry and PDM Data

In this section the use of the PAK-system will be described. On basis of an example, it will be shown how the experience of an enterprise can be provided to a designer, using the knowledge stored both in geometry files and a PDM-system. For the designer, first of all geometric similarity is relevant. In this case, a classification system concerning geometric principles is needed. But in different departments of an enterprise other aspects may be relevant, for example production engineering aspects. Therefore, more than one classification system may be needed in one enterprise. The PAK-system is able to manage at the same time as many classification systems as needed.

One established possibility for classification of geometric parts is the OPITZ classification system [3]. Figure 8 (E) shows a part of a classification system that is based on the OPITZ classification system, defined and modified with a module of the PAK-system, the so called PAKEditor. The other windows (A-E) show the specification of features, classes and rules.

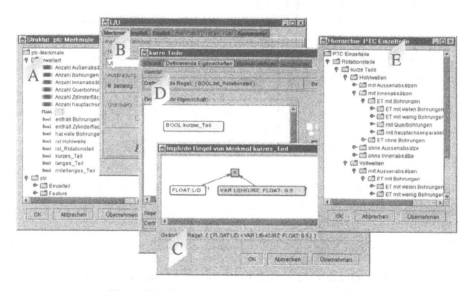

Figure 8: Definition of a classification system with PAK

The PAK-system is able to classify parts modeled in a 3D-CAD-system automatically according to the hierarchy based on the OPITZ classification system. Two modes are supported, the offline and the online mode. In the offline mode, several completely designed parts can be classified. In the online mode the PAK-system observes the designer during the design process and registers all changes made at the active part that is designed actually. Any change automatically results in a classification of the active part. If the PAK-system determines, that the part can be assigned to a class with no subclasses (a leaf of the hierarchy), the designer will be informed. All parts that were assigned to that class in former classification processes will be presented to the designer as geometrically similar design solutions (see Figure 9).

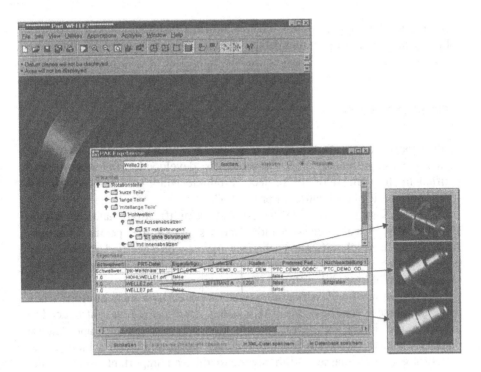

Figure 9: Geometrically similar design solutions found by the PAK-system

The designer now has to decide, whether one of the presented parts matches the requirements of his current design task, or if the development of a new part is necessary. Basically the re-use of existing design solutions is preferable, as each new design causes high cost in later stages of the product development process. In general, only geometric features are not sufficient to decide, which of the presented parts is suited best regarding cost incurred in later stages of the product development process. Therefore, the designer needs more information about existing design solutions, for example about standardized parts, extern parts, cost, if the part is in store, if it is a preferred part, and so on. Such kind of information is normally stored in EDM/PDM/PPS-systems. The PAK-system is able to compute classifications across system borders, that means, all relevant information about an object is considered, even if it is stored in different IT-systems. The result is also shown in Figure 9, where a couple of additional information for each part is presented to the designer.

In this way, it is possible to provide all relevant information to the designer that he needs for optimal usage of experience. The effort the designer has to take is minimal, as the initiative comes completely from the PAK-system. Therefore, the designer's debt to inform himself about existing design solutions has been changed to a debt of the classification system to inform the designer.

CONCLUSIONS

In this contribution, a concept for a system for automatic classification of products was presented. This concept was developed, implemented and verified in the context of the project "Produkte Automatisch Klassifizieren (PAK, automatic classification of products)".

The PAK-system is able to compute classifications and encodings automatically, based on classification schemes that can be specified flexibly. At present, these classification schemes have to be specified manually by using features, feature combinations and calculation rules. The data to be classified can be stored in different IT-systems.

The creation of a classification system is primarily a manual effort when using the present version of the PAK-system. An approach was introduced to automate the identification of classification relevant features and feature combinations, to create suggestions for class definitions, and a hierarchical structuring of this classes. With regard to this, an important aspect is the comprehensiveness of the rules that are created. The approach is based on methodologies of data mining and cluster analysis that will be used for classification tasks in the context of mechanical engineering in this way.

This approach will be elaborated, implemented and verified in a future research project.

REFERENCES

[1] Hain, K.: Automatische Gewinnung von Merkmalen und Klassifizierungseigenschaften für Produkte auf Basis eines integrierten Produktmodells, Dissertation TU Karlsruhe, Shaker Verlag, 1997

[2] Opitz, H.: Werkstückbeschreibendes Klassifizierungssystem (Teil 1), Verschlüsselungsrichtlinien und Definitionen zum werkstückbeschreibenden Klassifizierungssystem (Teil 2). Essen, Girardet, 1971

[3] Zimmermann, D.: Eine allgemeine Formenordnung für Werkstücke Gestaltung, Handhabung und Rationalisierungserfolg; Technischer Verlag Günter Grossmann GmbH, 1967 Stuttgart-Vaihingen

[4] Grabowski, H.; Weißkopf, J.: Metadatenkozept zur strukturierten Erfassung von Produktmerkmalen für die automatische Produktklassifikation; 9. Symposium, Fertigungsgerechtes Konstruieren, 15. und 16. Oktober 1998 in Schnaittach

[5] Wiendahl, H.-P.; Heuwing, F.-W.: Methode zur Klassifizierung von produktunabhängigen Baugruppen, Betriebstechnische Reihe RKW/REFA, Beuth Vertrieb GmbH, 1973

[6] Hahn, R.; Kunerth, W.; Roschmann, K.: Die Teileklassifizierung –Systematik und Anwendung im Rahmen der betrieblichen Nummerung". RKW Handbuch der Rationalisierung, Gehlsen, 1970

[7] K.M. Decker, S. Focardi : Technology Overview: A Report on Data Mining, TechReport TR-95-02, CSCS-ETH, 1995

CHAPTER 6

Data Mining for Knowledge Acquisition in Engineering Design

Yoko Ishino
okinaka@usc.edu
IMPACT Laboratory, Denney Research Building, Suite 101, University of
Southern California, Los Angeles, CA90089-1111

Yan Jin
yjin@usc.edu
IMPACT Laboratory, Denney Research Building, Suite 101, University of
Southern California, Los Angeles, CA90089-1111

ABSTRACT

Recently knowledge capturing in the design process using data mining
techniques has attracted attentions from researchers. In this paper, we focus
on how designers apply their knowledge and find ways to proceed with their
design in given design situations. This design *know-how* is the knack for a
design to be successful. In order to automatically capture the design *know-
how* through design processes without overburdening designers, we 1) -
introduced an object-oriented CAD system for data gathering, 2) - proposed a
three-layer design process model to represent generic design processes, and 3)
- developed a method, called *Extended Dynamic Programming* (EDP), to
extract *know-how* knowledge from the gathered design process data. The
effectiveness of the proposed approach has been demonstrated by a gear
design prototype system Gear-CAD.

D. Braha (ed.), Data Mining for Design and Manufacturing, 145–160.
© 2001 *Kluwer Academic Publishers. Printed in the Netherlands.*

INTRODUCTION

Designing modern engineering artifacts such as car and aircraft needs vast amounts of information and knowledge. Generally, design knowledge is acquired through experiences. Since the knowledge has its own life cycle, we always make an effort to capture new knowledge and keep it fresh in order to maintain the quality of design. Though the value of capturing and utilizing design knowledge has long been recognized and research has been carried out to date, capturing engineering knowledge is still a challenge since forcing designers to record their knowledge would interrupt their natural thinking process and become an unacceptable burden for them. Moreover, design knowledge is tacit and embedded in the design process in most cases. It is difficult for a designer to express his or her design knowledge fully and explicitly. Developing ways to acquire engineering knowledge without disturbing designers' normal design process is the key to achieve successful knowledge capturing.

We focus on *know-how*, which is the knack of a design in a design process. Our goal is to automatically capture how designers proceed with their design in various design situations during design to accomplish their design goals. This captured design process knowledge not only shows how a design is done, but also provides a basis for us to explore why the designers did their design in certain ways.

One major problem we face in achieving our goal is that design processes are often ill-structured and *ad hoc*, and vary greatly depending on design situations. To address this problem, we need to develop a general yet powerful design process model to represent design processes. This model should serve as a basis for identifying and gathering the useful data, and generating *know-how* knowledge through data mining.

We have introduced a way for data gathering through an object-oriented CAD system, and proposed a *three-layer design process model*. And finally, we have developed a method called *Extended Dynamic Programming* (EDP) to acquire *know-how* knowledge through data mining. Our knowledge acquisition framework has been successfully tested in a case study of designing a double-reduction gear system.

OBJECT-ORIENTED CAD SYSTEM

Our objective of research is to automatically capture *know-how* knowledge. In order not to interfere with a designer's normal design process, we take an action-based knowledge capturing approach and focus on the data that can be obtained through observing design activities using a CAD system. To make it possible to collect needed data, the CAD system we use must be an object-oriented system in which every geometric component has its own information as an object. For example, when a rectangle is registered as a robot arm in the CAD, it has several logical features such as geometric location, material, x-length, y-length and so on. While a designer does his or her design through such an object-oriented CAD system, all actions he or she takes during design process can be recorded by following the change in the features of objects.

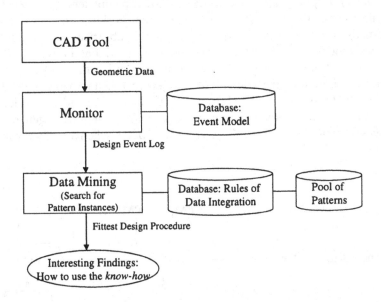

Figure 1: Architecture of Data Mining for Know-How Capturing

This approach alleviates the problem of interfering in the design process, but creates a new problem in managing the large volume of information recorded. The method to elicit meaningful chunks of knowledge from a pool of enormous data must be devised.

Our system architecture for knowledge capture is depicted in Figure 1. The monitor component receives the report from the object whose feature is changed, and matches it with the event ID from the database that stores every design event. Data-Mining component searches the similar pattern for *know-hows* that are registered beforehand. Finally, the knowledge, how to proceed with design, is clarified. The mechanism of the Data-Mining module is described in a later section.

ENGINEERING DESIGN PROCESS

Three-layer Design Process Model

Based on our observation of designers' behaviors, most designers make tentative design decisions and create a prototype design at first. They then repeatedly adjust and refine certain design parameters to meet specific requirements. Finally they reach the final design after evaluating all alternatives that have been explored. From this observation, we found that - design processes can generally be viewed as trial-and-error - and some primitive actions of design often occur in a specific order to fulfill a certain aim, *e.g.*, shortening the x-length of B-object and then moving down B-object are made to decrease the abrasion of B-object. This idea has led us to introducing a *three-layer design process model* to represent general design processes. The *three-layer design process model* is schematically illustrated in Figure 2.

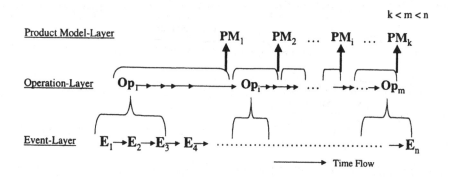

E: each event caused by a designer,
   e.g., "Change length A from 15 to 30," "See document B"

Op: meaningful operation which is a cluster of plural events,
   e.g. "Decrease the weight of the object," "Increase the strength of the arm"

PM: Product Model (design prototype) which represents design alternatives

Figure 2: Three-Layer Design Process Model

The *three-layer design process model* represents generic design processes based on three layers of information, namely, Event-Layer, Operation-Layer, and Product Model-Layer. Event-Layer captures primitive level design events that are generated by designers through operating a CAD system. For example,

"Change length A from 15 to 30" or "See document B," illustrated as "E" in Figure 2 is an event occurred at Event-Layer. Operation-Layer represents higher-level design operations that reflect meaningful design actions. Elements at Operation-Layer are design operations that can be generated by reasoning based on multiple design events found at Event-Layer. "Decrease the weight of the object" and "Increase the strength of the arm" illustrated as "Op" in Figure 2 are examples of design operations. The elements in Product Model-Layer are design alternatives, which are generated from multiple design operations, and are illustrated as "PM" in Figure 2. An element in Product Model-Layer is called a design alternative.

Based on our model of design processes, the goal of designers can be considered as to create a final product model. To do so, designers intentionally plan and perform sequences of operations ("Op"s). Although the sequences of operations a designer performed cannot be observed directly, we can capture the events ("E"s) that were generated while the designer performing the operations. We consider that how to proceed with design can be substituted for how operations appear in the design process. We regard knowing the sequences of operations as obtaining *know-how* knowledge. Especially, how to appear of the key operations should be the knack of successful design.

Data Mining of *Know-how*

Designers (or analysts) can produce in advance some sequences of design events that are considered to be key operations and have important roles for the design. Finding how the sequences appear in the design process can help us understand the design knack. By treating the sequences as templates, we found the similarities between this problem and DNA matching problem in the biological field, as depicted in Figure 3.

DNA consists of four kinds of base, A, T, G, and C. Searching the most matching spot with a DNA fragment is a typical problem of a pattern matching. We can use this analogy to capture the knowledge. In the case of design process, design events are primitive actions that are reported by a monitor module through a CAD system. For example, "Change length A from 15 to 30" or "See document B," illustrated as "E" in Figure 3 is an event. Design operation represents higher-level design operations that reflect meaningful design actions, which can be generated by gathering and analyzing design events. "Decrease the weight of the object" or "Increase the strength of the arm" that is illustrated as "Op" in Figure 3 is an example of design operations.

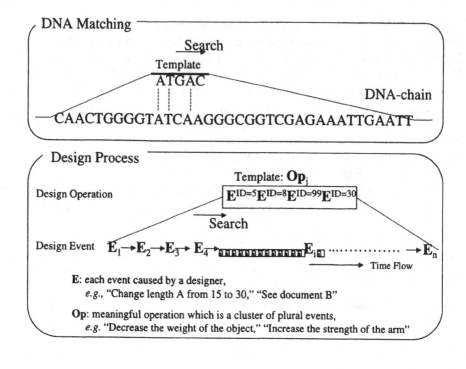

Figure 3: Analogy of Data Mining

Here, several problems exist to analogically apply a usual method of pattern matching to the design problem. The first problem is that real data of design event log has lots of noise that disturb the pattern matching. The second problem is that more than one template often compete to one another. Then, we have developed a method called *Extended Dynamic Programming* (EDP) to address these problems.

EXTENDED DYNAMIC PROGRAMMING APPROACH (EDP)

EDP for Capturing *Know-how*

Originally, Dynamic Programming (DP) is an approach to solving sequential decision problems developed by Richard Bellman in 1957. The simplest dynamic programming context involves an *n*-step decision-making problem, where the states reached after *n* steps are considered terminal states and have known utilities. The main concept of DP is that choosing the best[1] state in each step leads to the optimum final state. Recently, DP has been applied to solving various pattern matching problems, *e.g.*, speech recognition (Sakoe and Chiba 1990), image recognition (Chikada, *et al.* 1999), and bio-informatics (Krogh 1994). EDP is a method based on DP theory and modified to be suitable for this design problem.

When more than one prospective solutions exist for a pattern matching, EDP is able to detect the most reasonable solution. Here, "the most reasonable solution" means the solution that has the highest value for a user based on a given objective criterion. Moreover, EDP enables us to detect not only the same sequence as the template but also approximately similar ones. It allows sequential errors to some extent, *e.g.*, deletion, insertion or exchange of elements.

In EDP, every template has two elements. One is the definition of a one-to-one relationship between a typical event-sequence to grasp and an operation ID, and the other is the *value-score* of the template for a user. The value-score is determined by a user's subjective thought on a design problem. The examples of templates are shown in Table 1.

Table 1: Template Examples for EDP Approach

Template	Rules	Operation	Contents of Operation
T151	If it makes a combination (n=2) of {E91,E92} under the condition F=1, Then Op151, where Value-Score=0.6	Op 151	Move Z1, Formation=1
T152	If it makes a combination (n=2) of {E91,E92} under the condition F=0, Then Op152, where Value-Score=0.6	Op 152	Move Z1, Formation=0
T153	If it makes a combination (n=2) of {E97,E98} under the condition F=1, Then Op153, where Value-Score=0.6	Op 153	Move Z4, Formation=1
T154	If it makes a combination (n=2) of {E97,E98} under the condition F=0, Then Op154, where Value-Score=0.6	Op 154	Move Z4, Formation=0
T155	If it makes a combination (n=2, 4) out of {E91,E93,E95,E97}, Then Op155, where Value-Score = 0.2 * n	Op155	Adjust Y-Position of Gears (n)
T156	If it makes a combination (n=2, 4) out of {E92,E94,E96,E98}, Then Op156, where Value-Score = 0.2 * n	Op156	Adjust X-Position of Gears (n)

Algorithm of EDP

The EDP is schematically illustrated in Figure 4. In order of value-score, the pattern matching is executed. Figure 5 describes the algorithm of EDP.

The algorithm shown in Figure 5 works as follows. The main function of EDP is **CLUSTERING_BY_ EDP**, which receives the delivered event-sequence and the set of templates as its input and returns the corresponding operation-sequence as its output. First of all, the template that has the highest *value-score* of all templates is chosen and put into the variable, *probe*, by UPDATE-HIGHEST-VALUE-TEMPLATE function. Then, EDP-MANIPULATION function, which is discussed below, extracts the event-sequence whose mismatching score is less than the threshold by applying *probe* to *events*, and puts the result event-sequence into the variable, *extractant*. UPDATE-TEMPLATES function omits *probe* from *templates*. If *extractant* is not null, then TRANSLATE function obtains a corresponding operation ID in *operation* by referring *probe* and *extractant*, and this *operation* is added to *operation-list*, and *events* are updated where UPDATE-EVENTS function omits *extractant* from *events*. The above-mentioned actions from UPDATE-HIGHEST-VALUE-TEMPLATE to UPDATE-EVENTS are repeated until *events* or *templates* is empty. Finally, a set of operations can be acquired.

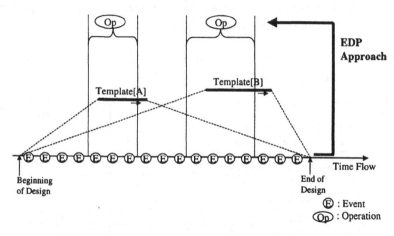

Template[A] has *value-score* **a**, and template[B] has *value-score* **b**.
If **a >= b**, then the search using template[A] has priority.
If **b > a**, then the search using template[B] has priority.

Figure 4: Extended Dynamic Programming Approach (EDP)

The algorithm of EDP-MANIPULATION, the center function in CLUSTERING_BY_ EDP works as follows. To begin with, the notation, $S = \text{EDP} - \text{MANIPULATION}\,(T, P)$, indicates that the sequence S, which is approximately similar to the sequence P, is extracted out of the sequence T. Since the elements in a sequence are arranged in a time series order, from left to right, their order is retained. Then the following definitions are introduced.

```
function CLUSTERING_BY_ EDP(events, templates)    return set of operations
    inputs: events ; sequence of events
            templates ; set of operation-templates, each template has two elements,
                        event-operation rule and its value-score
    variables: probe ; a specific template
            extractant ; an extracted sequence of events
            operation ; an operation risen from the event sequence
            operation-list ; a list of operation

    loop do until templates or events is empty
            probe  ⟵  UPDATE-HIGHEST-VALUE-TEMPLATE(templates)
            extractant  ⟵ EDP-MANIPULATION(events, probe)
            templates  ⟵ UPDATE-TEMPLATES(templates, probe)
            if extractant is not null then
                    operation  ⟵  TRANSLATE(extractant, probe)
                    operation-list  ⟵  add operation
                    events  ⟵  UPDATE-EVENTS(events, extractant)
    end
    return operation-list
```

Figure 5: Algorithm of EDP

Definitions:

- $T = [T_0, T_1, T_2, \ldots, T_i, \ldots, T_I]$, $P = [P_0, P_1, P_2, \ldots, P_j, \ldots, P_J]$

  T and P symbolize sequential strings. T_i and P_j symbolize an element of each sequence. The bracket means elements in it are ranged in a time series order, from left to right. I and J are the number of elements of T and P, respectively.

- $d(i, j)$ represents the distance between T_i and P_j.

- $D(T, S)$ represents the distance between T and P.

- $g(i, j)$ represents the distance between the string $[T_0, \ldots, T_i]$ and the string $[P_0, \ldots, P_j]$.

- $g(I, J) = D(T, S)$

- $B(i, j)$ represents the number by which the string $[T_{B(i, j)}, \ldots, T_i]$ matches the string $[P_0, \ldots, P_j]$ the best.

- p: Cost of exchange between two events

- q: Cost of deletion of an event
- r: Cost of insertion of an event

Lastly, the following manipulations are conducted.

Procedure of **EDP-MANIPULATION:**

1) Initialization
   $g(0,0) = 0$, $B(i, j) = 0$
   For $i = 1$ to $i = I$
   $\quad g(i,0) = 0$, $B(i,0) = i$
   For $j = 1$ to $j = J$
   $\quad g(0, j) = g(0, j-1) + r$, $B(0, j) = 0$

2) Iteration
   For $i = 1$ to $i = I$
   $\quad$ For $j = 1$ to $j = J$

$$g(i, j) = \min \begin{cases} g(i-1, j) + q & (1) \\ g(i-1, j-1) + d(i, j) & (2) \\ g(i, j-1) + r & (3) \end{cases}$$

where, If $T_i = P_j$, Then $d(i, j) = 0$
$\quad\quad$ Else $d(i, j) = p$

If (1)=(2)=(3), Then $B(i, j) = \min \begin{cases} B(i-1, j) \\ B(i-1, j-1) \\ B(i, j-1) \end{cases}$

Else (1)=(2)<(3), Then $B(i, j) = \min \begin{cases} B(i-1, j) \\ B(i-1, j-1) \end{cases}$

Else (2)=(3)<(1), Then $B(i, j) = \min \begin{cases} B(i-1, j-1) \\ B(i, j-1) \end{cases}$

Else (3)=(1)<(2), Then $B(i, j) = \min \begin{cases} B(i-1, j) \\ B(i, j-1) \end{cases}$

Else (1)<(2) and (1)<(3), Then $B(i, j) = B(i-1, j)$
Else (2)<(1) and (2)<(3), Then $B(i, j) = B(i-1, j-1)$
Else (3)<(1) and (3)<(2), Then $B(i, j) = B(i, j-1)$

3) Finalization
   For $i = 1$ to $i = I$
   If $g(i, J) \leq Threshold$, Then $Si = [T_{B(i, j)}, ..., T_i]$

By comparing Si, final S is/are determined in order to obtain the longest strings possible and avoid overlapping with each other. More than one S are sometimes found without overlapping.

Merits of EDP

Our proposed EDP has several advantages. First, this method has flexibility on template matching, and a tolerance to noise. EDP makes it possible to detect not only the same sequence as the template but also approximately similar ones. Second, since every template has a value-score reflecting the user's subjective belief, prioritized search for pattern instances is executed.

CASE STUDY

Double-Reduction Gear System

Our proposed methods were evaluated in a case study, the design of "Double-Reduction Gear System." The double-reduction gear system is composed of four gears, three shafts, bearings, and a case. Basically, the number of teeth in the gears determines the speed reduction rate. Since the power of the revolution makes the torque and the bending moment, gears and shafts are designed to stand up to the force. We developed the CAD system called "Gear-CAD." Gear-CAD is an integrated design environment that allows designers to access all the information they need and to use the tools they need. Gear-CAD has all domain knowledge about this double-reduction gear system, supports the design, and simultaneously records the entire designer's log, *i.e.*, all events he or she generated during design process. Figure 6 shows an example of the Gear-CAD screens.

The Design mission is below.

Requirement:
1) All design components are determined in detail, *i.e.*, size and position.
2) Required reduction ratio is 10:1.
3) Lighter, smaller, and cheaper is better on the assumption of using the equipment in outer space.

Conditions:
1) Spur gears that have teeth with a 20-degree pressure angle are utilized in this system.
2) The input power and speed of rotation are 10.0 kW and 500 rpm, respectively.

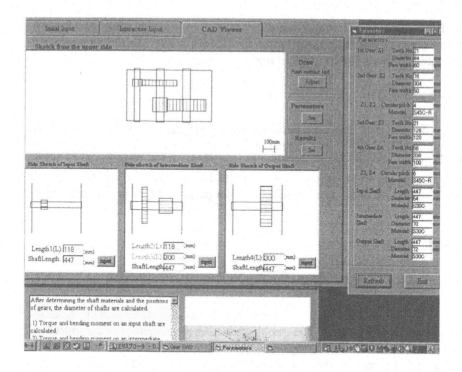

Figure 6: Example of Gear-CAD Screen

Our monitoring system including Gear-CAD, and knowledge capturing system were developed on Windows98 OS. The demo system was written in Visual Basic 6.0.

Results

A user who has enough knowledge on this problem designed the double-reduction gear system using Gear-CAD. In this design process, the user created eight alternatives, called product model (PM), and finally select the last PM as the best.

By using EDP, 52 operations corresponding to the templates (in Table 1) were captured from 472 events, as depicted in Figure 7. For example, in the event-sequence from event No.35 to No.43, S_0 = [E93, E94, E96, E91, E93, E94, E96, E95, E97], S_1 = [E93], S_2 = [E94, E96], S_3 = [E91, E93, E94, E96, E95, E97] were captured by EDP, as [Op155: Adjust Y-Positions of Z1 and Z2 Gears], [Op156: Adjust X-Positions of Z2 and Z3 Gears], and [Op155: Adjust Y-Positions of All Gears], respectively. In this case, the template P_1 = [E91, E93] for S_1, P_2 = [E94, E96] for S_2, and P_3 = [E91, E93, E95, E97] for

S_3 were applied. Since the longer template had the higher value-score, S_3 was found at first, and then S_1 and S_2 were found out of the rest. It turned out that EDP was flexible against a deletion, an insertion, and an exchange of elements, by comparing templates and obtained sequences.

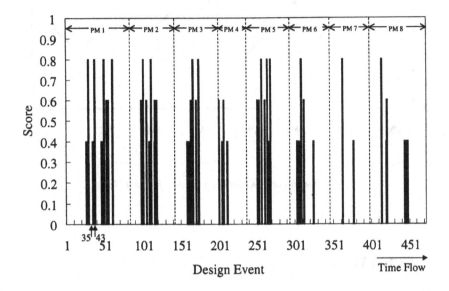

Figure 7: Appearance of Know-hows

The target operations, which were captured using templates in Table 1, were distributed in the design process as shown in Figure 7. Since templates in Table 1 represent some kinds of way to move the gears, Figure 7 shows where in the design process the location of gears were examined. The merged score at each product model was illustrated in Figure 8. According to Figure 8, as the design proceeded, we could see the downward trend in the frequency of examining the location of gears. Figure 7 and Figure 8 suggests that it be important to examine the location of gears in the early stage of the design process.

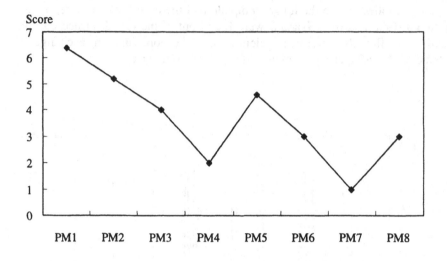

Figure 8: Appearance of Know-hows at every Product Model

RELATED WORK

Recently knowledge acquisition using data mining techniques has attracted attentions from many researchers in various fields. In the field of molecular biology, systems have been developed for finding patterns in molecular structures (Conklin 1993) and in genetic data (Holder 1994). In the business world, advanced data mining methods have been used for analysis and selection of stocks (Barr and Mani 1994).

In the engineering, machine learning techniques are used to predict the replacement of aircraft components (Letourneau 1999), and some classification learning techniques are utilized by astronomers to automatically identify stars and galaxies in a large-scale sky survey (Fayyad 1996).

Although many data mining researches have been done for manufacturing by using the data produced through manufacturing processes, data mining for design knowledge capturing in design process received little attention. That is because design processes are often ill-structured and *ad hoc*, and vary greatly depending on design situations. To deal with this problem, introducing an adequate model or representation for design process is very important. The quality of the model governs how data should be acquired and how data mining should be done to retrieve useful knowledge.

Despite the difficulty, some researchers point its importance out and tackle to capture knowledge in design processes by using data mining technique. KICAD (Knowledge Infrastructure for Collaborative and Agent-based Design) was proposed for the agent-system to capture and manage process knowledge in collaborative design (Jin, Zhao *et al.*, 1999) (Jin and Zhou, 1999), and the importance of capturing *know-how* knowledge was referred in it.

Myers *et al.* (Myers *et al.*, 1999) proposed the framework to automatically capture design rationale from a general CAD data. They use *design metaphor* in order to discover meaningful activities of a designer. However, their design metaphor means simple pattern matching using predefined rigid rules.

CONCLUSIONS

In this paper, we focus on *know-how*, which is the knack of a design in a design process. In order to automatically capture how designers do their design in a design process to accomplish their design goals, we have introduced a way for data gathering through an object-oriented CAD system at first, and then, we have proposed a *three-layer design process model* to represent generic design processes. Finally we have developed a method called *Extended Dynamic Programming* (EDP) for knowledge elicitation. The effectiveness of the proposed approach has been demonstrated by a gear design prototype system Gear-CAD. In an addition of capturing design process knowledge, our proposed method also enables us to understand and study a designers' behavior.

The limitation of our approach is that the knowledge we can capture must be pre-registered in advance. We plan to explore ways to identify new patterns that may emerge from design processes. We also plan to develop mechanisms to extract rationales behind the *know-how* knowledge being captured so that we will know why a designer did his or her design in a certain way.

NOTES

[1] Here, *the best* state means the state with the highest utility.

ACKNOWLEDGEMENTS

This research was supported in part by a NSF CAREER Award under grant DMI-9734006. Additional support was also provided by industrial sponsors. The authors are grateful to NSF and industrial sponsors for their support.

REFERENCES

Barr, D. and Mani, G., "Using Neural Nets to Manage Investments," AI Expert, 16-21, February 1994.

Chikada, T. and Yoshimura, M., *et al.*, "An off-line signature verification method based on a hidden Markov model using column images as features," in Proc. 9th Biennial Conference of the International Graphonomics Society (IGS'99), pp. 79-82, 1999.

Conklin, D., Fortier, S. and Glasgow, J., "Knowledge Discovery in Molecular Databases," IEEE Transactions on Knowledge and Data Engineering, 5(6), pp.985-987, Dec 1993.

Fayyad, U. M., Djorgovski, S. G. and Weir, N., "Automating the Analysis and Cataloging of Sky Surveys," in Advances in Knowledge Discovery and Data Mining, pp. 471-493, AAAI Press/ The MIT Press: California, 1996.

Holder, L., Cook, D. and Djoko, S., "Substructure Discovery in the SUBDUE system," in Proceedings of KDD-94: the AAAI-94 Workshop on Knowledge Discovery in Databases, pp. 169-180, 1994.

Jin, Y. and Zhou, W., "Agent-based knowledge management for collaborative engineering," in Proc. Design Engineering Technical Conferences (DETC'99) in ASME, 1999.

Jin, Y., Zhao, L. and Raghunath, A., "ActivePROCESS: A process-driven and agent-based approach to collaborative engineering," in Proc. Design Engineering Technical Conferences (DETC'99) in ASME, 1999.

Krogh, A. and Brown, M., *et al.*, "Hidden Markov models in computational biology: Application to protein modeling," Journal of Molecular Biology, Vol. 235, pp.1501-1531, 1994.

Letourneau, S., Famili, F. and Matwin, S., "Data Mining to Predict Aircraft Component Replacement," in Data Mining: A Long-Term Dream, *IEEE Intelligent Systems*, pp. 59-66, November/ December 1999.

Myers, K. L., Zumel, N. B. and Garcia, P., "Automated capture of rationale for the detailed design process," in Proc. The 11th Conference on Innovative Applications of Artificial Intelligence. (IAAI'99), 1999.

Sakoe, H. and Chiba, S., "Dynamic programming algorithm optimization for spoken word recognition," Readings in Speech Recognition, (Waibel, A. and Lee, K., Eds.) pp. 159-165. Morgan Kaufmann, San Mateo, California, 1990.

CHAPTER 7

A Data Mining-Based Engineering Design Support System: A Research Agenda

Carol J Romanowski
cfr@acsu.buffalo.edu
Department of Industrial Engineering, State University of New York at
Buffalo, 342 Bell Hall, Buffalo, New York 14260-2050

Rakesh Nagi
nagi@acsu.buffalo.edu
Department of Industrial Engineering, State University of New York at
Buffalo, 342 Bell Hall, Buffalo, New York 14260-2050

ABSTRACT

Currently, designers do not have access to product life cycle information and
other feedback, causing costly design iterations and increased time-to-market.
Our research proposes using data mining to incorporate this heterogeneous
and distributed information into the beginning stages of design - thereby
reducing iterations and lowering the cost of product design.

D. Braha (ed.), Data Mining for Design and Manufacturing, 161–178.
© 2001 *Kluwer Academic Publishers. Printed in the Netherlands.*

INTRODUCTION

Engineering design is a multi-disciplinary, multi-dimensional, non-linear process. Once performed by a solitary engineer wielding drafting tools and vellums, today's design and development is likely to be a team process – concurrent engineering – performed with CAD programs, interdisciplinary meetings, conference calls, and e-mail.

Although some design task characteristics have changed, others remain constant. The design begins with requirements – customer or otherwise – that are often ill-defined. The resulting concept is subject to constraints that are often conflicting. The final product is prey to unforeseen consequences of design choices – as well as to possible end-user abuse – with accompanying legal ramifications.

So how can we make the design task easier and more efficient? How can we take into account all the possible combinations of requirements, choices, constraints, and consequences? Usually, our first thought is to automate the process as much as possible, using the computer's ability to perform complex tasks quickly and efficiently. In this case, however, some aspects may be better left to human capabilities – since design so often requires innovation and a depth of knowledge that is difficult to articulate or encode.

Many researchers believe that computational approaches to design should enhance, not replace, human practice (Duffy and Duffy 1996; Reich et al. 1993). However, the difficulties inherent in capturing and maintaining the contextual knowledge needed for good design have hindered the application of machine learning systems to this process (Reich 1994).

Another difficulty involves the re-use of previous design knowledge in the form of archival design documents, testing and analysis reports, and life-cycle information such as warranties, product reviews, and sales returns. These types of information are distributed as well as heterogeneous, making integration into computer-assisted design systems a challenge. With computerized acquisition of this distributed information – and cheap data storage – the sheer amount of data available can easily overwhelm anyone attempting to make use of it. Therefore, critical feedback on design successes and failures is not considered early in the design process. Omitting this essential information results in high design costs from redesigns, increased time-to-market, and product recalls.

Data mining, a related machine learning field, is a promising answer to the problem of incorporating life cycle information and integrating previous design knowledge into the beginning design stages. The algorithms and methodologies used in data mining are tailored for large, heterogeneous databases, searching for "nuggets" of knowledge that would otherwise be overlooked or unused. This research proposes a data mining approach to such information integration, with the goal of reducing design iterations and thereby lowering the cost of product design.

OVERVIEW OF ENGINEERING DESIGN

Hazelrigg (Hazelrigg 1996) describes design as "the effective allocation of resources" and "a process of synthesis and integration". Pahl and Beitz (Pahl and Beitz 1999) see engineering design as "a creative activity", "the optimization of objectives within partly conflicting constraints", and "an essential part of the product life cycle". Probably the most comprehensive definition is given by Dym and Levitt (Dym and Levitt 1991), who define engineering design as the "systematic, intelligent generation and evaluation of specifications for artifacts whose form and function achieve stated objectives and satisfy specified constraints."

Dym and Levitt note that, unlike other design domains, engineering design does not directly result in a physical product – but a set of specifications to construct or fabricate the product (Dym and Levitt 1991). Therefore, mapping design objectives and constraints to specifications is an important design issue.

The engineering design process

Several models attempt to describe the process of designing; Pahl and Beitz (Pahl and Beitz 1999) present a model of systematic design with specific tasks and outputs; their model incorporates a more general model proposed by French (French 1992). Another systematic model, proposed by the German technical society of professional engineers, (Verein Deutscher Ingenieure, or VDI) delineates seven stages and the outputs resulting from completion of each stage.

Although nearly all of these models show feedback from successive stages to earlier stages, none of the models explicitly includes feedback from life cycle information once a product is turned over to manufacturing and sales. Information-processing-based design models, such as that proposed by Ullman et al. (Dym 1994), do include external information – files, handbooks, manuals, etc. – in their mappings of the design state. However, since they deal primarily with cognitive processes those models are outside the scope of this research and will not be discussed.

Design tasks can be classified into at least three major types (Dym 1994):
1. Routine design, where the domain is well understood, the process is easily decomposed into tasks, and information about constraints and possible failure modes is extensive.
2. Variant design, where the amount of domain knowledge is substantial but how to effectively apply that knowledge is not well understood.
3. Creative, or unique, design, where both domain knowledge and an effective problem-solving strategy are unknown.

Pahl and Beitz (Pahl and Beitz 1999) use the term "original design" to denote new or existing design tasks solved using new solution principles; "adaptive" design uses known solution principles to adapt the design to changed requirements. This type of design may also include original design of some components. "Variant" design refers to designing parts and assemblies within constraints – such as dimensions – set by previous structure designs. In this research, we will follow the definitions of routine, variant, and original design as proposed by Dym.

The Product Life Cycle (PLC)

Every product goes through a series of stages from the initial concept to obsolescence and disposal. Nahmias' model (Nahmias 1993) has four stages: startup, rapid growth, maturation, and decline. Hazelrigg (Hazelrigg 1996) defines seven stages: research to proof-of-concept; engineering and design; test and evaluation; manufacturing; distribution and sales; operations, maintenance and repair; and disposal. Dym and Levitt (Dym and Levitt 1991) define stages similar to Hazelrigg, but separate the disposal stage into three states: energy recovery, recycling, and disposal. Nahmias' view is concerned mainly with the volume of units sold, while both Hazelrigg and Dym and Levitt define the stages in terms of the product state.

Information collected during a product's life cycle provides feedback on a product's performance that can be used to assess the quality of the design (Ertas and Jones 1996). For instance, products are generally conceived to meet a market need or appeal to a certain demographic group. Sales information, readily available as POS (point of sale) transaction data, can analyze the effectiveness of the design in reaching the target group. Consumer reviews point out a product's strengths and shortcomings. Merchandise returns and warranty information reveal flaws in both form and function. Process plans and revisions, bills of material, and routings all contain useful information that may enable designers to generate more efficient and optimal designs.

Many design support systems attempt to optimize some of the total cost of design, but because post-production information is difficult to integrate, indirect costs such as the cost to customers and society are largely ignored. Studies show that such indirect costs can be as much as 4-5 times the amount of direct labor and material costs (Prasad 2000). Product, process, and field or warranty measures assess the impact of these costs on the overall product design.

OVERVIEW OF MACHINE LEARNING IN DESIGN

Many routine design tasks can be performed by computers, such as computation and computer-aided drawing (CAD). As computers became more powerful, the focus turned to assisting the designer in more complex tasks. Initially, machine learning in design (MLinD) research focused on solving simple, simulated routine design tasks with existing techniques(Reich 1998). One such system, BRZYDL1 (Arciszewski et al. 1987) induced decision rules that classified structural beam configurations as either feasible or infeasible. Human experts examined the generated rules and submitted corrective examples to increase the accuracy of the system. McLaughlin and Gero (McLaughlin and Gero 1987) used ID3, a decision tree algorithm, to infer characteristics of Pareto-optimal and non-optimal building designs. Both of these systems dealt with small datasets and only two classes. Lu and Chen (Lu and Chen 1987) simulated a face milling operation to generate design examples, then applied a clustering algorithm to group the examples into classes. An inductive program found generalized rules to describe each class.

MLinD research has shifted from these simple, simulated studies to more complex issues in applying ML techniques to real-life problems(Reich 1998). As this shift took place, several dichotomies characterizing the MLinD role were identified. For instance, Reich (Reich 1998) describes two functional roles for machine learning in design:

1. **The performance role**, where the "best" ML program – whose constructed model is statistically validated as superior to other ML program models – learns examples from large data in a structured, algorithmic process; the output is typically a prediction. These systems often act as black boxes.

2. **The understanding role**, where humans employ different ML programs to better understand the data. The resulting models are not required to be of good statistical quality, as their purpose is to provide comprehensible insight into the data. According to Reich, data mining fills this role in MLinD.

Another MLinD dichotomy states that computer-assisted design systems have two purposes (Sim and Duffy 1998):

1. **Acquiring knowledge of a product and associated design concepts from past designs.** Examples of such product-knowledge systems include BRIDGER (Reich 1993), NETSYN (Ivezic and Garrett 1994), PERSPECT (Duffy and Duffy 1996), IM-RECIDE (Gomes et al. 1998), and FORESEE (Borg et al. 1999). FORESEE is of particular interest in this research, as it is a feature-based design system that attempts to incorporate life cycle consequences in a design knowledge base. The system, written in Clips, produces rules that relate materials, forms, and processes to

possible failure modes or design for X concerns. Completed design solutions are evaluated by their impact on time, cost, and quality for 6 life cycles phases (design, manufacture, assembly, use, service, and disposal). In testing this deterministic system, Borg et al. noted the difficulty of obtaining relevant knowledge as well as the cognitive load generated by considering consequences over multiple life phases.

2. **Acquiring knowledge of the design process** by recording the decisions made at each step, and the actions and consequences resulting from those decisions. This knowledge can be captured two ways: as a sequence of dependent actions, called a *plan*, or as condition-action pairs on a blackboard in a blackboard system. ADAM, BOGART, ARGO, and DONTE are planning systems; DDIS is a blackboard system (Sim and Duffy 1998).

INFORMATION INTEGRATION IN ENGINEERING DESIGN

In 1986 the National Science Foundation sponsored a workshop to identify critical areas for research in engineering design. In the committee's report (Rabins et al. 1986), information integration and management was designated as a supporting discipline important to the growth of the engineering design field. In the committee's view, information integration and management includes:

- the study of usage of information in design
- development of engineering data models
- characterization of design information
- identification of design process dynamics
- identification of interactions of information with design tasks
- development of engineering databases
- development of engineering database techniques
- identification of design environments
- development of ways to collect design information
- development of ways to present design information
- study of the feedback from lifecycle to design

This view of information integration defines information as "interpreted data" – including sources, dependencies, authorizations, uses, and release of data, and sees design information as dynamic, growing in volume and complexity as the design progresses. However, simply providing access to this information is not enough to boost efficiency; approximately half of designers' time is spent organizing and managing data (Rabins et al. 1986; Rychener 1988). Truly assistive design systems would provide the designer with knowledge – understanding of the stored information in relation to the design task – as well as access to the interpreted data.

Court (Court et al. 1998) studied the information needs of designers in both the UK and the US. He found that designers first turned to their personal files – usually hard-copy – and then searched computer records for design information. US designers were more likely than their UK counterparts to access computerized files, such as CAD drawings, reflecting the emphasis on computerized drafting aids in US engineering curricula. The reliance on personal files points out the need to increase the value of computerized information: unless information is "accurate, available, accessible, applicable, and in sufficient quantity" (Court et al. 1997), the value of that information to the designer is low. Court's studies showed that designers will proceed with decisions – potentially sub-optimal or incorrect – if the information they seek is not supplied or found in a timely manner.

OVERVIEW OF DATA MINING

Data mining (DM) can be defined as the application of computer algorithms to discover useful knowledge in large databases. The process of mining a database is known as *knowledge discovery in databases* (KDD); data mining is one step in this process. DM uses statistical and visualization techniques as well as machine learning algorithms (both supervised and unsupervised) to analyze and describe the dataset.

A data mining approach can be either symbolic or non-symbolic; predictive or classifying; but it is always interactive and iterative(Fayyad et al. 1996). Domain experts – who understand the meaning of the data being mined – and data experts – who understand the structure of the database – are necessary to the success of any mining task.

Most data mining algorithms are drawn from the fields of statistics or machine learning. Neural networks, decision trees, rule induction, and case-based reasoning are the predominant methods used in DM, while data visualization, Bayesian belief nets, support vector machines, and text mining are newer additions to the miner's toolbox. Other technologies, such as genetic algorithms, fuzzy sets, and rough sets, are usually used in conjunction with neural nets, trees, or rules.

Applications of data mining to engineering design

Many application of ML to learning design problems exist, but to our knowledge no research exists that uses data mining methodology in the sense of knowledge discovery and use of that knowledge. Most MLinD research emphasizes learning from a knowledge base, but that knowledge base is built "by hand" – thus incorporating incomplete human knowledge and biased opinion about what is important. Also, assumptions may be hidden; for example, Reich et al. (Reich et al. 1993) points out that bridges designed in the US must conform to a different standard than bridges designed in the UK – thus influencing the design decisions on which structural elements to use in a particular situation.

Moczulski (Moczulski 1998) used case-based reasoning to design anti-friction bearing systems. His knowledge base came from text sources, whose quantitative values were converted into qualitative ones by hand, and thus susceptible to the bias of the person doing the coding.

Yang et al. (Yang et al. 1998) used singular value decomposition to build in-process and post-process thesauri from electronic design notebooks and informal design documents to aid in design document retrieval. They found that domain knowledge improved the accuracy of the machine-generated thesaurus compared with a human-generated thesaurus, but added more human effort to the process. Their research also found that the differences in terminology between in-process design documents and final documents significantly impacted the recall precision of thesauri generated from post-process documentation.

THE NEXT STEP: A DATA MINING-BASED DESIGN SUPPORT SYSTEM

The demands of the current business climate require companies to be agile and proactive, moving products quickly from concept to market. On the other hand, increasingly cheap data storage has resulted in huge amounts of critical data that is not being effectively used. Additionally, current engineering information systems do not capture or maintain data generated during the design process.

We know these things about designers:
1. Designers first look at their own files – usually hardcopy notes and documents
2. Designers spend 20-30% of their time looking for information
3. Designers spend 20-30% of their time handling information

Therefore, any design support system must first incorporate the designers' own files; make information easy to find; and make information easy to use. Ferguson et al. (Ferguson et al. 1998) identify three important aspects of design support systems research related to solving these problems:
- How do we determine what data and information is relevant to the design task?
- How do we structure the system to allow designers easy access to this data and information?

- How do we present information in a meaningful, understandable, and useful form?

We can add a few more questions:

- How do we obtain, and efficiently and quickly mine, relevant data?
- How do we incorporate the mined knowledge into the knowledge base?
- How do we deal with changing knowledge?

Data mining can enhance different areas of the design support system (DeSS). Initially, DM methods can be used to mine archival data and assist in determining relevant and critical attributes for inclusion in a knowledge base. Additionally, mining archival data also produces design rules and constraints to populate the knowledge base. Each use requires a different DM algorithm and different data preparation.

Because no single algorithm can handle every type of data, a data mining-based design support system needs to contain several algorithms to deal with the heterogeneous nature of product life cycle data. These algorithms need to be fast and efficient, with understandable output. Because the amount of information is vast, we need algorithms that learn incrementally but are insensitive to the order in which data is presented. We need a method of resolving conflicts between information "nuggets" and a way of updating "old" rules. If a legacy knowledge base or decision support system exists, we need to integrate the systems. We need to know how to add to and maintain, and how often to update, the knowledge base. We also need to provide the capability to interface with CAD or analysis software.

Several other problems make designing a useful, efficient design support system challenging. For instance:

- much of the data is in hardcopy, not digital, and not necessarily in scannable form
- the electronic data is distributed and heterogeneous
- no mapping exists from design concepts to product life cycle concepts
- rules aren't enough – the designers need explanations for the rules and also access to the original documents.
- in general, designers are not overwhelmed with the quality and functionality of existing support systems, which means they may be resistant to using yet another system

The next logical step in integrated design support systems must make use of enterprise-wide and engineering data to determine life cycle costs arising from design decisions at each stage of the design process. Completely answering all the questions posed – and meeting all the needs of designers and collaborative design work teams – would be a huge task that would take a very long time. But, that doesn't mean we can't envision how such a complete system could be structured.

The heart of any support system is CAD, coupled with CAM, CAE, a component library, and optimization capabilities. Hopefully, the CAD system is built to support the varying styles and skills – as well as the non-linear process – of designers. Supporting CAD are the information modules that gather, integrate, mine, and present relevant product life cycle information to the designer.

Information modules necessary for integrated data mining and information retrieval include a data preparation module, a data mining manager, and an information retrieval engine. Each of these modules contains subsystems that perform various tasks. An imaged-document warehouse could be built to house scanned-in hardcopy legacy data.

Within the information retrieval engine is a series of mediators to retrieve and integrate data from heterogeneous, distributed sources throughout the enterprise, including an engineering data warehouse.

The data mining manager contains a conflict resolution module that mediates the data mining output and selects the most relevant rule or itemset based on the data.

As designers select components from the CAD component library, relevant mined rules and total cost of the design life cycle at that point in the process – as well as hyperlinks to the data tuples referencing the rules – can be presented on the screen. Cost parameters can be transferred to the optimization module for design cost analysis and testing. Figure 1 shows a general framework for a data mining-based support system.

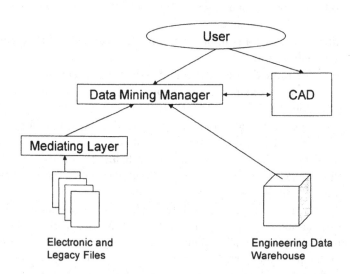

Figure 1: A framework for a data mining-based support system

Enabling technologies for a data mining-based design support system

Engineering data warehouse

Engineering data by its very nature is different from business or enterprise data. Therefore, building an efficient, easily accessible engineering data warehouse is the first step in an integrated design support system. Typically, data in a warehouse expires after a certain amount of time and is removed from the system. However, we may want to keep historical design data far longer than the usual transaction warehouse data. Data compression, using lossless methods with low memory requirements and fast retrieval capabilities, will be a necessary component as the amount of storage space taken up by legacy files, images, and original documentation increases. The support system must include capabilities to compress the original files using methods that allow searches by the information retrieval engine and mining by the data mining engine.

Making data access easy and transparent to the user requires an understanding of the types of queries users are apt to make. In some cases, queries may require multiple SQL or other types of query language statements to generate the desired output. The user should have the ability to perform ad hoc queries as well as pre-defined queries.

Imaging hardcopy files

Existing OCR algorithms cannot produce consistent electronic versions of this legacy information, but some promising new data mining algorithms such as SubdueCL (Cook and Holder 2000) are being applied to imaged files with good results. These algorithms have not yet been used in a mixed image/text environment such as an engineering drawing, logbook, or parts catalog. A vital research issue is the effective mining of such mixed-media documents, including a methodology for handling the different data types. One possible approach to this problem, using existing algorithms, is to mine images and text separately, subsequently integrating the initial results and then applying a classification algorithm. Other methods could include development of an algorithm that would be able to mine both images and text at the same time.

Compression of mixed text and image files at this time relies on heuristic methods that are not particularly robust (Witten et al. 1999). The current approach to these types of documents is to determine and correct orientation, removing any skew from the scanning process; segmenting the document into textual, graphics, and halftone regions; classifying the regions, and applying appropriate compression techniques to the classified regions.

Data Mining Manager

An important consideration in the use of data mining algorithms available in the Data Mining Manager is the capability for incremental learning as the amount of stored data increases, and the speed of the algorithms in cases where incremental learning schemes are not possible. Dynamic data mining algorithms need to be developed that can adapt the existing mined output to new data. Wang's (Wang 1997), Raghavan's (Raghavan and Hafez 2000) and Zhou's (Zhou et al. 2000) work in clusters and association rules is promising; but we need an effective dynamic *rule* mining algorithm as well. As the information in the data warehouses increases, the probability of rules changing drastically also increases; therefore, a mechanism to capture these changes and keep track of versions is necessary, such as Raghavan's memory approach.

Conflict resolution is a seldom-discussed area in data mining. Most often, some sort of weighted scheme is used to decide between rules, or the miner relies on support and confidence measures based on the available data. Support is defined as the percentage of records in the data containing a certain relationship. Confidence is determined by how much one attribute is dependent on another. If a particular combination of attributes and values is rare but important, support and confidence measures will be misleading. We need to evaluate the different conflict resolution schemes and develop methods that reflect the importance of a particular concept. Domain knowledge supplies the criteria for interestingness and importance, thus underscoring the need for close collaboration with industrial designers in building a design support system.

Design index/thesaurus development

Yang et al. (Yang et al. 1998) showed that design concept terminology changed as the design process progressed. Therefore, the mapping between terms at each stage of the design process – as well as the life cycle – is critical to understanding how a product evolves throughout its lifetime, and how life-cycle measures and metrics relate to design concepts. This mapping contributes directly both to efficient information retrieval of original documentation, but also to the effective mining of manufacture-to-disposal data.

Information retrieval

Because the document collections in an enterprise are often distributed, the problem of effective and fast distributed retrieval becomes significant. Several different models for ranked queries exist, but developing fast, accurate ranking methods is still an open area of research. Interactive retrieval, another open area of study, allows the user to refine queries over a retrieved set of documents without requerying the collection, and performs relevance feedback and probabilistic rankings based on the acceptance or rejection of retrieved document summaries.

Component library

In variant design, a component library is essential to exploring the many different options available. Group technology (GT), product data models, and generic bills of material (GBOMs) are enabling technologies for building these libraries.

In the component library, the knowledge representation chosen will need to include relevant product life cycle consequence data metric values that will be used to calculate the total design cost, and hyperlinks to the information retrieval engine to access the original documentation.

Optimization

Prasad (Prasad 2000) identified at least 21 life-cycle cost drivers impacting the total cost of a product from conception to disposal. Several of these costs are indirect (not labor or materials) and therefore often overlooked in optimization efforts. This module would provide the designer with metric values that measure direct costs and indirect costs, based on mining enterprise-wide data. Where no information exists about a particular metric, default values can be user-defined, or set using a loss function. Quantifying these indirect costs can be a difficult task; however, analysis of customer returns and warranty claims may give us some insight into their expected values and ranges.

Knowledge representation

Exploration of hybrid schemes that link the languages of engineering design – text, graphical representations, mathematical models, numerical values, rules, and object models – and measure both their ability to convey design information and their effects on cognitive loading is necessary. Borg et al. (Borg et al. 1999) noted in their work on FORESEE that designers expressed concern about the mental work needed to keep track of multiple life cycle consequences, yet still focus on individual issues. We are adding another level of complexity to the design task by requiring designers to consider future, possibly indirect effects in addition to satisfying constraints and adhering to fundamental engineering principles. The way product life cycle information is presented is a non-trivial consideration in building a design support system. Merely stating IF-THEN rules is insufficient for complete understanding of design consequences, and the limitations of rules alone do not allow for the richness of design models to be sufficiently expressed. Object models are promising, but may require numerous instantiations to cover all the possible combinations of components, parts, and assemblies. In addition, certain engineering design languages cannot be incorporated into an object model.

Acceptance of an aiding system such as a design support system is predicated on the user's perception that the system does not attempt to usurp authority or restrict options (Dong and Agogino 1998), and also that the knowledge representation matches the users' mental model of the problem (Madni 1988). One of the problems Borg et al.'s evaluators pointed out was the disconnect between the designers' graphical visualization of the solution and the text-based representation offered by FORESEE (Borg et al. 1999). Incorporating an intelligent CAD system provides graphical capabilities; the major research issue then becomes how the relevant product life cycle consequences are presented to the designer.

Evaluation of the design and the system

As Siemieniuch and Sinclair have pointed out (Siemieniuch and Sinclair 1993), new systems of working require a change in the organizational structure. They also require evaluation. Use-value analysis (Pahl and Beitz 1999), utility analysis (Thurston 1990), various fault metrics (Anglade and MacRae 1987) and relative or absolute comparisons (Diteman and Stauffer 1992; Hyde and Stauffer 1990) are examples of ways designs have been evaluated either during the process or after production has begun. Reich noted the absence of rigorous evaluation procedures or even research for machine learning applications in design (Reich et al. 1993).

For any system or process, we need to objectively and thoroughly look at not only the solution produced, but also the impact of the system on the users. Who uses the system – either directly or indirectly? The most elegant technical solutions can be rendered completely useless if no one can understand, manipulate, or evaluate them.

Computer-based design evaluation capacity should be incorporated into the design support system. Hyperlinks can access forms for comparisons or analyses, with the relevant numerical information exported directly to the forms. Evaluating the various components of the system, and the system's usefulness itself, requires analysis of the human-computer interface as well as algorithmic accuracy and efficiency.

CONCLUSION

Companies today are increasingly concerned with getting their products to the market faster than their competitors, yet they must still meet the high expectations of customers. Mass customization manufacturing systems are becoming the norm, and are generating mountains of data. This research accomplishes two industrial goals: making use of critical data that would lower design and product costs and realization times, and developing methods to integrate that data into the design process. The results are extendable from a single designer to a concurrent engineering design group.

The benefits of using product life cycle information reach far beyond reducing product costs. Knowledge generated from mining enterprise-wide data results in a better understanding of the consequences of decisions made at all levels of the company. Understanding the target audience, for example, influences form, function, materials, maintenance, disposal, and even the tone and complexity of operating manuals.

The fields of concurrent engineering design, information integration, and data mining offer an exciting opportunity to merge the technologies and create a design support system that truly assists designers. This type of system can help reach the concurrent engineering goals of getting the design right the first time and reducing the time-to-market interval. Incorporating data mining and information integration into design allows designers to access critical product life cycle information.

At least three important results can be achieved by implementing a data mining-based design system. Firstly, design engineers will develop a better understanding of the consequences their designs have on a product from concept to disposal;. Secondly, integrating product life cycle information into all design stages will lower overall life cycle cost by reducing design iterations, engineering changes, product failures, warranty claims, and time-to-market. And, thirdly, research into data mining and knowledge representation issues specific to engineering will result in a methodology and framework for using these important tools effectively in engineering domains.

ACKNOWLEDGEMENTS

Carol Romanowski acknowledges the support of the Engineering Research Program of the Office of Basic Energy Sciences at the Department of Energy. Rakesh Nagi acknowledges the support of the National Science Foundation under career grant DMI-9624309.

REFERENCES

Anglade, E., and MacRae, A. U. , "End-product hardware design quality metrics - definitions," in International Conference on Communications, pp. 1182-1187, 1987.

Arciszewski, T., Mustafa, M., and Ziarko, W., "A methodology for design knowledge acquisition for use in learning expert systems," International Journal of Man-Machine Studies, 27, pp. 23-32, 1987.

Borg, J. C., Yan, X-t., and Juster, N. P., "Guiding component form design using decision consequence knowledge support," Artificial Intelligence for Engineering Design, Analysis and Manufacturing, 13, pp. 387-403, 1999.

Cook, D. J., and Holder, L. B., "Graph-based Data Mining," IEEE Intelligent Systems, (March/April), pp. 32-41, 2000.

Court, A. W., Culley, S. J., and McMahon, C. A., "The influence of information technology in new product development: observations of an empirical study of the access of engineering design information," International Journal of Information Management, 17(5), pp. 359-375, 1997.

Court, A. W., Ullamn, D. G., and Culley, S. J., "A comparison between the provision of information to engineering designers in the UK and the USA," International Journal of Information Management, 18(6), pp. 409-425, 1998.

Diteman, M., and Stauffer, L., "Usability Analysis of a Computer Tool for Evaluating Design Concepts." Design Theory and Methodology, 42, pp. 41-44, 1992.

Dong, A., and Agogino, A. M., "Managing design information in enterprise-wide CAD using 'smart drawings'," Computer-Aided Design, 30(6), pp. 425-435, 1998.

Duffy, S. M., and Duffy, A. H. B., "Sharing the learning activity using intelligent CAD," Artificial Intelligence for Engineering Design, Analysis and Manufacturing, 10, pp. 83-100, 1996.

Dym, C. L., Engineering Design: A Synthesis of Views. Cambridge: Cambridge University Press, 1994.

Dym, C. L., and Levitt, R. E., Knowledge-Based Systems in Engineering. New York: McGraw-Hill, 1991.

Ertas, A., and Jones, J., The Engineering Design Process. New York: John Wiley & Sons, 1996.

Fayyad, U. M., Piatetsky-Shapiro, G., Smyth, P., and Uthurusamy, R., Advances in Knowledge Discovery and Data Mining. Menlo Park, CA: AAAI Press/The MIT Press, 1996.

Ferguson, C.-J., Lees, B., MacArthur, E., and Irgens, C., "An Application of Data Mining for Product Design," IEE Colloquium Digest, 434, pp. 5/1-5/5, 1998.

French, M. E., Form, Structure and Mechanism. London: MacMillan, 1992.

Gomes, P., Bento, C., and Gago, P., "Learning to verify design solutions from failure knowledge," Artificial Intelligence for Engineering Design, Analysis and Manufacture, 12, pp. 107-115, 1998.

Hazelrigg, G. A., Systems Engineering: An Approach to Information-Based Design. Upper Saddle River, NJ: Prentice-Hall, Inc., 1996.

Hyde, R. S., and Stauffer, L., "The comparison of the reliability of three psychometric scales for measuring design quality," in Design Theory and Methodology, 40, pp. 349-354, 1990.

Ivezic, N., and Garrett, J. H., "A neural-network-based machine learning approach for supporting synthesis," Artificial Intelligence for Engineering Design, Analysis and Manufacture, 8, pp. 143-161, 1994.

Lu, S., and Chen, K., "A machine learning approach to the automatic synthesis of mechanistic knowledge for engineering decision-making," Artificial Intelligence for Engineering Design, Analysis and Manufacturing, 1(2), pp. 109-118, 1987.

Madni, A. M., "The Role of Human Factors in Expert Systems Design and Acceptance," Human Factors, 30(4), pp. 395-414, 1988.

McLaughlin, S., and Gero, J. S., "Acquiring expert knowledge from characterised designs," Artificial Intelligence for Engineering Design, Analysis and Manufacturing, 1, pp. 73-87, 1987.

Moczulski, W., "Inductive Learning in Design: A Method and Case Study Concerning Design of Antifriction Bearing Systems," in Machine Learning and Data Mining (Michalski, R., Bratko, I., and Kubat, M., Eds.), pp. 203-219, John H. Wiley & Sons, Ltd.: Chichester, 1998.

Nahmias, S., Production and Operations Analysis. Burr Ridge IL: Irwin, 1993.

Pahl, G., and Beitz, W., Engineering Design. London: Springer-Verlag, 1999.

Prasad, B., "Survey of life-cycle measures and metrics for concurrent product and process design," Artificial Intelligence for Engineering Design, Analysis and Manufacturing, 14, pp. 163-176, 2000.

Rabins, M. J., Ardayfio, D., Balzar, R., Fenves, S., Nadler, G., Richardson, H., Rinard, I., Roth, B., Seireg, A., and Wozny, M., "Goals and Priorities for Research in Engineering Design: A Report to the Design Research Community," New York, NY: American Society of Mechanical Engineers, 1986.

Raghavan, V., and Hafez, A., "Dynamic Data Mining," in Intelligent Problem Solving-Methodologies and Approaches: Proc. of Thirteenth International Conference on Industrial Engineering Applications of AI & Expert Systems, pp. 220-229, 2000.

Reich, Y., "The development of BRIDGER: A methodological study of research on the use of machine learning in design," Artificial Intelligence in Engineering, 8, pp. 217-231, 1993.

Reich, Y., "Layered models of research methodologies," Artificial Intelligence for Engineering Design, Analysis and Manufacturing, 8, pp. 263-274, 1994.

Reich, Y., "Learning in design: From characterizing dimensions to working systems," Artificial Intelligence for Engineering Design, Analysis and Manufacturing, 12, pp. 161-172, 1998.

Reich, Y., Konda, S., Levy, S., Monarch, I., and Subrahmanian, E., "New roles for machine learning in design," Artificial Intelligence in Engineering, 8, pp. 165-181, 1993.

Rychener, M. D., "Research in Expert Systems for Engineering Design," in Expert Systems for Engineering Design, pp. 1-33, Academic Press, Inc.: Boston, 1988.

Siemieniuch, C. E., and Sinclair, M. A., "Implications of concurrent engineering for organizational knowledge and structure: a European, ergonomics perspective," Journal of Design and Manufacturing, 3, pp. 189-200, 1993.

Sim, S. K., and Duffy, A. H. B., "A foundation for machine learning in design," Artificial Intelligence for Engineering Design, Analysis and Manufacture, 12, pp. 193-209, 1998.

Thurston, D. L. "Subjective Design Evaluation with Multiple Attributes," in Design Theory and Methodology - DTM '90, pp. 355-361, 1990.

Wang, K., "Discovering patterns from large and dynamic sequential data," Journal of Intelligent Information Systems, 9(1), pp. 33-56, 1997.

Witten, I. H., Moffat, A., and Bell, T. C., Managing Gigabytes: Compressing and Indexing Documents and Images. San Francisco, CA: Morgan Kaufmann Publishers, 1999.

Yang, M. C., Wood, W. H., and Cutkosky, M. R., "Data mining for thesaurus generation in informal design information retrieval," in Congress on Computing in Civil Engineering, pp. 189-200, 1998.

Zhou, A., Jin, W., Zhou, S., Qian, W., and Tian, Z., "Incremental mining of the schema of semi-structured data," Journal of Computer Science and Technology, 15(3), pp. 241-248, 2000.

CHAPTER 8

Data Mining for High Quality and Quick Response Manufacturing

Jang-Hee Lee
janghlee@major.kaist.ac.kr
Department of Industrial Engineering, Korea Advanced Institute of Science and Technology (KAIST), 373-1 Kusong-dong, Yusong-ku, Taejon 305-701, Republic of Korea

Sang-Chan Park
sangpark@kaist.ac.kr
Department of Industrial Engineering, Korea Advanced Institute of Science and Technology (KAIST), 373-1 Kusong-dong, Yusong-ku, Taejon 305-701, Republic of Korea

ABSTRACT

As the manufacturing industry becomes more and more competitive, both intelligent process control and fast manufacturing cycle time are more crucial than ever. Recently, semiconductor manufacturing has become increasingly complex due to device size reduction and consequently, those objectives become necessities in the survival and can be achieved through optimal sampling strategy utilizing inspection resources effectively without incurring a loss in quality or output. We propose a new and better application of data mining in developing an optimal measurement sampling method for process parameter monitoring in a wafer fab and illustrate the effectiveness of proposed sampling method using actual fab data. The results indicate that if the sampling chip locations and their size are chosen rationally by data mining, that sampling can provide a good sensitivity of 100% wafer coverage and defect detection for high quality and quick response manufacturing in spite of smaller sampling size.

D. Braha (ed.), Data Mining for Design and Manufacturing, 179–205.

INTRODUCTION

The effect of increasing international competition in business is relentless pressure on manufacturing organizations to lower product costs and increase product quality. For manufacturing organization, these pressures translate into a continuous effort to reduce process variability and maintain a stable manufacturing operation. This is usually accomplishes through tighter process control and more frequent process observations. As the semiconductor industry becomes more and more competitive, both fast high yield learning and sustaining high yield by detecting any process abnormality as early as possible are more crucial than ever.

Moreover, in recent years, semiconductor manufacturing has become increasingly complex due to device size reduction and consequently, the manufacturing cycle time, also called *turn around time* (TAT), which is defined as the time required from wafer input through probing test, becomes longer year by year. The delay between the occurrence of process defects and their detection caused by longer more complex process has become a significant problem, as production of defective wafers continue from defect occurrence until the defect is detected and eliminated. Thus it is important to reduce the time from defect occurrence to their detection at probing test, which requires reduction of fab TAT.

One of the main vehicles available to achieve those intelligent process control and cycle time reduction is an economic and effective sampling strategy for in-line process parameter monitoring and control. Process sampling is perceived as being "non-value add" step, generally for the purpose of taking some sort of measurements and so these metrology operations are frequently the primary focus for further monitor reduction and/or elimination to reduce operational times. As the process matures, the value of process sampling becomes unclear. An economic and effective sampling in semiconductor manufacturing represents a significant opportunity for cost, or time reduction and improvement in operational efficiency without incurring a loss in quality or output through efficient allocation of the inspection resources.

During the wafer fabrication process, it is quite common to perform a five-location per wafer sample-averaged measurement of the parameters. In most cases, equal-size and even-spaced locations sample measurements within the wafer are taken at equal-spaced time epochs and this sampling scheme does not change in spite of process change over time. In current industrial practice three different layouts of measurement locations as in Figure 1 are commonly used.

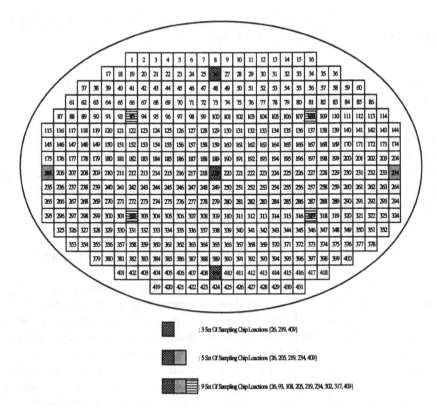

Figure 1. Three common layouts of measurement sampling locations of process parameter

These locations are chosen approximately evenly across the wafer, in order to have all regions of the wafer equally well represented, but they are not adequate if process-related defective chips are distributed with spatial pattern within the wafer. In practice, it is common for defective chips to occur in clusters or to display other systematic patterns. The spatial clustering phenomenon has been discussed extensively in the IC-fabrication literature (Mallory et. al., 1983).

In this study we present a hybrid approach of data mining techniques that generates an economic, efficient in-line measurement sampling method of process parameter based on the spatial information of defect in the *wafer bin map* (WBM) data. The WBM data show the classification results of chips within wafer into good and defective bins in probe test. The sampling method specifies the chip locations within the wafer to be measured and the optimal size of measured chip locations per wafer in order to represent a good sensitivity of 100% wafer coverage and defect detection for intelligent process control and cycle time reduction.

CYCLE TIME AND PROCESS CONTROL IN SEMICONDUCTOR FABRICATION

There is wide consensus within the semiconductor industry that TAT is an important measure of manufacturing performance. Stalk and Hout (1990) discuss the many potential benefits of TAT as a metric of manufacturing performance. The three main benefits of TAT reduction for memory device manufacturers can be summarized as follows:

1) Faster ramp up of yield due to faster process improvement.

2) Improved yield due to reduction of the time wafers are in the fab.

3) Cost savings due to reduction of *work in process* (WIP) inventory.

For defects can only be identified at probe, TAT reduction reduces the time that elapses between process defect occurrence and detection because it allows wafer lots to reach probe sooner. If a problem is detected at probe, a shorter TAT reduces the number of wafers that may have been exposed to the problem, and permits corrective action to be taken sooner.

In the operation of any semiconductor fabrication facility, process control is a crucial ingredient. The techniques and methodologies used for process control determine the effectiveness of each piece of equipment and influence the organization and operation of a facility. Figure 2 illustrates a generic model for a manufacturing process and related equipment with two types of inputs: the inputs that we have control over, and the disturbances that we have no control over (Montgomery, 1991).

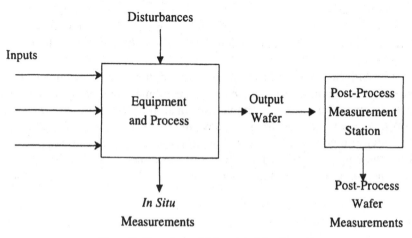

Figure 2. Generic model for manufacturing process

The inputs that we have control over include mask geometry, equipment settings, and incoming work material. We can alter the output of a process by changing these controllable inputs. Disturbances are those inputs which are subject to unintended and undesired variations. They include variations in the properties of incoming work material, variations in the equipment settings, and variations in other factors affecting the process. There are two types of outputs: the output wafer itself and *in situ* state. The *in situ* state is measured inside the equipment itself while the process is taking place resulting in *in situ* measurements. The output wafer is often measured post-process at a separate station resulting in post-process wafer measurements. Examples of post-process measurements include thickness measurements on deposited films, and linewidth measurements.

SPATIAL VARIATION AND IN-LINE SAMPLING

As device and interconnect dimensions continue to shrink and wafer sizes increasing, maintaining process uniformity is increasing in importance and difficulty. An understanding of variation, particularly spatial variation, is essential to both control the process and to design manufacturable high-performance circuit. Spatial variation across each wafer results from equipment or process limitations and variation within each chip may be exacerbated further by complex pattern dependencies. Variation in semiconductor manufacturing appears largely at four different scales in time and space: lot-to-lot, wafer-to-wafer, within wafer (wafer-level), and intra-chip (chip-level).

Lot–to-lot variation is the tendency of the lot mean of a device or process parameter (e.g. the mean of channel length computed over the entire lot) to vary from one lot to the next. Lot-to-lot variation is often monitored using statistical process control and may be compensated for using run by run or other feedback control approaches, e.g. (Friedman et. al., 1993). Wafer-to-wafer variation may be either temporal or spatial in nature. Temporal wafer-to-wafer variation is generally caused by drift in process equipment operation from one wafer to the next. This variation is increasing in importance as single-wafer processing equipment expands in use. Spatial wafer-to-wafer variation may also result from non-ideal process equipment, e.g. due to different positions of wafers in a boat during a batch furnace step.

Wafer-level variation is generally caused by additional equipment nonuniformity and other physical effects such as thermal gradients and loading phenomena. It often exhibits symmetrical properties such as radial (or "bull's eye") patterns or slanted planes. Intra-chip variation is often caused by layout and topography interaction with the process and key examples include pattern planarization in chemical mechanical polishing (Telfeyan et. al., 1996).

These spatial variations in semiconductor manufacturing finally cause the spatial pattern of defects on a wafer because of a design's intolerance of them and the analysis of spatial defect pattern will give valuable insight into the root cause. For example, a radial pattern, with defects in the pattern of "spokes" will point to a spin type process, such as resist coating or spin-on dielectric. Very little work has been done, however, to systematically exploit this information for yield and process improvement. Most of the existing literature is focused on examining the effect of clustering on yield prediction (Cunningham, 1990). Albin et. al. (1989), for example, proposed control charts for monitoring yield in the presence of spatial clustering. Their work is based on the fitting "overdispersion" models such as the negative binomial or Neyman-Type A dispersions to account for the excess variation and using these to develop appropriate control limits. In this study, we use that spatial pattern of defective chips on the whole wafers to design an optimal sampling method for process parameter monitoring and control.

The objectives of inspection sampling for in-line process monitoring are grouped into the following three categories:
1) Excursion Monitoring and Control
The objective here is to monitor the process at frequent intervals so that any process deviations are caught and the causes for the process excursion are fixed.
2) Yield Prediction and Wafer Start Adjustment
The objective here is solely sampling for yield prediction of lots and wafers so as to adjust the fab releases (and sometimes even lot sequencing at process and machine level) to meet demand schedules (to meet delivery dates or minimize penalty and inventory holding costs).
3) Defect Detection and Learning
The main objective of sampling here is to learn about the different defect types and their causal mechanisms; killer rates, the rate of sampling has to enable defect detection at a rate that is matched to that of root-cause analysis and problem fixing.

The sampling strategy could be significantly different for each of these objectives and in our work we focus on excursion monitoring and control of process parameter. Using the fact that the yield loss is increasing as a function of the number of defects found on the wafer, many researches about in-line defect sampling plan focused on the monitoring of defects on the wafer. Integrating the traditional cost-based approach with declustering, and using the random defect data, Raman et al. (1996) presented a comprehensive cost-based defect sampling plan to extend the existing models and the framework for developing defect sampling methodologies for different objectives such as yield learning. Little work has been done, however, to sampling plan for process-dependent measurements such as CD (*Critical Dimension*), thickness. Cost-based approaches to sampling plan design for *statistical process control* (SPC) have appeared in the literature as early as 1931 (Shewart, 1931). Lorenzen and Vance (1986) have unified all the past approaches into a common framework.

DEVELOPMENT OF OPTIMAL SAMPLING METHOD

We propose the methodology for generating optimal sampling method of post-process measurements by applying optimal data mining techniques such as SOM (*Self-Organizing Map*) neural network, statistical homogeneity test, to WBM data within a certain time period. The framework of our methodology is presented in Figure 3.

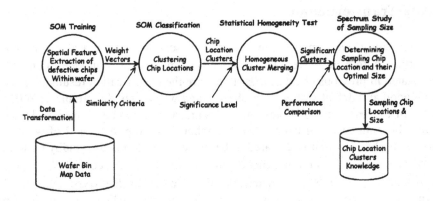

Figure 3. Framework of optimal sampling generation methodology

First of all, we focus on existing spatial patterns of historical WBM data that are sensitive enough to explain the complex spatial signatures left by the various process steps. And so we extract features of defect pattern on all chip locations within the wafer through SOM training of them as the first step of the sampling plan development. The SOM was developed in its present form by Kohonen (1982) and it has demonstrated its efficiency in various engineering problems (Kohonen et. al., 1984) including clustering, the recognition of patterns, the reduction of dimensions and the extraction of features. More in Data Mining veneer SOM has been used in visualization of complex process and systems and discovery dependencies and abstractions from raw data (Kaski, 1997).

In the second step, we construct the set of chip location clusters that have similar feature through SOM classification process and then merge homogeneous clusters into significant clusters through statistical homogeneity test on their observed defect data in the third step. We determine the sampling chip locations set that consists of the matched chip location to the resulted significant cluster through SOM classification in the third step and optimal sampling size by the performance comparison through spectrum analysis of sampling size within the possible inspection capability in the final step. These mined features, chip location clusters and sampling chip location results are accumulated in the chip location cluster knowledge of data mining system to use them later.

Data Transformation

In this study, we consider the WBM data scheme with l wafers, m chip locations within the wafer, and n defective bin classification (see Figure 4). As stated in previous chapter, WBM data are the classification results of chips into good and defective bins in probe test and we can see the sample of WBM data in the left hand of Figure 4. In Figure 4, the numbers on chip denote those bin numbers. For example, the number 3 on chip location 2 of wafer 1 means by bin 3 which is declared to be defective in terms of standby current. The number of m and n depends on the type of device but l is not deterministic. One way to determine l is to fix as the fixed number of fabricated wafers and a knowledge expert for the application domain can determine based on his or her knowledge of application.

We shall refer to i as the wafer number, and j as the chip location number within the wafer, and k as the defective bin number. Each chip on wafer i is subjected to a series of n bin tests and if jth chip is declared to be defective in kth test, then we display $Xijk$ as 1, otherwise $Xijk$ as 0. Under certain conditions, the set of total kth defect frequency at chip location j across the l wafers $\left\{ X_{\bullet jk} = \sum_{i=1}^{l} X_{ijk}, 1 \leq k \leq n \right\}$ will follow a multinomial distribution. We can transform the raw WBM data as in Figure 4.

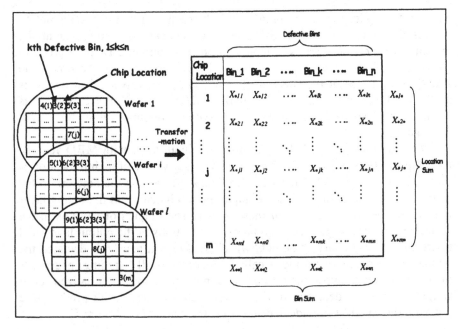

Figure 4. Wafer Bin Map data and their transformation

Multinomial Bin Distribution Feature Extraction through SOM Training

In order to extract the existing *multinomial bin distribution* (MBD) features on chip locations within the fabricated wafers, we are going to start SOM training. When starting the SOM training, for all chip locations of within the wafer, the MBD vectors $X_j = (X_{\bullet j1}, X_{\bullet j2}, \ldots\ldots, X_{\bullet jk}, \ldots\ldots, X_{\bullet jn}), 1 \leq j \leq m$, is first normalized so that they have values between 0 and 1 inclusive. Normalization equalizes the scale for computing distances. In our case, we use the normalized bin vector $\hat{X}_j$, which is obtained by dividing the bin distribution vector of each chip location by the corresponding chip location sum, as normalized input vector of SOM. We intend to compare all chip locations within the wafer in terms of all bins portion.

The next stage is to develop a neural architecture in accordance with MBD data. The SOM network is made up of two layers. *n* neurons in the input layer is chosen, that is to say, the same as the number of test bins, and P^2 neurons in the output layer arranged in a *P*P* square grid in order to adequately accommodate the *m* input patterns (vectors), the same as the number of chip locations within the wafer. If the size of neurons in the output layer is chosen to be too small, SOM network can not be trained to map all the existing MBD characteristics of input patterns onto output 2-D space properly. Otherwise, if the size is too lavishly large, the network becomes too inefficient in terms of the speed or the computation involved. However, it is preferable to construct a little large-sized SOM network and to merge its clusters into small-sized network to avoid oversimplifying the input distribution.

SOM training proceeds by presenting the input layer with MBD vectors, one chip location at a time. When the MBD input vector of a chip location is presented to the network, the values flow forward through the network to the neurons in the output layer. The net input into each neuron in the output layer is equal to the weighted sum of these values. The outputs of neurons in the output layer will compete through lateral connections in the output layer, and the neuron with the largest output "wins". This output neuron is declared the winner in the competition and is called winner-takes-all (winning) neuron. If the dot products between the normalized input vector and a normalized set of weight vectors are determined, the neuron with the largest dot product (the one with the smallest Euclidean distance) is the winner (see Figure 5).

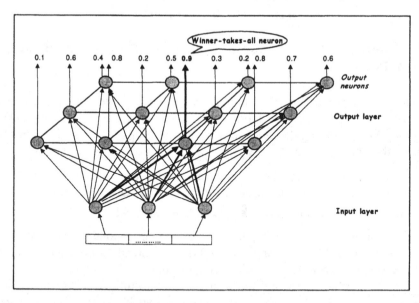

Figure 5. Winner-takes-all neuron and its path

Consider a normalized set of weight vectors lying around a unit circle, with the normalized input vector displayed graphically on the same circle (Caudill, 1987), as shown in Figure 6. As learning involves adjustment of weight vectors, only the neurons within neighborhood of the winning neuron are allowed to learn with a presented input vector: adjusting their weights closer to the input vector carries out learning for the neurons within the neighborhood as shown in Figure 6. The neighborhood is defined by physical proximity of the neurons to the winning neuron. The size of the neighborhood is initially chosen to be fairly large, and may include all neurons in the output layer. Hence, a large number of neurons have their weights adjusted at the start. However, as learning proceeds, the size of the neighborhood is progressively reduced to a pre-defined limit. Thus during the latter stages, fewer neurons have their weights adjusted closer to the input vector. Lateral inhibition of weight vectors that are distant from the input vector may also be carried out as shown in Figure 6.

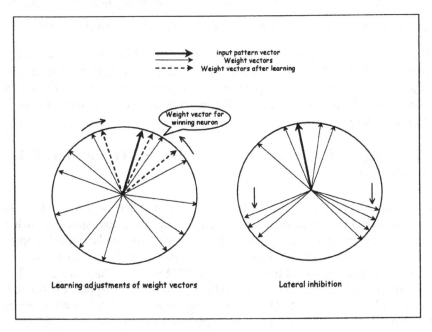

Figure 6. Competitive learning between neurons

Kohonen (1982) provided the following formula for updating the weights of the winning neuron q^* and the neurons within its neighborhood $Nbd(q^*)$ at discrete time t:

$$\hat{\mathbf{W}}_q(t+1) = \hat{\mathbf{W}}_q(t) + \alpha(t)[\hat{\mathbf{X}}(t) - \hat{\mathbf{W}}_q(t)], q \in Nbd(q^*) \qquad (1)$$

where $\hat{\mathbf{X}}$ is the (presented) normalized input vector, $\hat{\mathbf{W}}_q$ is the normalized weight vector from the input layer to output neuron q and $\alpha(t)$ is the corresponding value of the learning rate parameter.

There is no theoretical basis for the selection of $\alpha(t)$ and it is usually determined by a process of trial and error. Nevertheless, Kohonen (1989) provide a useful guide of it. In particular, during the first thousand iterations or so $\alpha(t)$ should begin with a value close to unity; thereafter, $\alpha(t)$ should decrease gradually, but staying above 0.1. The effect of the update equation (1) is to move the weight vector $\hat{\mathbf{W}}_q$ of the winning neuron $q*$ and the weight vectors of its neighboring neurons $Nbd(q*)$ toward the presented MBD vector $\hat{\mathbf{X}}$. Upon repeated presentations of the MBD vector, the weight vectors tend to follow the distribution of the input vectors due to the neighborhood updating. These processes are repeated until the map formation is completed. Once it is completed, the weights are fixed and the network is ready to used.

Clustering Chip Locations within the wafer through SOM Classification

SOM tries to project the multidimensional input space, which in our case could be MBD information of chip location, into the output space in such a way that the input patterns whose variables present similar values appear close to one another on the map which is created. For the clustering of chip locations, two-dimensional trained map in SOM training phase are used. The main question to be answered is: Which part of the mapped distribution corresponds best to the MBD data of a chip location? In other words, where the chip locations MBD patterns are located on the trained map. The maps have a regular structure, made out of an array of neurons. Every neuron can be described as a n-dimensional bins vector, calling weight vector.

Once the SOM has completed its learning session, the weights associated with each neuron can be fixed, the trained MBD patterns of all chip locations within the wafer can be re-presented for classification by checking the winner-takes-all neurons. If θ is the intersection angle between $\hat{\mathbf{X}}_j$ for some j , $1 \le j \le m$ and $\hat{\mathbf{W}}_q$ in the update equation (1), $\hat{\mathbf{X}}_j \cdot \hat{\mathbf{W}}_q^{t} = \left|\hat{\mathbf{X}}_j\right|\left|\hat{\mathbf{W}}_q\right| \cos\theta$ indicates the similarity measure between $\hat{\mathbf{X}}_j$ and $\hat{\mathbf{W}}_q$. We define it as "similarity score".

The similarity score plays an important role in clustering of chip locations within the wafer. The trained SOM network classifies the MBD pattern of a chip location by its similarity score to the trained weight vector associated with each output neuron. The higher the similarity score the more likely an input vector will pass the threshold described below and be allocated to a cluster. If the similarity score of an input is larger than the threshold, the input vector will be recognized as one of the trained weight vector patterns.

In our methodology we redefine winner-takes-all neuron as all neurons that are activated by overcoming threshold; we call that threshold "similarity criteria". In other words, the winner-takes-all neuron of the MBD pattern of a chip location is the neuron that has higher similarity score than the predefined similarity criteria, and can be several output neurons; the maximum number of that is the number of output neurons. And, the similarity criteria is the lowest level of similarity to be recognized as a cluster, and in the range between 0 and 1.

The SOM classification is introduced to divide all chip locations within the wafer into the set of (redefined) winner-takes-all neuron based on their similarity scores. And these results of all chip locations can be visualized efficiently via the chip location clusters map which offers information of correlations between all chip locations within the wafer and of the cluster structure of those. Because of the important role that humans have in Data Mining, visualization is essential in reporting results, or creating knowledge (Fayyad et. al., 1996) and that visualization consolidates the knowledge extracted by SOM training and classification.

Merging the Homogeneous Chip Location Clusters

The clusters constructed by SOM classification contain the chip locations that have the similar characteristics each other. The neighboring clusters have the property of topology-preserving, roughly meaning that similar patterns are gathered into a part of output SOM neurons and the initial size of clusters was required to be large enough to represent the characteristics of MBD on all chip locations within the wafer into the output layer of SOM network. So, we adopt the second cluster merging process that reduces the number of clusters.

The cluster merging process is carried out by statistical hypothesis testing for homogeneity of MBD between two clusters of all pairs of constructed clusters after clustering the chip locations through SOM classification, based on the MBD observation of all chip locations belonging to those clusters. In general, we have the following hypothesis for homogeneity test: Null Hypothesis H_0: $(P_{11}, P_{12},..., P_{1n}) = (P_{21}, P_{22},..., P_{2n})$ and Alternative hypothesis H_1: $(P_{11}, P_{12},..., P_{1n}) \neq (P_{21}, P_{22},..., P_{2n})$. Here, the P_{ij} is the proportion belonging to jth bin of the cluster i among two clusters in pair. Hypothesis testing is used together to ascertain, with some confidence, if the MBDs of two clusters are homogeneous or not. If the test rejects H_0, then we say that the two clusters are not homogeneous and consider as different clusters; otherwise we merge the two clusters and consider as one cluster that has similar MBD characteristics.

Determining Sampling Chip Locations and their optimal size through Spectrum Study of sampling size

For a given similarity criteria and significance level setting, we constructed the set of significant clusters through SOM classification and cluster merging phase. Hence, we prepared the phase of determining sampling chip locations and their optimal size. In this chapter we present procedures to determine answers to questions on which chips within the wafer need to be measured, and how many chips within the wafer need to be measured for a good sensitivity of 100% wafer coverage and defect detection.

The procedures are based on the simple idea that the sampling chip size for all significant clusters increases with the possession of each cluster within the wafer. If the overall coverage within the wafer can be explained by the sum of significant cluster coverage and null-cluster coverage, then the sampling size based on the limited inspection capability can be reallocated between significant clusters based on their respective coverage. For example, process A had a capacity of sampling 10 chip locations per wafer with a total of 100 chip locations. Based on the SOM classification and merging results we found that four significant clusters exist within the wafer and the number of chip locations belonging to each cluster is 10, 20, 30, and 40 respectively. In this model every chip location should belong to only one cluster that has the highest similarity score among significant clusters based on SOM classification results. By allocating 10 sampling chip locations per wafer in these cluster member ratios, we find that the sampling plan needs to select one chip location from the cluster 1, two from the cluster 2, three from the cluster 3, and four from the cluster 4.

A spectrum study of the sampling chip size within the possible inspection capacity will help us to determine the optimal sampling chip size. In other words, we choose the size with having a better performance in terms of the accurate representation of total distribution and defect detection among all possible sampling size. Since the common sampling methods have the maximum sampling size as 9 in current industrial practice like Figure 1, we consider the spectrum study of sampling size with range from one to ten in our work.

The next question is which chip locations within each cluster need to be inspected. Based on SOM classification results, we select only the corresponding size of chip locations having the highest similarity score in order among chip locations belonging to that cluster. In the above example, for cluster 1, we select one chip location having the highest similarity score among the chip locations belonging to cluster 1. For cluster 2 we select two chip locations having the highest and the second highest similarity score among the chip locations belonging to cluster 2. For cluster 3 we select three chip locations having the highest, the second, and the third highest similarity score among the chip locations belonging to cluster 3. For cluster 4 we select four chip locations having the highest, the second, the third, and the fourth highest similarity score among the chip locations belonging to cluster 4. It is clear that the similarity score of selected chip locations is higher than predefined similarity criteria.

ILLUSTRATIVE EXAMPLE

Example Data

In this study reported below, WBM data with a total of 431 chip locations within the wafer and 11 probe test bins were used as example. In general, for different product types and different companies, the number of bin classification and designated bin number differ. In case of our example, 32 bin classification levels are reserved, but only 11 bin levels among them are actually used. We consider 200 product wafers in a week from fab A which is in mature production phase. There is 9 sampling chip locations (number 26, 93, 108, 205, 219, 234, 302, 317, and 409) set up that are chosen approximately evenly across the wafer like Figure 1 in its fabrication line for the excursion monitoring purpose and it is still used.

MBD Feature Extraction through SOM Training

A neural network with 11 neurons is chosen in the input layer, and 9 neurons in the output layer arranged in a 3*3 square grid in order to limit resulting sampling size to nine chip locations for practical usage and performance comparison with the current sampling size. The MBD vectors $\{\mathbf{X}_j, 1 \leq j \leq 431\}$ of all chip locations within the wafer are first normalized by the way described in previous SOM training chapter and then fed into the SOM. A training algorithm of a SOM adjusts the weights using the 431 training patterns through the update equation (1). In our case, the initial neighborhood size and learning rate are set to 1 and 0.9 respectively. After the training of

SOM network is finished, we obtained the weight vectors which appear as Table 1. Here come 9 weight vectors out.

Table 1. The resulting weight vectors through SOM Training

cluster\Bin	Bin3	Bin4	Bin5	Bin6	Bin7	Bin8	Bin10	Bin16	Bin17	Bin18	Bin31
cluster1	0.0000	0.0000	0.0000	0.0000	0.0000	0.0000	0.0000	0.0000	0.1246	99.8754	0.0000
cluster2	0.0000	0.0000	0.0000	0.0000	0.0000	0.0000	0.0000	0.0000	4.0056	91.5007	4.4937
cluster3	1.9002	39.7599	6.6996	0.0745	9.9727	6.7318	33.4230	0.0585	0.1070	0.6989	0.5739
cluster4	1.0008	8.8337	7.8471	0.8333	0.0000	0.0000	0.0000	0.0000	0.0000	30.1652	51.3200
cluster5	1.5000	1.0000	4.0000	6.0000	0.0000	0.0000	0.0000	0.0000	0.0000	81.5000	6.0000
cluster6	4.1889	62.5421	17.0456	0.6691	1.0159	0.6386	6.3292	0.0521	0.4857	2.6561	4.3768
cluster7	3.1956	36.2996	12.1910	0.5991	3.7033	2.2527	12.9630	0.0502	0.8041	3.5980	24.3432
cluster8	10.0050	6.7567	66.6356	1.1249	0.0000	0.0000	0.0000	0.0000	0.5005	5.7413	9.2361
cluster9	2.2589	57.2801	5.5995	0.1698	5.6239	4.0583	23.3852	0.0565	0.1444	0.6349	0.7883

Clustering Chip Locations within the wafer through SOM Classification

Table 2 summarizes all the main results of SOM classification by two different similarity criteria setting, such as 0.9 and 0.95.

*Table 2. The SOM clustering results of all chip locations by
two different similarity criteria setting*

Cluster	Similarity Criteria	
	0.9	0.95
1	6 Chip Locations {1,2,326,353,420,431}	4 Chip Locations {1,326,353,431}
2	6 Chip Locations {1,2,326,353,420,431}	2 Chip Locations {2,420}
3	156 Chip Locations {20,21,.....,423,427}	92 Chip Locations {20,22,.....,423,427}
4	1 Chip Location {421}	1 Chip Location {421}
5	1 Chip Location {38}	1 Chip Location {38}
6	119 Chip Locations {5,6,.....,418,430}	21 Chip Locations {5,6,7,.....,403,418}
7	7 Chip Locations {115,174,206,327,328,425,426}	2 Chip Locations {425,426}
8	3 Chip Locations {325,354,379}	1 Chip Location {325}
9	235 Chip Locations {7,10,11,.....,424,428,429}	126 Chip Locations {7,10,14,......,405,411,429}

In case of SOM classification with 0.95 similarity criteria setting, 431 chip locations are divided to 9 chip location clusters. Table 3 shows the similarity score of chip locations obtained by SOM classification with similarity criteria setting of 0.95.

Table 3. The similarity score of chip locations in 0.95 similarity criteria setting

Cluster	Chip Location No.	Similarity Score	Cluster	Chip Location No.	Similarity Score	Cluster	Chip Location No.	Similarity Score
Cluster 1	326	0.9988	Cluster 4	421	0.9571	Cluster 9	373	0.9727
	353	0.9988	Cluster 5	38	1		153	0.9726
	431	0.9988	Cluster 6	116	0.9824		228	0.9725
	1	0.9962		403	0.9739		108	0.9724
Cluster 2	2	0.9951		342	0.9714		389	0.972
	420	0.9909		367	0.9706		315	0.9714
Cluster 3	89	0.9975		381	0.9702		348	0.971
	46	0.9968		62	0.9671		183	0.9706
	48	0.996		6	0.9651		198	0.9692
	82	0.9942		221	0.9646		137	0.9685
	27	0.9923		336	0.9609		7	0.9679
	180	0.9915		341	0.9585		304	0.9674
	141	0.9904		310	0.9572		429	0.9673
	239	0.9902		5	0.9563		217	0.967
	260	0.9886		252	0.9549		282	0.9666
	83	0.9884		251	0.9546		399	0.9664
	44	0.9875		279	0.9539		128	0.9659
	81	0.9872		429	0.9538		235	0.9658
	25	0.9866		314	0.9527		374	0.9656
	92	0.9866		418	0.952		162	0.9644
	63	0.9864		7	0.9517		284	0.9639
	121	0.9858		334	0.9512		199	0.9627
	119	0.9854		220	0.9508		218	0.9613
	179	0.9853	Cluster 7	426	0.9872		346	0.9612
	20	0.9852		425	0.9596		103	0.9609
	55	0.9851	Cluster 8	325	0.9667		166	0.9609
	85	0.9844	Cluster 9	372	0.9988		165	0.9605
	320	0.9841		197	0.998		318	0.9603
	52	0.984		219	0.9978		136	0.9602
	123	0.9836		222	0.9976		10	0.96
	47	0.9828		400	0.9964		73	0.9599
	93	0.9825		133	0.9959		366	0.9598
	110	0.982		265	0.9954		288	0.9592
	292	0.9819		164	0.9951		360	0.9571
	112	0.9818		127	0.9946		411	0.957
	323	0.9815		134	0.994		102	0.9568
	178	0.9809		213	0.9929		220	0.9568
	150	0.9799		274	0.9924		99	0.9561
	43	0.9796		312	0.992		314	0.956
	67	0.9795		194	0.9918		295	0.9559
	152	0.979		289	0.9918		357	0.9556
	410	0.979		163	0.9917		405	0.9554
	120	0.9786		352	0.9915		361	0.9544
	210	0.9776		332	0.9907		246	0.9538
	321	0.9775		77	0.9904		396	0.9532
	172	0.9772		305	0.9903		161	0.9529
	54	0.9764		193	0.99		259	0.9518
	34	0.9756		138	0.9898		336	0.9518
	237	0.9752		125	0.9894		392	0.9518
	118	0.9734		316	0.9893		160	0.9517
	40	0.9733		224	0.9885		342	0.9512
	32	0.9727		196	0.9875		279	0.951
	208	0.9726		126	0.9872		19	0.9508
	28	0.9724		368	0.9869		238	0.9508

Table 3. The similarity score of chip locations in 0.95 similarity criteria setting (continued)

Cluster	Chip Location No.	Similarity Score	Cluster	Chip Location No.	Similarity Score
Cluster 3	91	0.972	Cluster 9	338	0.9867
	209	0.9719		104	0.9864
	22	0.9716		186	0.9861
	181	0.9715		129	0.9857
	49	0.9713		243	0.9857
	147	0.9694		175	0.9855
	31	0.969		216	0.9855
	201	0.9689		157	0.9854
	84	0.9676		317	0.9842
	423	0.9674		214	0.984
	261	0.9667		159	0.9839
	41	0.9659		370	0.9826
	412	0.9656		225	0.9824
	232	0.9655		130	0.9822
	299	0.9653		135	0.9821
	233	0.9646		404	0.9813
	268	0.9643		313	0.9811
	117	0.9639		365	0.9811
	51	0.9633		363	0.9808
	35	0.9626		358	0.9807
	406	0.9624		167	0.9805
	171	0.9622		106	0.9803
	142	0.9619		362	0.9803
	139	0.9615		98	0.9795
	149	0.9609		155	0.9794
	23	0.9603		205	0.9793
	291	0.9596		78	0.979
	58	0.9586		168	0.9789
	148	0.9582		254	0.9789
	350	0.9579		333	0.9789
	290	0.9578		14	0.9787
	68	0.9571		190	0.9781
	264	0.9569		303	0.9777
	53	0.9565		105	0.9774
	122	0.9555		107	0.9773
	203	0.9544		371	0.9769
	427	0.9543		345	0.9752
	80	0.9531		369	0.9749
	113	0.9528		188	0.9742
	33	0.9526		364	0.9736
	202	0.9526		131	0.9735
	39	0.9522		287	0.9734
	140	0.9519		242	0.9732
	59	0.95		258	0.9731

Note that 243 chip locations among a total of 431 chip locations belong to only one cluster that has the highest similarity score. For example, chip location 7 belongs to the cluster 9 that has the highest similarity score as 0.9679 rather than the cluster 6 that has a similarity score as 0.9517 (refer to Table 3).

Figure 7 shows the visualization which displays spatial patterns of chip location clusters within the wafer with similarity criteria setting of 0.95.

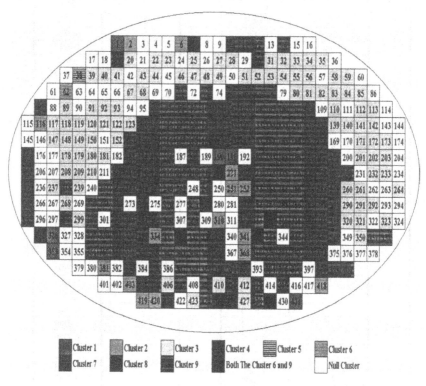

Figure 7. The chip location cluster map with similarity criteria setting 0.95

Note that there are many chip locations belonging to more than one cluster and in case of 0.9 similarity criteria setting, the cluster 1 and 2 have the same chip locations. The cluster 1 and 2 are homogeneous clusters having the same MBD characteristics.

Merging the Homogeneous Clusters

We found that in case of 0.9 similarity criteria setting, the cluster 1 and 2 are identical and we can merge those clusters into one cluster labeled C_{12}. So, in case of 0.9 similarity criteria setting we evaluate 28 cluster pairs of test for homogeneity, where 28 is made out of $_8C_2$, and in case of 0.95, 36 cluster pairs, where 36 out of $_9C_2$. Table 4 shows the observed MBD data which are obtained by summing the corresponding MBD vector of all chip locations belonging to each cluster, and Table 5 shows the results of homogeneity test for all cluster pairs in similarity criteria setting of 0.95.

Table 4. The observed bin distribution of all clusters in similarity criteria setting 0.95

Cluster_Observed_Sum	Count	Bh3	Bh4	Bh5	Bh6	Bh7	Bh8	Bh10	Bh16	Bh17	Bh18	Bh31	Sum
Cluster1_Observed_sum	4	0	0	0	0	0	0	0	0	1	799	0	800
Cluster2_Observed_sum	2	0	0	0	0	0	0	0	0	16	366	18	400
Cluster3_Observed_sum	92	370	7261	1245	19	1866	1266	6117	13	22	135	86	18400
Cluster4_Observed_sum	1	2	1	36	0	0	0	0	0	0	71	90	200
Cluster5_Observed_sum	1	3	2	8	12	0	0	0	0	0	163	12	200
Cluster6_Observed_sum	21	140	2604	459	20	109	93	660	3	13	57	42	4200
Cluster7_Observed_sum	2	8	145	39	1	11	8	76	1	16	4	91	400
Cluster8_Observed_sum	1	26	27	135	0	0	0	0	0	2	5	5	200
Cluster9_Observed_sum	126	535	14424	1395	56	1448	1020	5922	18	17	149	216	25200

Table 5. The result of homogeneity test for all pairs
of clusters in 0.95 similarity criteria setting

Cluster Pair	Homogeneity Test Results	
	Chi-Square Statistics	P-value
(1,2)	65.974	2.6382E-10
(1,3)	16265.27	0
(1,4)	294.023	2.8498E-57
(1,5)	151.091	2.2229E-27
(1,6)	4591.773	0
(1,7)	1174.794	3.937E-246
(1,8)	962.165	2.635E-200
(1,9)	21732.628	0
(2,3)	12878.133	0
(2,4)	263.563	7.5947E-51
(2,5)	58.401	7.2645E-09
(2,6)	3722.001	0
(2,7)	689.751	9.953E-142
(2,8)	551.032	5.39E-112
(2,9)	17064.611	0
(3,4)	6501.001	0
(3,5)	8739.309	0
(3,6)	1268.356	2.579E-266
(3,7)	2424.532	0
(3,8)	1263.578	2.769E-265
(3,9)	1540.313	0
(4,5)	125.873	3.2394E-22
(4,6)	2166.414	0
(4,7)	281.745	1.1152E-54
(4,8)	237.118	2.7592E-45
(4,9)	6217.486	0
(5,6)	2736.568	0
(5,7)	480.404	6.7757E-97
(5,8)	317.483	3.1118E-62
(5,9)	11031.13	0
(6,7)	724.135	4.129E-149
(6,8)	623.21	1.868E-127
(6,9)	460.639	1.1225E-92
(7,8)	293.149	4.3597E-57
(7,9)	2105.611	0
(8,9)	1541.427	0

We have used chi-square test statistics χ^2 with the degrees of freedom 10 and a *P*-value of 0.05 or more will be necessary to validate that the MBD characteristics between two clusters in pair are statistically homogeneous. We can infer from Table 5 that all 9 clusters are significantly different each other.

Generation of the Proposed Sampling Method with 9 Sampling Size and Its Performance

For comparative study of performance with the *current sampling method* (CSM) which has 9 sampling chip locations, we assumed the sampling chip size of proposed sampling method is 9. By allocating 9 sampling chip locations in the number of chip locations ratios of each cluster, such as 4:2:92:1:1:21:2:1:126, we find that the sampling plan needs to select three chip locations from the cluster 3, one from the cluster 6, and five from the cluster 9. The number 3 of cluster 3 is calculated as:

$$\frac{92}{243} \times 9 = 3.41 \cong 3$$

and the number 1 of cluster 6 is calculated as:

$$\frac{17}{243} \times 9 = 0.63 \cong 1$$

and the number 5 of cluster 9 is calculated as:

$$\frac{123}{243} \times 9 = 4.56 \cong 5$$

The next question is which chip locations need to be inspected for the above clusters. From Table 3, we can select chip locations in similarity score order, such as 89, 46, and 48 among chip locations belonging to the cluster 3, and select chip location 116 in similarity score order among chip locations belonging to the cluster 6, and select chip locations in similarity score order, such as 372, 197, 219, 222, and 400 among chip locations belonging to the cluster 9. Therefore, we can finally generate the *proposed sampling method* PSM = {46, 48, 89, 116, 197, 219, 222, 372, 400} (see Figure 8).

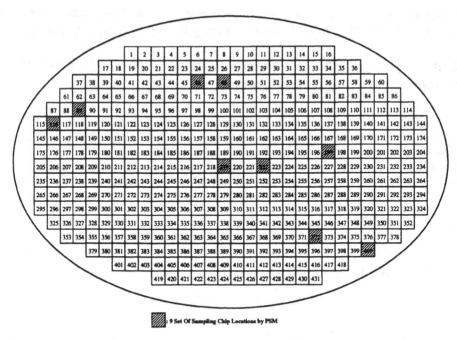

Figure 8. Generation of 9-set of sampling chip location by the proposed sampling method

After obtaining the observed MBD data from all sampling chip locations generated by the PSM (see Table 6), the performance was compared under two criteria, namely the accurate representation of the total MBD and the number of detected bins on its sampling locations.

Table 6. The observed multinomial bin distribution

Sampling Method	Bin3	Bin4	Bin5	Bin6	Bin7	Bin8	Bin10	Bin16	Bin17	Bin18	Bin31
PSM	52	932	145	8	102	90	434	3	2	20	12
CSM	36	867	120	0	137	99	514	1	2	14	10
Total Observation	1996	42606	6380	178	5668	3936	21263	47	162	2278	1686

The detailed comparison results are shown in Table 7 and 8. Here the P-value of homogeneity test between the observed bin distribution of each method and total MBD within the wafer is chosen to measure accurate representation of the total distribution of that method. Note that if P-value is higher, the sampling method has a better performance in terms of the accurate representation of total distribution.

Table 7. The results of representation accuracy by each method

Sampling Method	Homogeneity Test Results	
	Chi-Square Statistics	**P-value**
PSM	48.86	**4.3283E-07**
CSM	65.37	**3.4435E-10**

Table 8. The results of bin detection by each method

Sampling Method	Number of Detected Bins
PSM	11
CSM	10
Total Distribution	11

As the result suggests, the PSM performs the best in the accurate representation and bin detection. Compared to the CSM, the PSM gave 1257 times higher *P*-value than that of the CSM, where 1257 is made out of 4.3283E-07 ÷ 3.4435E-10, and one more bin detection, such as bin 6(refer to Table 8). However, considering relatively much lower than the 0.05 significance level, the PSM does not represent the total MBD well.

Determining Optimal Sampling Method through Spectrum Study of Sampling Size

Table 9 and 10 show the spectrum study results of sampling size with range from one to ten and performance results of three common types of CSM respectively.

Table 9. Spectrum study results of sampling size

Sampling Size	Proposed Sampling Chip Locations	Performance			
		Chi-square Statistics	P-value	Number of Detected Bin	Missing Bin
1	{372}	15.43	0.117	6	bin6,16,17,18,31
2	{89,372}	19.30	0.087	7	bin6,16,17,31
3	{89, 197, 372}	26.59	0.003	10	bin6
4	{46, 89, 197, 372}	36.49	6.929E-05	10	bin6
5	{46, 89, 197, 219, 372}	37.96	3.853E-05	10	bin6
6	{46, 89, 116, 197, 219, 372}	33.27	2.450E-04	11	
7	{46, 48, 89, 197, 219, 222, 372}	56.40	1.727E-08	11	
8	{46, 48, 89, 116, 197, 219, 222, 372}	44.36	2.836E-06	11	
9	{46, 48, 89, 116, 197, 219, 222, 372, 400}	48.86	4.328E-07	11	
10	{46, 48, 82, 89, 116, 197, 219, 222, 372, 400}	45.07	2.116E-06	11	

Table 10. Performance results of three types of current sampling method

Sampling Size	CSM			
	Chi-square Statistics	P-value	Number of Detected Bin	Missing Bin
3	21.66	0.017	10	bin 6
5	42.618	5.819E-06	10	bin 6
9	65.37	3.444E-10	10	bin 6

As the results suggest, the PSM with one sampling size performs the best in the accurate representation of total distribution and the PSMs with more than six sampling size perform the best in the bin detection. So, we can finally conclude that the PSM with six sampling size performs better than the other PSMs in terms of both side and it is optimal sampling method. In spite of three less chip locations than the CSM with 9 sampling size, it performs better in terms of both sides and accordingly, the use of it will present an intelligent process control and cycle time reduction.

CONCLUSIONS

Sample measurement inspecting for a process parameter is a necessity in semiconductor manufacturing because of the prohibitive amount of time involved in 100% inspection while maintaining sensitivity to all types of defects and abnormality. Yet, manufacturers does not know how to design an optimal sampling method to reduce the current cycle time and simultaneously provide an accurate representation of the true process parameter and defect population on the whole wafers.

The results presented in our work indicate that sampling can provide an good representation of the total bin distribution and existing bin of probe test if the sampled chip locations and their size are chosen carefully by data mining application to WBM data. From the results of the example using actual fab data, it is found that the proposed sampling method can be very effective in spite of smaller sampling size.

REFERENCES

Albin S. L., and Friedman D. J., "The Impact of Clustered Defect Distribution in IC Fabrication, Management Science," 35, 1066-1078, 1989

Caudill M., "Neural networks primer, part I," AI Expert, December, pp. 46-52, 1987

Cunningham J. A., "The Use and Evaluation of Yield Models in Integrated Circuit Manufacturing," IEEE Trans. Semiconduct. Manufact., 3, 60-71, 1990

Duncan A. J., "The economic design of X-bar charts used to maintain current control of a Process," J. Amer. Statistical Assoc., 51, 228-242 , 1956

Fayyad U., and Piatetssky -Shapiro G., et al., Advances in Knowledge Discovery and Data Mining, AAA Press/ MIT Press, California, 1996

Friedman D. J., and Hansen H., et al., A Method for Characterizing Defects in Integrated Circuits, U.S. Patent 5,240,866 , 1993

Kaski S., Data Exploration Using Self- Organizing Maps, PhD thesis, Helsinki University of Technology, 1997

Lorenzen T. J. and Vance L. C., "The Economic Design of Control Charts: A Unified Approach," Technometrics, 28, 3-10 , 1986

Mallory C. L., and Perloff D. S., et al., "Spatial Yield Analysis in Integrated Circuit Manufacture", Solid State Technology, 26, 121-127, 1983

Montgomery D. C., Introduction to Statistical Quality Control. New York: Wiley, 1991

Kohonen T., "Self-organized formation of topologically correct feature maps," Biological Cybernetics, vol. 43, pp. 59-69, 1982

Kohonen T., and Oja E., et al., "Engineering applications of the self-organizing map," Proceedings of the IEEE, 1984

Raman K., Ram Akella, and Andrzej J. Strojwas, "In-Line Defect Sampling Methodology in Yield Management: An Integrated Framework," IEEE Trans. Semiconduct. Manufact., 9, 506-517 , 1996

Shewart W. A., Economic Control of Quality of Manufactured Product. New York: Van Nostrand, 1931

Stalk G. and Hout T. M., Competing Against Time. New York: Free Press, 1990.

Telfeyan R., and Moyne J., et al., "A Multilevel Approach To The Control Of A Chemical-Mechanical Planarization Process", J. Vac. Sci. Tech. A, Vol. 14, No 3, 1907-1913, 1996

CHAPTER 9

Data Mining for Process and Quality Control in the Semiconductor Industry

Mark Last
mlast@bgumail.bgu.ac.il
Department of Information Systems Engineering, Ben-Gurion University,
Beer-Sheva 84105, Israel

Abraham Kandel
kandel@csee.usf.edu
Department of Computer Science and Engineering, University of South
Florida , 4202 E. Fowler Avenue, ENB 118, Tampa, FL 33620 USA

ABSTRACT

Like in any other industry, manufacturing departments of semiconductor plants are evaluated by their ability to meet the delivery schedules. However, the final quantities and the flow times of individual semiconductor batches are affected by multiple uncertain factors, like material quality, process variability, equipment condition, and others. Thus, the tasks of predicting the batch quality (measured by yield) and its total flow time are an important part of the production planning activities. Beyond prediction, the plant management is interested in identifying the main causes of yield excursion and process delays.

In this paper, we are applying several methods of data mining and knowledge discovery to WIP (Work-in-Process) data, collected in a semiconductor plant. The information on each manufacturing batch includes its design parameters, process tracking data, line yield, etc. The data is prepared for data mining by converting a sequential dataset into a relational format. Classification models for predicting line yield and flow times are built from the pre-processed data by using the Info-Fuzzy Network (IFN) methodology. Fuzzy-based techniques of automated perception are used for post-processing the data mining results. We conclude the paper with a critical evaluation of the discovered knowledge and the methods used.

D. Braha (ed.), Data Mining for Design and Manufacturing, 207–234.
© 2001 Kluwer Academic Publishers. Printed in the Netherlands.

INTRODUCTION

The outgoing *yield* of manufactured batches is the basic measure of profitability in the semiconductor industry. Overall, or line yield of the manufacturing process is defined as the ratio between the number of good parts (microelectronic chips) in a completed batch and the maximum number of parts, which can be obtained from the same batch. Since capitalization costs constitute the major part of manufacturing costs in the semiconductor industry, the *direct cost* of producing a single batch is almost fixed. However, the *income* from a given batch is proportional to the number of good chips. Thus, there is a direct relationship between the yield and the profits of semiconductor companies, which usually treat the yield performance as one of their top commercial secrets.

The expected yield of every batch is also an important parameter for the production planning and control. An "optimistic" estimate of the outgoing yield may cause delays in the order delivery due to insufficient quantities produced. On the other hand, "overbooking" in the number of batches designated for a given order may lead to a waste of precious resources, like technicians, machines, and electricity. Unnecessary batches may also cause delays in the production of other, more critical batches. Consequently, this is the primary interest of the planning personnel to have an accurate prediction of the actual yield.

The problem of yield prediction is closely related to another problem of planning the supply of orders, namely the problem of predicting the *flow times* of individual batches. Though the net time of each production step in the semiconductor industry is nearly fixed, the waiting times between the operations are very hard to predict, due to such variable factors as equipment downtimes, process failures, and line congestions.

Controlling and preserving the yield is a complex engineering problem. Both new and mature semiconductor products suffer from variability of yield within and between individual batches and even on specific wafers of the same batch. Improved understanding of this variability can save significant manufacturing costs by focusing on problematic processes and taking appropriate actions, whenever excursion of yield is expected for a given batch, wafer, etc.

Although the amount of manufacturing data collected by semiconductor companies is constantly increasing, it is still hard to identify the most important parameters required for yield modeling and prediction. As indicated by Tobin et al. (2000), the amount of data generated is exceeding the yield engineer's ability to effectively monitor and correct unexpected trends and excursions. Consequently, there is a strong need for automated yield management systems, which will be able to explain and predict yield excursions by using sophisticated data management and data mining tools.

The use of standard data mining methods for the yield and the flow time prediction in the semiconductor industry faces several difficulties. First, such common methods as association rules (see Agrawal et al., 1996) and decision trees (see Quinlan, 1993) are aimed at analyzing structured data, organized in two-dimensional tables, where all the variables relevant to the target are located in a single row (record) of one table. In a real-world situation, the factors affecting the final yield of a manufacturing batch are spread across many data tables, which may be even stored in separate database systems. Moreover, multiple records of parametric data, collected in the course of the manufacturing process, may be related to the outcome of a single batch. Important information may be lost if the timing of those records is ignored by the data mining tools.

Second, the basic assumption of most data mining methods is that all the data stored in a database is complete and correct. The actual semiconductor data may not comply with the assumption of data correctness for several reasons including erroneous data entry, inaccurate measurements, etc. Since it is not practical to ensure the 100% data quality, the methods used should be robust to the presence of errors in the mined data. Another common problem of data quality is the absence of values for certain attributes due to either incomplete data entry process or non-existence of the values themselves (e.g., missing measurements of a scraped wafer). In both cases, a data mining method should deal with a missing value without ignoring the values of the other attributes in the same record.

Finally, the abundance and the diversity of automatically collected data pose another potential problem to data mining. Most of the available attributes may be completely irrelevant to the target (e.g., the final yield). Since most data mining methods (especially those based on the Bayesian approach) are trying to incorporate as many attributes as possible in the model, the learning process may be misled by the irrelevant attributes, resulting in over complex and inaccurate representations of discovered knowledge. The continuous nature of the target (dependent) variables, like the yield and the flow time, precludes them from being used directly by *classification* techniques (e.g., decision trees and Bayesian methods), which assume the target to take a limited number of distinct values.

The process of knowledge discovery in semiconductor databases, presented in this paper, is based on a novel method for mining real-world data, termed IFN for Info-Fuzzy Network (see Maimon and Last, 2000). The method builds upon the principles of Shannon's information theory (see Cover, 1991) and fuzzy logic (see Klir and Yuan, 1995). It is applicable to databases of mixed nature containing quantitative (continuous), qualitative (nominal), and binary-valued attributes. The relevant features are selected automatically in the process of constructing the prediction model (IFN). Ranked association rules between the selected factors and the target attribute (e.g., final yield) can be extracted from the network structure. The network output can also be used for evaluating reliability of target data. In (Maimon and Last, 2000), the IFN method is shown to have several benefits, including: built-in feature selection, compact and interpretable representation of extracted knowledge, reasonable predictive accuracy (compared to other methods), stability of obtained models, and robustness to noisy and incomplete data.

Section 2 of this Chapter describes the information-theoretic fuzzy approach to knowledge discovery in databases. The process of rule fuzzification and reduction is presented in Section 3. In Section 4, we proceed with a detailed case study of knowledge discovery in manufacturing data. The Chapter is concluded by Section 5, which evaluates the potential of the information-fuzzy approach for data mining in the semiconductor industry.

THE IFN METHODOLOGY FOR KNOWLEDGE DISCOVERY AND DATA MINING

The *Info-Fuzzy Network* (IFN) methodology, presented by us in (Maimon and Last, 2000), is a novel and unified approach to automating the process of *Knowledge Discovery in Databases* (KDD). The main stages of the KDD process handled by IFN include discretization of continuous attributes, feature selection, prediction and classification, extraction of association rules, and data cleaning. The method is aimed at maximizing the *mutual information* (see Cover 1991) between input (predicting) and target (dependent) attributes. The following sub-sections describe the data model used by IFN, the general structure of an info-fuzzy network, the network construction algorithm, and the procedure for extracting association rules from the network structure.

The Data Model

The IFN method distinguishes between the following types of attributes in a database:

1) O - a subset of *target* ("output") attributes ($O \subset R$, $|O| \geq 1$). The information-theoretic network is constructed to predict the values of target attributes, based on the values of *input* attributes (see below).

2) C - a subset of *candidate input* attributes ($C \subset R$, $|C| \geq 1$). This is a subset of attributes, which *can be* used to predict the values of *target* attributes.

3) I_i - a subset of *input* attributes selected by the network construction procedure for predicting the value of the target attribute i ($\forall i: I_i \subset C$).

A database is assumed to satisfy the following conditions:

1) 40 $\forall i : I_i \cap O = \varnothing$ (An attribute cannot be both input and a target). This means that no cyclic relationships between attributes may be revealed by the method.

2) $\forall i : I_i \cup O \subseteq R$ (Some attributes may be neither input, nor target). Usually, the key attributes are not used in the knowledge discovery process, since they have no practical meaning, except for the purpose of identifying individual records.

The Info-Fuzzy Network Structure

An Info-Fuzzy Network (see Figure 1 below) has the following components:

1) $|I_i|$ - total number of hidden (internal) layers in a network of a target attribute i. The network in Figure 1 has two internal layers (No. 1 and No. 2). Each internal layer is uniquely associated with an input attribute by representing the interaction of that attribute and the input attributes of the previous layers. Layer No. 0 includes only the root node and is not associated with any input attribute. The layers of the network differ from the decision-tree structure used by CART (Breiman et al., 1984) and C4.5 (Quinlan, 1993) in the following aspects: only one input attribute is used to split the nodes of the same layer, multiple splits of continuous attributes are allowed, and the partitioning of continuous attributes is identical at all the split nodes.

2) L_l - a subset of nodes z in a hidden layer No. l . Each node represents a conjunction of values of the first l input attributes, which is similar to the definition of an internal node in a standard decision tree. In Figure 1, the first input attribute has three values, represented by nodes no. 1,2, and 3 in the first layer, but only nodes no. 1 and 3 are split due to the statistical significance testing (see next sub-section). The second layer has four nodes standing for the combinations of two values of the second input attribute with two split nodes of the first layer.

3) K_i - a subset of distinct target nodes V_{ij} in a network of the target attribute i (the target layer). $|K_i| = M_i$. Each target node is associated with a value in the domain of the target attribute i. This layer is missing in the standard decision-tree structure. In our example, the target attribute has three values, represented by three nodes in the target layer.

4) w_z^{ij}- an information-theoretic weight of the connection between a terminal (unsplit) node z and a target node V_{ij}. Each connection represents an association between a conjunction of input attribute-values and a value of the target attribute. In Figure 1, there are 15 connections between the five terminal (unsplit) nodes and the three target nodes. The calculation of the weights is explained in the next sub-sections.

The connectionist nature of our system (each terminal node may be connected to every target node) resembles the structure of multi-layer Neural Networks (see Mitchell, 1997). Consequently, we define our system as a *network* and not as a *tree*.

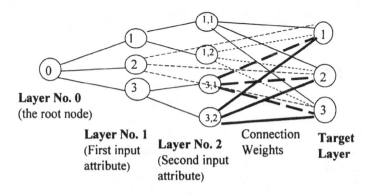

Figure 1 IFN: An Example

Network Construction Procedure

Without loss of generality, we present here an algorithm for constructing an information-theoretic network of a *single* target attribute A_i. In a general case, the network should be re-built for every target attribute defined in a database. The pseudocode of the network construction procedure is shown in Table 1 below.

Table 1 The IFN Construction Algorithm

Inputs

A_i is the target attribute

C is a subset of candidate input attributes

S is a subset of training examples

α is the significance level used by the algorithm

Procedure **IFN** (A_i, C, S, α)

$I_i = \varnothing$ // *initialize the set of input attributes to an empty set.*

$|L_0| = 1$ // *Initialize layer 0 to the root node*

Define *the layer of target nodes*

For $l = 1$ to $|C|$ // *repeat for the maximum number of layers (number of candidate input attributes)*

> For $i' = 1$ to $|C|$ // *repeat for every candidate input attribute*
>
>> If $A_{i'} \notin I_i$ // *if an attribute is not an input attribute*
>>
>>> $MI\ (A_{i'}; A_{i'}) = 0$ // *Initialize the mutual information between the attribute $A_{i'}$ and the attribute A_i to zero.*
>>>
>>> For $z = 1$ to $|L_{l-1}|$ // *repeat for every node of the final hidden layer*
>>>
>>>> Calculate $MI\ (A_{i'}\ ;\ A_i\ /\ z)$ // *calculate conditional mutual information of a candidate input attribute i' and a target attribute i, given a node z (see below)*
>>>>
>>>> If $MI\ (A_{i'}\ ;\ A_i\ /\ z)$ is greater than zero at the significance level α (see below), $MI\ (A_{i'}; A_{i'}) = MI\ (A_i; A_{i'}) + MI\ (A_{i'}\ ;\ A_i\ /\ z)$

> If max $MI\ (A_{i'}; A_i\ ) = 0$, **Stop and** return the set of input attributes (I_i)
> i'

> $i^* = \arg$ max $MI\ (A_{i'}; A_i\ )$ // *find the best input attribute*
> i'

> $I_i = I_i \cap A_{i^*}$ // *update the set of input attributes*
>
> Update the network structure with a new layer of hidden nodes

Calculate the information-theoretic weights of the input-target connections (see below)

Return the set of input attributes (I_i)

The network construction procedure starts with a single-node network representing an empty set of input attributes. A node in the network is split if it provides a statistically significant decrease in the *conditional entropy* of the target attribute. As indicated in (Cover, 1991), conditional entropy measures the uncertainty of a random variable Y, given the values of other variables X_1, ..., X_n. A decrease in the conditional entropy of a random variable is termed *"conditional mutual information."* The conditional mutual information of a candidate input attribute i' and a target attribute i, given a node z ($MI\ (A_{i'}\ ;\ A_i\ /\ z)$), is estimated by the following formula (based on Cover, 1991):

$$MI\ (A_{i'}; A_i\ /\ z) = \sum_{j=0}^{M_i-1} \sum_{j'=0}^{M_{i'}-1} P(V_{ij}; V_{i'j'}; z) \bullet \log \frac{P(V_{i'j'}^{ij}\ /\ z)}{P(V_{i'j'}\ /\ z) \bullet P(V_{ij}\ /\ z)}$$

where

M_i / $M_{i'}$ - number of values of the target attribute i /candidate input attribute i'. This formula assumes that all continuous attributes are discretized to a finite number of intervals.

P $(V_{i'j'}/z)$ - an estimated conditional (*a posteriori*) probability of a value j' of the candidate input attribute i', given the node z.

P (V_{ij}/z) - an estimated conditional (*a posteriori*) probability of a value j of the target attribute i, given the node z.

P $(V_{i'j}{}^{ij}/z)$ - an estimated conditional (*a posteriori*) probability of a value j' of the candidate input attribute i' and a value j of the target attribute i, given the node z.

P $(V_{ij};\ V_{i'j'};\ z)$ - an estimated joint probability of a value j of the target attribute i, a value j' of the candidate input attribute i' and the node z.

The statistical significance of the estimated conditional mutual information, is evaluated by using the likelihood-ratio statistic (based on Attneave, 1959):

$$G^2\ (A_{i'}\ ;\ A_i\ /\ z) = 2 \bullet (ln2) \bullet E^*(z)\ \bullet MI\ (A_{i'}\ ;\ A_i\ /\ z)$$

Where $E^*(z)$ is the number of tuples associated with the node z.

The null hypothesis (H_0) of the likelihood-ratio test is that the conditional mutual information is zero (which means that the attributes are conditionally independent, given the node). If H_0 holds, then the likelihood-ratio statistic G^2 $(A_{i'};\ A_i\ /\ z)$ is distributed as chi-square with $(NI_{i'}\ (z) - 1) \bullet (NT_i\ (z) - 1)$ degrees of freedom, where $NI_{i'}\ (z)$ is the number of values of a candidate input attribute i' at node z and $NT_i\ (z)$ is the number of values of a target attribute i at node z (based on Rao and Toutenburg, 1995). The default significance level (*p-value*), used by the information-theoretic algorithm, is 0.1%. We have found empirically that the higher values of the *p-value* tend to decrease the generalization performance of the network.

A new input attribute is selected to maximize the total significant decrease in the conditional entropy as a result of splitting the nodes of the last layer. The nodes of a new hidden layer are defined for a Cartesian product of split nodes of the previous hidden layer and the values of the new input attribute. According to the chain rule (see Cover, 1991), the *mutual information* between a set of input attributes and the target (defined as the overall decrease in the conditional entropy) is equal to the sum of drops in conditional entropy at all the hidden layers. If a candidate input attribute significantly decreasing the conditional entropy of the target attribute cannot be found, the network construction stops.

The conditional entropy of the target attribute can only be calculated with respect to attributes taking a finite number of values. The algorithm performs discretization of continuous attributes "on-the-fly" by using an approach, which is similar to the information-theoretic heuristic of Fayyad and Irani

(1993): recursively finding a binary partition of an input attribute that minimizes the conditional entropy of the target attribute. However, the stopping criterion we are using is different. Rather than searching for a *minimum description length* (minimum number of bits for encoding the training data), we make use of a standard statistical *likelihood-ratio test* (see above). The search for the best partition of a continuous attribute is *dynamic*: it is performed each time a candidate input attribute is considered for selection. Detailed descriptions of the algorithm steps, including the dynamic discretization procedure, are provided in (Maimon and Last, 2000).

The IFN construction procedure is a highly scalable algorithm. As shown in (Maimon and Last, 2000), its run time is quadratic in the number of candidate input attributes. It is also linear in the number of records and the number of values taken by each candidate input / target attribute. The dynamic discretization procedure increases the run time per each continuous attribute by the factor of $m \ log \ m$, where m is the total number of distinct attribute values (bounded by the number of data records).

Rule Extraction and Prediction

Each connection between a terminal node and a node of the target layer represents an association rule of the form *if conjunction of input values, then the target value is likely / unlikely to be...* These are not *prediction rules*, like the rules extracted by the C4.5 algorithm (Quinlan, 1993), since multiple rules may be associated with the same terminal node. An information-theoretic weight of an association rule between a terminal node z and a target value V_j is given by:

$$w_z^j = P(V_j; z) \bullet \log \frac{P(V_j / z)}{P(V_j)}$$

Where $P \ (V_j; z)$ is an estimated joint probability of the value V_j and the node z, $P \ (V_j / z)$ is an estimated conditional (*a posteriori*) probability of the value V_j, given the node z, and $P \ (V_j)$ is an estimated unconditional (*a priori*) probability of the value V_j.

According to the information theory (see Cover, 1991), the above weight represents a contribution of a node-pair to the total mutual information between the input attributes and the target attribute. The weight is positive if the conditional probability of a target attribute value, given the node, is higher than its unconditional probability and negative otherwise. A zero weight means that the target attribute value is independent of the node value. Thus, each positive connection weight can be interpreted as the *information content* of an appropriate rule of the form *if node, then target value is....*

Accordingly, a negative weight refers to a rule of the form *if node, then target value is **not**... .*

Since IFN represents a disjunction of conjunctions of the input attribute values, each record in a relational data table having the same schema like the training data set can be associated with a single terminal node in the network. The target attribute in that record can be assigned a predicted value j^* by the following *maximum a posteriori* (MAP) rule:

$$j^* = \arg\max_j P(V_j / z)$$

For discrete target attributes, V_j stands for an actual attribute value. If a target attribute is continuous, V_j represents a discretized interval, which is converted by IFN into a continuous predicted value by using the mean of the corresponding interval.

The IFN method of prediction and classification is based on the Bayesian approach to learning from data (see Mitchell, 1997). According to the Bayesian reasoning, a *consistent* learning algorithm (i.e., an algorithm that outputs an error-free model over noiseless training data) has to use a MAP hypothesis for prediction.

Fuzzification and Reduction of Association Rules

The number of rules extracted from IFN may be quite large. It is bounded by the product of the number of terminal nodes and the number of target nodes and the previous applications of the algorithm show that this bound is relatively sharp. Although every rule is important for the predictive accuracy of the network, the user may find it difficult to comprehend the entire set of rules and to interpret it in natural and actionable language. As we have shown in (Last and Kandel, 2001 and Last, Klein, and Kandel, 2001), the *fuzzification* of the information-theoretic rules provides an efficient way for reducing the dimensionality of the rule set, without losing its actionable meaning. The process of rule reduction includes the following stages:

   Stage 1 - Fuzzifying the information-theoretic rules
   Stage 2 – Reducing the set of fuzzified rules by conflict resolution
   Stage 3 – Merging rules from the reduced set

Fuzzifying Association Rules

Although the boundaries of the discretized intervals, determined in the process of network construction (see sub-section 0 above), are aimed at minimizing the uncertainty of the target attribute, the user may be more interested in the linguistic descriptions of these intervals, rather in their precise numeric boundaries. For example, the user is more interested in the rules of the form "If current is high, then the yield is low", which is closer to the human way of reasoning, rather than "If current is between A and B then the yield is between X and Y". People tend to "compute with words" rather than with precise numbers. In addition, the total number of rules, extracted from a typical dataset may be much larger than the number of rules generally used by people in their decisions.

As indicated by (Klir and Yuan, 1995), the "linguistic ranges" of continuous attributes may be expressed as lists of terms that the attributes can take ("high", "low", etc.).   The user perception of each term may be represented by fuzzy membership functions. According to (Zadeh 1999), this is the first stage in an automated reasoning process, based on the Computational Theory of Perception (CTP), which can directly operate on perception-based, rather than measurement-based, information. Subsequent CTP stages include constructing the initial constraint set (ICS), goal-directed propagation of constraints, and creating a terminal constraint set, which is the end result of the reasoning process.

Sometimes the "crisp" rules cannot be presented to outsiders, because they contain some sensitive information. This may be an obstacle to open exchange of technological information in forums like professional conferences, multi-company consortia, etc. According to (Shenoi 1993), fuzzification of numeric attributes in a real-world database may be used for an additional purpose: *information clouding*. The user may be unwilling to disclose the actual values of some critical performance indicators associated with manufacturing, marketing, sales, and other areas of business activity. In many cases, data security considerations prevent results of successful data mining projects from being ever published. The application part of this chapter deals with highly sensible data obtained from a semiconductor company. Direct presentation of rules extracted from this data could provide valuable information to the company competitors. However, we are going to "hide" the confidential context of the rules by presenting them in their fuzzified form only.

The terms assigned to each simple condition and to the target (consequence) of the association rule are chosen to maximize the membership function at the middle point of the condition / consequence interval. Thus, we convert a crisp rule into a *fuzzy relation* (Klir and Yuan, 1995). Since a complex condition is a conjunction of simple conditions, an algebraic product is used to find the fuzzy intersection of the simple conditions. Fuzzy implication of Mamdani type (see below) is applied to each rule. Mamdani implication is more appropriate for the fuzzification of the information-theoretic rules due to the local nature of these rules. The informativeness of each fuzzified rule is represented by weighting the implication grade by the information-theoretic weight of the original crisp rule (see sub-section 0 above). If the weight is positive, the rule is stated as "*If <conjunction of terms assigned to rule conditions>, then <term assigned to the rule target >*". If the weight is negative, the rule will be of the form "*If <conjunction of terms assigned to rule conditions>, then **not** <term assigned to the rule target >*". The expression for calculating the weighted membership grade of an association rule is given below.

$$\mu_R = w \bullet [\prod_{i=1}^{N} \max_{j}\{\mu_{A_{ij}}(V_i)\}] \bullet \max_{k}\{\mu_{T_k}(O)\}$$

Where

    w – information-theoretic weight of the crisp rule

    N – number of simple conditions in the crisp rule

    V_i – crisp value of the simple condition i in the crisp rule (middle point of the condition interval)

    O – crisp value of the rule target (middle point of the target interval)

    $\mu_{A_{ij}}(V_i)$ - membership function of the simple condition i w.r.t. term j

    $\mu_{T_k}(O)$ - membership function of the target value O w.r.t. term k

Conflict Resolution

Since an information-theoretic ruleset includes association rules between conjunctions of input values and all possible target values, several rules may have the same IF parts, but different THEN parts. Moreover, the rule consequents may differ in their numeric values, but be identical in their linguistic values. This means that the set of fuzzy rules, produced above, may be *inconsistent*. To resolve this conflict, we calculate the grade of each distinct fuzzy rule by summing up the grades of all identical fuzzified rules and choose from each conflict group the target value that has a maximum grade. A similar approach is used by (Wang and Mendel, 1992) for resolving conflicts in fuzzy rules generated directly from data.

    In our procedure, there is no explicit distinction between positive and negative rule grades. For example, the fuzzified rules of the form "If A then B" and "If A then not B" are associated with the same consequent in the same distinct rule. However, their combined grade will be equal to the *difference* of their absolute grades, giving a preference to one of possible conclusions (**B** or **not B**). Eventually, the target value with the maximum *positive* grade will be chosen by the above procedure. This closely agrees with the interests of the database users, who need to estimate *positively* the value of the target attribute.

Merging Reduced Rules

In the previous sub-section, we have shown a method for handling rules having *identical antecedents* and *distinct consequents*. However, the resulting set of conflict-free rules may be further reduced by merging the rules having *distinct antecedents* and *identical consequents*. Thus, any two rules (I) and (II) having the form:

1. *If a is A and b is B and c is C, then t is T*
2. *If d is D and e is E and f is F, then t is T*

   can be merged into a single rule of the following form:

3. *If a is A and b is B and c is C or d is D and e is E and f is F, then t is T*

Using the above approach, we can create a rule base of a minimal size, limited only by the number of distinct target values. However, this approach may produce a small number of long and hardly useable rules (like the rule 3 above). Therefore, we perform the merging of disjunctive values *for the last rule condition only*. The procedure of merging fuzzy conjunctive rules (see Last and Kandel, 2001) is based on the assumption that each fuzzy rule has the same partial ordering of input attributes, which is true for any rule extracted from a given IFN (see sub-sections 0 and 0 above). The grade of the merged rule is calculated by using a fuzzy union ("max" operation). The resulting rule base can be considered a terminal constraint set in a CTP process (Zadeh 1999).

KNOWLEDGE DISCOVERY IN SEMICONDUCTOR DATA

In this section, we are applying the information-theoretic fuzzy approach to a real-world data set provided by a semiconductor company. The semiconductor industry is a highly competitive sector, and the original data used in our analysis, as well as any conclusions made from that data are considered highly sensitive proprietary information. Consequently, we were forced to omit or change many details in the description of the target data and the obtained results. As indicated in sub-section 0 above, fuzzification of continuous attributes has helped us to "hide" the proprietary information from the unauthorized (though, probably, curious) reader. As part of the "information clouding" effort, we also refrain here from mentioning the name of the company that has provided the data.

Data Description

The company has provided us with the data on the manufacturing batches that completed their production in the beginning of 1998. The data was derived from the company database in the form of a single table ("view" in the database terminology), which is shown schematically in Figure 2 below.

Record_ID	Batch_ID	Spec_ID	Priority	Oper_ID	Date_Fin	Qty_Trans	Qty_Scrap

Figure 2 The Original Data Table

The obtained table includes about 110,000 records. A short explanation about each attribute in the table and its relationships with other attributes is given below.

- *Record_ID*. This is the primary key attribute of the table, which uniquely identifies a specific manufacturing operation applied to a given batch.
- *Batch_ID*. This is the identification number of each manufacturing batch.
- *Spec_ID*. This is a specification (part) number of a batch No. *Batch_ID*. It specifies the manufacturing parameters of the batch, like current, voltage, frequency, chip size, etc.
- *Priority*. This is the priority rank of a batch, assigned by the marketing department.
- *Oper_ID*. The ID number of a specific operation applied to the batch No. *Batch_ID*. To preserve the confidentiality of the data, we have converted the actual operation codes into meaningless serial numbers.
- *Date_Fin*. The date when the operation *Oper_ID* was completed. After completion of an operation, the batch is transferred to the next fabrication step on its routing list.
- *Qty_Trans*. The quantity of good chips transferred to the next step. If a batch consists of wafers (before they are cut into individual chips), the number of good chips is calculated automatically from the number of wafers.
- *Qty_Scrap*. This is the number of chips scraped at the operation *Oper_ID*. It is equal to the difference between the number of chips transferred from the previous operation and the number of chips transferred to the next step. If an entire wafer is scraped, the number of scraped chips is calculated automatically by the maximum number of good chips, which can be obtained from a wafer.

By directly applying a data mining algorithm to the above records, one can easily obtain some basic statistical results like the distribution of the number of scraped chips at each operation as a function of the *Spec_ID*. This type of analysis is performed routinely by process and quality engineers. However, much more important and less obvious information may be hidden *between* the records. To discover those "nuggets" of knowledge, some pre-processing of data is required. The process of data preparation is described in the next sub-section.

Data Preparation

The original dataset included batches from a variety of microelectronic products, each having a different set of functional parameters and requiring a different sequence of operations ("routing"). Rather than trying to build a single data mining model from all the records, we have decided to focus our analysis on a group of 1,635 batches related to a single product family. The batches of this family have three main parameters (chip size, capacitance, and tolerance) and their manufacturing process requires about 50 operations. The selected batches were represented by 58,076 records in the original dataset.

As indicated by (Pyle, 1999), the process of data preparation strongly depends on the specific objectives of knowledge discovery. Here, we are interested to predict the following two parameters: the *yield* and the *flow time* of each manufacturing batch. The process of preparing the selected dataset for the data mining included the following steps:

- *Data normalization.* The original data table does not comply with the *third normal form* of relational tables (see Korth and Silberschatz, 1991), since the attributes *Spec_ID* and *Priority* are fully functionally dependent on the attribute *Batch_ID*, which is a non-key attribute. Consequently, we have moved *Spec_ID* and *Priority* from the original table (which was named *Batch_Flow*) to a new table *Batches*, where *Batch_ID* was defined as the primary key attribute.

- *Calculating batch parameters.* The product parameters (chip size, capacitance, and tolerance) were extracted from the attribute *Spec_ID* by using the available metadata on the attribute's encoding schema and stored in the *Batches* table. Chip size (*Size*) and tolerance (*T_Code*) are nominal attributes, which take a limited number of values, while capacitance is a continuous attribute. In our presentation or results (see below), we have replaced the actual size and tolerance codes with meaningless letters and numbers.

- *Calculating the yield.* The line yield of each batch can be found from dividing the value of the attribute *Qty_Trans* in the last operation of the batch manufacturing process by its value in the first operation. However, the line yield is difficult to analyze, since during the first part of the manufacturing process, the batches are transferred and scraped as *wafers* rather than individual chips. Consequently, the overall line yield is a combination of two yields: the so-called *wafer yield* and the *pieces yield*. A loss in the pieces yield is more expensive than a loss in the wafer yield, since it means that the defects are discovered in one of the final tests, after the manufacturing costs have already been incurred. Our objective here is to predict the pieces yield (*Yield_P*), which is calculated as the ratio between *Qty_Trans* in the last operation and its value immediately after the wafers are cut ("died") into chips. The attribute *Yield_P* is stored in the *Batches* table.

- *Calculating the flow times.* Another target attribute, the flow time of a batch, was derived as the difference between the values of the attribute *Date_Fin* in the last and the first operation.

- *Discretization of target attributes.* Since the information-theoretic algorithm of sub-section 0 above cannot be applied directly to continuous targets, the attributes *Yield_P* and *Flow_Time* were discretized into ten intervals of approximately equal frequency.

- *Storing completion dates.* The completion dates of all the operations applied to each batch were retrieved from the *Batch_Flow* table and stored in the *Batches* table. This required defining a new attribute for each operation occurring at least once in the routings of the selected batches. This step has created certain amount of redundancy in the database, but it was necessary to eliminate the need of accessing the relatively large *Batch_Flow* table by the data mining algorithm. The *Batch_Flow* table has about 58,000 records vs. 1,635 records only in the *Batches* table.

Constructing the Info-Fuzzy Networks

The information-fuzzy networks related to the target attributes *Yield_P* and *Flow_Time* were constructed by using the algorithm of sub-section 0 above. To predict *Yield_P*, the network construction algorithm was trained only on the records of 1,378 batches, where the number of pieces immediately after the wafers were cut into chips was reported into the system. In Table 2 below, we show the four input attributes included in the network of *Yield_P*. The selected attributes are chip size, product capacitance, and the completion dates of two operations (No. 32 and 38). The column "Mutual Information" shows the cumulative association between a subset of input attributes, selected up to a given iteration inclusively, and the target attribute. Since the mutual information is defined as the difference between unconditional and conditional entropy (Cover 1991), it is bounded by the unconditional entropy of *Yield_P*, which is 3.32 (log_2 10). The estimated net increase in the mutual information, due to adding each input attribute, is presented in the column "Conditional MI". The last column "Conditional Entropy" shows the difference between the unconditional entropy (3.32) and the estimated mutual information.

Table 2 Selected Attributes (Target: Yield_P)

Iteration	Attribute Name	Mutual Information	Conditional MI	Conditional Entropy
0	Size	0.111	0.111	3.212
1	Date_32	0.237	0.126	3.086
2	Capacitance	0.286	0.05	3.036
3	Date_38	0.306	0.02	3.017

From the engineering point of view, it is not surprising that the product parameters *Size* and *Capacitance* are related to the yield value. A more interesting and potentially useful result is the selection of completion dates for two operations (32 and 38) as additional factors that affect the yield. This finding may be interpreted as an indication of process instability: batches that went through these operations during certain periods of time had lower quality than batches manufactured in other periods. The boundaries of each period were determined automatically by the dynamic discretization procedure of the algorithm. To find the real causes of yield excursion, the process engineers should compare between the tool conditions in these operations during the "good and the "bad" periods. Other time-dependent factors (e.g., staffing) may be examined as well.

The attributes included in the network of *Flow_Time* are shown in Table 3 below. The selected attributes represent the three main parameters of the product in question: capacitance, chip size, and tolerance. This means that different products experience different kinds of delays in the production line. Some of these delays are probably caused by product-specific manufacturing problems, which may be revealed from the routing documentation of the relevant batches.

One attribute is conspicuous by its absence in Table 3. This attribute is *Priority*, which represents the priority rank assigned to every batch in the beginning of the manufacturing process. Surprisingly enough, the data mining algorithm showed that the batch flow time is *not* affected by its priority. This strange result can be partially understood from a close look at the data: about 95% of all batches in the dataset have the same priority rank, which, of course, undermines the basic idea of prioritizing. Anyway, it indicates a failure to impose a pre-defined schedule on the order of batch production.

Table 3 Selected Attributes (Target: Flow_Time)

Iteration	Attribute Name	Mutual Information	Conditional MI	Conditional Entropy
0	Capacitance	0.107	0.107	3.204
1	Size	0.266	0.159	3.045
2	T_Code	0.354	0.088	2.958

Rule Extraction and Prediction

The Yield Network

The information-fuzzy network built for the *Yield_P* target attribute has four layers (corresponding to four input attributes), total of 23 hidden nodes, 16 terminal (unsplit) nodes, and ten target nodes representing the ten intervals of the discretized target attribute. Thus, the network can have up to 16*10=160 connections between its 16 terminal nodes and the ten nodes of the target layer. The actual number of connections having non-zero information-theoretic weights is 125. Each connection represents an association rule of the form

If Size = V [and Date_32 is between A and B],[and Capacitance is between C and D], [and Date_38 is between E and F], then Yield_P is [not] between G and H

where *V* represents a valid value from the domain of the corresponding nominal attributes (*Size*) and *A, ..., H* stand for the boundaries of intervals in discretized continuous attributes (*Capacitance, Date_*, and Yield_P*). The rules having the highest positive and the smallest negative connection weights are given below (confidential information was replaced by meaningless letters).

- **Rule No. 77**: If Size is Z and Date_32 is between 02-Dec-97 and 10-Feb-98, then Yield_P is more than C (weight = 0.0671).
- **Rule No. 27**: If Size is W and Date_32 is between 02-Dec-97 and 10-Feb-98, then Yield_P is not more than C (weight = -0.0163).

The predictive accuracy of the above information-fuzzy network was estimated by holding out one third of the data as a validation set. Thus, we chose randomly 433 validation records and used the other 945 records for constructing the network. The IFN classification accuracy (the probability of identifying the correct interval out of ten) is 23.4% on the training set and 17.3% on the validation set. The 95% confidence interval for the validation accuracy is between 13.8% and 20.9%. The low accuracy of the information-fuzzy network results from the large number of target classes (10) and the inherent noisiness of the manufacturing data. Still, it appears to be higher than the validation accuracy of the C4.5 algorithm (15.9% only), which was applied to the same data, though the difference between the accuracy of the two algorithms is not statistically significant.   At the same time the classification model produced by C4.5 is much more complex than the IFN model: after pruning, the C4.5 tree includes 399 nodes vs. only 23 nodes in IFN. C4.5 uses 37 attributes for prediction, while IFN is satisfied with four input attributes only. Default settings were used for both C4.5 (see Quinlan, 1993) and IFN (see Maimon and Last, 2000).

The Flow Time Network

The information-fuzzy network of the *Flow_Time* target attribute consists of three layers (corresponding to three input attributes), total of 37 hidden nodes (including 28 terminal nodes), and ten target nodes representing the ten intervals of the discretized target attribute.   The network can have up to 28*10=280 connections between its 28 terminal nodes and the ten nodes of the target layer, but the actual number of connections having non-zero information-theoretic weights is only 191. Each connection represents an association rule of the form

   If Capacitance is between A and B [and Size is V_1], [and T_Code is V_2] then Flow_Time is [not] between C and D

   where V_i represents a valid value from the domain of the corresponding nominal attributes (*Size* or *T_Code*) and *A,..., D* stand for the boundaries of intervals in discretized continuous attributes (*Capacitance*, and *Flow_Time*). The rules having the highest positive and the smallest negative connection weights are given below (confidential information was replaced by meaningless letters).

- **Rule No. 135**: If Capacitance is between A and B and Size is W and T_Code is 2 then Flow_Time is between C and D (weight = 0.0282).
- **Rule No. 25**: If Capacitance is between A and B and Size is V then Flow_Time is not between E and F (weight = -0.0057).

   The predictive accuracy of the above information-fuzzy network was estimated by holding out one third of the data as a validation set. Thus, we chose randomly 533 validation records and used the other 1,102 records for constructing the network. The IFN classification accuracy (the probability of identifying the correct interval out of ten) is 23.9% on the training set and 18.6% on the validation set. The 95% confidence interval for the validation accuracy is between 15.3% and 21.9%. The low validation accuracy of the information-fuzzy network results again from the inherent noisiness of the manufacturing data. It is lower than the validation accuracy of the C4.5 algorithm (22%), which was applied to the same data, though the difference between the accuracy of the two algorithms is statistically significant at the 5% level only (not at the 1% level). Like in the previous case, the classification model produced by C4.5 is much more complex than the IFN model: after pruning, the C4.5 tree includes 168 nodes vs. 37 nodes only in IFN. C4.5 uses four attributes for prediction, while IFN is satisfied with three input attributes only.

Rule Fuzzification and Reduction

The Yield Rules

The "crisp" rules extracted from the information-fuzzy network that was described in sub-section 0 above cannot be presented here in their explicit form due to confidentiality of the information they contain. However, their explicit presentation to the users (process engineers) could be hardly useful either, since it is very difficult to generalize and interpret manually a set of more than 100 numeric rules. To make the rule set more compact, interpretable, and less explicit, we have fuzzified two numeric attributes included in the extracted rules: *Capacitance* and *Yield_P*. The following terms (words in natural language) were chosen by us for each fuzzified attribute:

- *Capacitance*: low, high.
- *Yield_P*: low, normal, high.

To convert the above attributes into linguistic variables, we have defined triangular membership functions associated with each term by using the frequency histograms of attribute values. Triangular functions are frequently used in the design of fuzzy systems (Wang 1997). The membership functions are shown in Figures 3 and 4 below without the values of the X-axis, to protect the confidentiality of the original data.

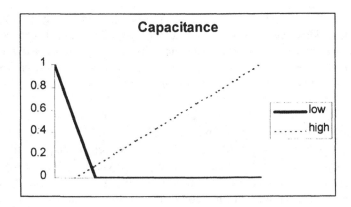

Figure 3 Membership Functions of Capacitance

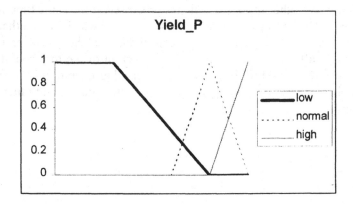

Figure 4 Membership Functions of Yield_P

Fuzzification of the "crisp" rules having the highest and the lowest connection weights (see sub-section 0 above), results in the following linguistic rules:

- **Rule No. 77**: If Size is Z and Date_32 is between 02-Dec-97 and 10-Feb-98, then Yield_P is normal (grade = 0.0387).
- **Rule No. 27**: If Size is W and Date_32 is between 02-Dec-97 and 10-Feb-98, then Yield_P is not normal (grade = - 0.0094).

In Table 4 below, we present the consistent set of fuzzy rules, extracted from the set of fuzzified rules by using the conflict resolution procedure of sub-section 0 above. The last column represents the number of original rules (crisp / fuzzified), associated with a given fuzzy rule. As one can see, the size of the fuzzy rule base has been significantly reduced from 125 original rules to 16 rules only (a decrease of almost 90%).

All the rules in shown in Table 4 are *conjunctions* of fuzzy and "crisp" conditions. However, some rules, like rules No. 2, 3, and 4, can be merged into a *disjunction*, since they have the same consequent (*Yield is normal*). The formal algorithm for merging fuzzy rules was described in sub-section 0 above. The resulting set of nine merged fuzzy rules is shown in Table 5 below.

The users (process engineers) would be particularly interested in the rules describing problematic situations, i.e., rules predicting the yield to be low. Thus, Rule 1 indicates that batches of size W, which passed the operation 32 before December 2, 1997, tend to have a low yield. This means that the engineers should look carefully both at the routing records of all batches that meet these criteria and at the condition of tools and machines used by operation 32 before the above date. Rules 6 and 7 intensify the suspicion that something went wrong with the equipment of operation 32 before Dec. 2, 1997, since two other groups of batches processed at the same time had yield problems. These groups include batches of size X, which have either high or low capacitance, given that the low capacitance batches were processed at operation 38 after Dec. 22, 1997. The inspection of the tool conditions and routing records related to the problematic period at operation 32 may lead to changes in maintenance guidelines, process control limits, and other working procedures.

Table 4 Yield: The Set of Consistent Fuzzy Rules

Rule No	Rule Text	Grade	Number of Crisp Rules
0	If Size is Y then Yield_P is normal	0.0084	7
1	If Size is W and Date_32 is between 21-Dec-95 and 02-Dec-97 then Yield_P is low	0.0237	10
2	If Size is W and Date_32 is between 02-Dec-97 and 10-Feb-98 then Yield_P is normal	0.0097	10
3	If Size is W and Date_32 is between 10-Feb-98 and 12-Mar-98 then Yield_P is normal	0.021	10
4	If Size is W and Date_32 is after 12-Mar-98 then Yield_P is normal	0.0087	10
5	If Size is X and Date_32 is between 10-Feb-98 and 12-Mar-98 then Yield_P is normal	0.0324	9
6	If Size is X and Date_32 is after 12-Mar-98 then Yield_P is normal	0.0096	10
7	If Size is Z and Date_32 is between 21-Dec-95 and 02-Dec-97 then Yield_P is normal	0.0051	4
8	If Size is Z and Date_32 is between 02-Dec-97 and 10-Feb-98 then Yield_P is normal	0.0338	7
9	If Size is Z and Date_32 is between 10-Feb-98 and 12-Mar-98 then Yield_P is normal	0.0073	5
10	If Size is Z and Date_32 is after 12-Mar-98 then Yield_P is normal	0.001	2
11	If Size is X and Date_32 is between 21-Dec-95 and 02-Dec-97 and Capacitance is high then Yield_P is low	0.0065	7
12	If Size is X and Date_32 is between 02-Dec-97 and 10-Feb-98 and Capacitance is low then Yield_P is normal	0.0185	10
13	If Size is X and Date_32 is between 02-Dec-97 and 10-Feb-98 and Capacitance is high then Yield_P is normal	0.0036	9
14	If Size is X and Date_32 is between 21-Dec-95 and 02-Dec-97 and Capacitance is low and Date_38 is between 18-Aug-97 and 22-Dec-97 then Yield_P is low	0.0094	10
15	If Size is X and Date_32 is between 21-Dec-95 and 02-Dec-97 and Capacitance is low and Date_38 is after 22-Dec-97 then Yield_P is normal	0.0055	5

Table 5 Yield: The Set of Merged Fuzzy Rules

Rule No	Rule Text	Target	Grade
0	If Size is Y then Yield_P is	normal	0.0084
1	If Size is W and Date_32 is between 21-Dec-95 and 02-Dec-97 then Yield_P is	low	0.0237
2	If Size is W and Date_32 is after 02-Dec-97 then Yield_P is	normal	0.0097
3	If Size is X and Date_32 is after 10-Feb-98 then Yield_P is	normal	0.021
4	If Size is Z and Date_32 is after 21-Dec-95 then Yield_P is	normal	0.0087
5	If Size is X and Date_32 is between 02-Dec-97 and 10-Feb-98 then Yield_P is	normal	0.0096
6	If Size is X and Date_32 is between 21-Dec-95 and 02-Dec-97 and Capacitance is high then Yield_P is	low	0.0324
7	If Size is X and Date_32 is between 21-Dec-95 and 02-Dec-97 and Capacitance is low and Date_38 is between 18-Aug-97 and 22-Dec-97 then Yield_P is	low	0.0051
8	If Size is X and Date_32 is between 21-Dec-95 and 02-Dec-97 and Capacitance is low and Date_38 is after 22-Dec-97 then Yield_P is	normal	0.0338

The Flow Time Rules

In addition to fuzzifying the attribute *Capacitance* (see sub-section 0 above), we have defined the following terms for the attribute *Flow_Time* (to avoid disclosing its real values): *short, medium,* and *long.* The triangular functions for these terms are shown in Figure 5 below.

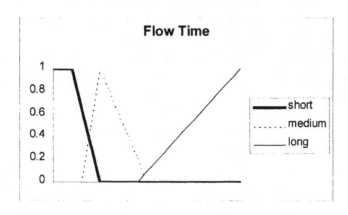

Figure 5 Membership Functions of Flow_Time

Fuzzification of the "crisp" rules having the highest and the lowest connection weights (see sub-section 0 above), results in the following linguistic rules:

- **Rule No. 135**: If Capacitance is low and Size is W and T_Code is 2 then Flow_Time is medium (grade = 0.0129).
- **Rule No. 25**: If Capacitance is low and Size is V then Flow_Time is not long (weight = -0.0057).

In Table 6 below, we present the consistent set of fuzzy rules, extracted from the set of fuzzified rules by using the conflict resolution procedure of sub-section 0 above. The last column represents the number of original rules (crisp / fuzzified), associated with a given fuzzy rule. As one can see, the size of the fuzzy rule base has been significantly reduced from 191 original rules to 12 rules only (a decrease of more than 90%).

The seven rules merged by the procedure of sub-section 0 above are shown in Table 7 below. It appears from the table that the company has more delays in the low capacitance batches than in the high capacitance ones. Special problems with flow times were experienced for sizes V and Y (see Rule 1) and for tolerance 1 of size W (see Rule 5). The engineers should study the routing records of the relevant products to find out, what were the leading delay factors of the batches in question.

Table 6 Flow Time: the Set of Consistent Fuzzy Rules

Rule No	Rule Text	Grade	Number of Crisp Rules
0	If Capacitance is high then Flow_Time is short	0.0082	16
1	If Capacitance is low and Size is V then Flow_Time is long	0.0176	25
2	If Capacitance is low and Size is X then Flow_Time is short	0.0056	18
3	If Capacitance is low and Size is Y then Flow_Time is long	0.0019	10
4	If Capacitance is low and Size is Z then Flow_Time is medium	0.0056	20
5	If Capacitance is low and Size is W then Flow_Time is medium	0.0149	20
6	If Capacitance is low and Size is W and T_Code is 0 then Flow_Time is medium	0.0075	10
7	If Capacitance is low and Size is W and T_Code is 1 then Flow_Time is long	0.009	14
8	If Capacitance is low and Size is W and T_Code is 2 then Flow_Time is medium	0.019	15
9	If Capacitance is low and Size is X and T_Code is 0 then Flow_Time is medium	0.0011	13
10	If Capacitance is low and Size is X and T_Code is 1 then Flow_Time is medium	0.0053	11
11	If Capacitance is low and Size is X and T_Code is 2 then Flow_Time is medium	0.004	19

Table 7 Flow Time: The Set of Merged Fuzzy Rules

Rule No	Rule Text	Target	Grade
0	If Capacitance is high then Flow_Time is	short	0.0082
1	If Capacitance is low and Size is V or Y then Flow_Time is	long	0.0176
2	If Capacitance is low and Size is X then Flow_Time is	short	0.0056
3	If Capacitance is low and Size is Z or W then Flow_Time is	medium	0.0019
4	If Capacitance is low and Size is W and T_Code is 0 or 2 then Flow_Time is	medium	0.0056
5	If Capacitance is low and Size is W and T_Code is 1 then Flow_Time is	long	0.0149
6	If Capacitance is low and Size is X and T_Code is 0 or 1 or 2 then Flow_Time is	medium	0.0075

CONCLUSIONS

In this chapter, we have presented a systematic approach to mining the process and quality data in the semiconductor industry. The construction of the data mining model is based on the *Information-Fuzzy Network* (IFN) methodology introduced in (Maimon and Last, 2000). Post-processing of the IFN output follows the Computational Theory of Perception (CTP) approach and it includes information-theoretic fuzzification of numeric association rules, removal of conflicting rules and merging of consistent rules. As demonstrated by the case study of an actual semiconductor database, the method results in a compact and reasonably accurate prediction model, which can be transferred into a small set of interpretable rules. The fuzzification of the rules can also be used for hiding confidential information from external users.

The sound theoretical basis of the information-fuzzy approach and the promising results obtained so far encourage us to further develop this methodology in several directions. These include mining very large and non-stationary datasets, using IFN as a feature selector in the knowledge discovery process, and combining multiple IFN models (see chapter by Maimon and Rokach in this volume).

ACKNOWLEDGMENTS

This work was partially supported by the USF Center for Software Testing under grant no. 2108-004-00.

REFERENCES

Agrawal, R., Mehta, M., Shafer, J., and Srikant, R., "The Quest Data Mining System," in Proceedings of KDD-96, pp. 244-249, 1996.

Attneave, F., Applications of Information Theory to Psychology, Holt, Rinehart, and Winston, 1959.

Breiman, L., Friedman, J.H., Olshen, R.A., & Stone, P.J., Classification and Regression Trees, Wadsworth, 1984.

Cover, T. M., Elements of Information Theory, Wiley, 1991.

Fayyad, U. and Irani, K., "Multi-Interval Discretization of Continuous-Valued Attributes for Classification Learning," in Proceedings of the 13th International Joint Conference on Artificial Intelligence, pp. 1022-1027, 1993.

Korth, H.F. and Silberschatz, A., Database System Concepts, McGraw-Hill, Inc., 1991.

Last, M. and Kandel, A., "Fuzzification and Reduction of Information-Theoretic Rule Sets," in Data Mining and Computational Intelligence, pp. 63-93, Physica-Verlag, 2001.

Last, M., Klein, Y., and Kandel, A., "Knowledge Discovery in Time Series Databases," IEEE Transactions on Systems, Man, and Cybernetics, 31 (1), 160-169, 2001.

Maimon, O. and Last, M., Knowledge Discovery and Data Mining, The Info-Fuzzy Network (IFN) Methodology, Boston: Kluwer Academic Publishers, 2000.

Mitchell, T.M., Machine Learning, McGraw-Hill, 1997.

Pyle, D., Data Preparation for Data Mining, Morgan Kaufmann, 1999.

Quinlan, J. R., C4.5: Programs for Machine Learning, Morgan Kaufmann, 1993.

Rao, C.R. and Toutenburg, H., Linear Models: Least Squares and Alternatives, Springer-Verlag, 1995.

Shenoi, S., "Multilevel Database Security Using Information Clouding," in Proceedings of IEEE International Conference on Fuzzy Systems, pp. 483-488, 1993.

Tobin, K. W., Karnowski, T.P., and Lakhani, F., "A Survey of Semiconductor Data Management Systems Technology," in Proceedings of SPIE's 25th Annual International Symposium on Microlithography, Santa Clara, CA, February 2000.

Wang, L.-X. and Mendel, J.M., "Generating Fuzzy Rules by Learning from Examples," IEEE Transactions on Systems, Man, and Cybernetics, 22 (6), 1414-1427, 1992.

Wang, L.-X., A Course in Fuzzy Systems and Control, Prentice-Hall, 1997.

Zadeh, L. A., "A New Direction in System Analysis: From Computation with Measurements to Computation with Perceptions," in New Directions in Rough Sets, Data Mining, and Granular-Soft Computing, pp. 10-11, Springer-Verlag, 1999.

CHAPTER 10

Analyzing Maintenance Data Using Data Mining Methods

Carol J Romanowski
cfr@acsu.buffalo.edu
Department of Industrial Engineering, State University of New York at
Buffalo, 342 Bell Hall, Buffalo, New York 14260-2050

Rakesh Nagi
nagi@acsu.buffalo.edu
Department of Industrial Engineering, State University of New York at
Buffalo, 342 Bell Hall, Buffalo, New York 14260-2050

ABSTRACT

Preventive maintenance activities generate information that can help determine the causes of downtime and assist in setting maintenance schedules or alarm limits. When the amount of generated data becomes large, humans have difficulty understanding relationships between variables. In this paper, we explore the applicability of data mining, a methodology for analyzing multi-dimensional datasets, to the maintenance domain. Using data mining methods, we identify subsystems responsible for low equipment availability; recommend a preventive maintenance schedule; and find sensors and frequency responses giving the most information about fault types. The data mining approach achieves good, easily understandable results within a short training time.

D. Braha (ed.), Data Mining for Design and Manufacturing, 235–254.
© 2001 *Kluwer Academic Publishers. Printed in the Netherlands.*

INTRODUCTION

When discussing ways to become more competitive in today's business environment, companies rarely mention reducing their maintenance costs. Yet, machine maintenance contributes 15-40% to the cost of production, and a third of this amount is spent on unnecessary or improperly performed maintenance activities (Mobley 1990). The increasing numbers of computerized maintenance management systems and businesses practicing condition-based maintenance show that companies are beginning to understand the importance of maintenance policies as a major cost driver.

Computerized data collection and asset management systems provide a wealth of information for the maintenance engineer; however, cheap computer storage and advanced monitoring capability often results in large, multi-dimensional databases. Sifting through and analyzing these mountains of data is a daunting task even for experts.

In response to the need for methods to deal with multi-dimensional, terabyte-sized databases, researchers have developed algorithms that cluster, classify, or predict based on the available information in the data. Popularly called "data mining", this practice of discovering knowledge in large databases deals with "big data" – many variables, many values, and many records. Instead of the traditional hypothesis-driven models, data mining models are data-driven.

In this research, we look at two types of maintenance data – a time-based and a condition-based maintenance domain – analyzed using both traditional methods and data mining, and compare the results. Our goal is to establish the applicability of data mining methods in the maintenance domain, and to show the benefits of using this approach within the maintenance field.

In Section 2, we briefly discuss three major preventive maintenance policies and give some general background on data mining, on decision trees in particular. Section 3 applies decision trees to scheduled maintenance data; Section 4 applies the method to condition-based data. Conclusions and recommendations for future work are contained in Section 5.

BACKGROUND

Preventive Maintenance Overview

Three main types of maintenance policies predominate in both theory and practice. Within each type, however, are sub-types that essentially are intermediate steps toward the next level of complexity. Figure 1 shows the relationships between the different maintenance policies.

Unplanned maintenance, or run-to-failure, is a reactive program that performs only emergency repairs. In this approach, inventory levels of spare parts must be kept high, since there is no prior knowledge of when breakdowns will occur. Maintenance costs are high also; emergency repairs take longer and cost more than planned downtimes (Mobley 1990; Hsu 1995). However, many companies that practice run-to-failure at least perform some limited preventive maintenance such as lubrication and visual inspections. Williams et al. (Williams 1994) makes a distinction between corrective maintenance – the repair of a non-serious failure – and emergency maintenance, where repair is needed immediately to avoid some dire consequences.

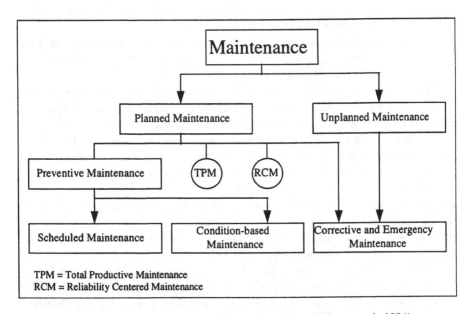

Figure 1: Maintenance policies (adapted from Williams et al., 1994)

Scheduled maintenance programs perform maintenance tasks at predetermined intervals. The entire machine may be overhauled whether necessary or not. Costs are generally less than in corrective maintenance programs, but production schedules may still be disrupted and unscheduled breakdowns may still occur frequently (Mobley 1990). The engineer must determine the parameters for scheduling activities, such as time from the last maintenance, amount of usage, etc.

Condition-based maintenance is the most comprehensive of the policies and generates large amounts of data. This approach measures the state of the machine either at intervals or continuously through vibration monitoring, tribology, visual inspection, and monitoring of process parameters. Monitoring data is analyzed to classify the varying states of the machine's health, and the extracted knowledge used to forecast when breakdowns are likely to occur.

Vibration analysis, the most common condition-based maintenance method, is composed of three techniques (Mobley 1990):

- Broadband trending gives relative vibration values at specific points on a machine train over time. The values are compared to baseline readings or vibration severity charts to determine the machine's condition.
- Narrowband trending monitors a specific bandwidth representing certain components or failure modes. The advantage to narrowband trending is the ability to monitor critical areas instead of the entire machine train.
- Signature analysis looks at each frequency component generated by the equipment. The signatures are matched against baselines and known failure modes.

Methods such as Fast Fourier Transforms and multiple narrowband filtering are used to decompose a vibration time series into frequency and amplitude components for frequency spectrum analysis.

Tribology studies the bearing-lubrication-rotor support structure with such techniques as lubricating oil analysis, wear particle analysis, and ferrography (Mobley 1990). These methods monitor the physical condition of lubricating fluids as well as the presence of debris from system components. Particulate contamination has been shown to be the major cause of fluid power machinery failure in industry (Williams 1994). As computers become more powerful and more integrated with manufacturing, these methods are increasingly available to industry.

Although vibration monitoring is the most common form of condition-based monitoring, a total predictive maintenance program includes other techniques tailored to the types of equipment found in the plant (Mobley 1990). When maintenance policies are integrated with other aspects of manufacturing, the result is increased productivity, reliability, and maintainability (Williams 1994).

Preventive Maintenance and Production

In the past, preventive maintenance approaches have generally been absent from studies of the production process. Perhaps the move toward JIT and flexible manufacturing has sparked an interest into integrating PM with traditional areas of production research such as job scheduling, safety stock, lot sizes, and economic manufacturing quantities. Several researchers have considered the link between production and maintenance, as shown by Kelly, Hsu, and Cheung (Hsu 1995; Cheung 1997; Kelly 1997).

As a non-value-added activity, maintenance is often ignored or downplayed by management. However, it is clear that maintenance policies have a great impact on the efficiency, quality and cost of a production process. Regardless of the policy or approach, massive amounts of useful data can be generated and stored in preventive maintenance databases. Using data mining techniques, we can analyze this data and use the discovered knowledge to optimize PM activities.

Data Mining Overview

Data mining deals with "big data" – many variables, many values, and many records. The two purposes of data mining, knowledge discovery and prediction, have different goals and methodologies. Regardless of the goal or method, data mining is an iterative process.   The methodology chosen depends in a large part on the particular data mining purpose. Unsupervised learning methods such as cluster analysis are appropriate when the data is unclassified and the purpose of the data mining task is knowledge discovery. When the task is predictive, supervised learning methods such as decision trees are used. The data mining goal influences the choice of algorithms within each method. The user, however, must also consider other factors such as CPU time, accuracy, and ease of interpretation.

Table 1, Common Data Mining Methods, summarizes major data mining methods and their characteristics according to Kennedy et al. (Kennedy 1998), Elder et al. (Elder and Pregibon 1996), and Berry et al. (Berry 1997). Genetic algorithms and rough sets are not included in the table because they are usually not used as stand-alone mining methods. Question mark entries indicate that no information was available for a particular characteristic.

Algorithm	Learning Type	Classification	Estimation	Association	Memory Requirements	Training Time (CPU seconds)	Testing Time (CPU seconds)	Ease of interpretation	Batch only	Type of Data	Accuracy	Scalability
Linear regression	S	N	Y	N	V. Low	Fast	V. Fast	Easy	Y	Num	Med	High
Neural nets	S	N	Y	N	Low-med	V. Slow	V. Fast	V. Hard	N	Num	High	Low
Rule induction	S	Y	N	N	High	Fast	V. Fast	Easy	Y	N/C	Med Low	High
Decision trees	S	Y	Y	N	Low	V. Fast	V. Fast	V, Easy	Y	N/C	Med Low	High
Case-based reasoning	S	Y	Y	N	High	V. Fast	Fast	Easy	N	N/C	?	High
Kth nearest neighbor	S	Y	Y	Y	V. High	V. Fast	Slow	Easy	Y	N/C	MedLow	V. Low
Self-organizing maps	U	N	N	Y	Low-med	Fast	NA	Hard	Y	N/C	NA	Low
Bayesian belief nets	U	Y	N	Y	?	Slow	NA	Hard	Y	N/C	?	?
Support vectors	S	Y	Y	N	Med	Slow	Med	Easy	Y	Num	High	?

Table 1: Common Data Mining Methods

KEY: S = supervised; U = unsupervised; Y = yes; N = no; Num = numerical;
N/C = numerical or categorical NA = not applicable

In all the hype that has surrounded data mining, we must not assume that the process is automatic and needs no human intervention. Expert knowledge is still needed to prepare the data, to choose the methodology, and, most importantly, to evaluate the results. Are the outcomes what we would expect? Did we find out anything new, something we did not see using traditional methods? What are the advantages of these methods over statistical techniques?

Mining with Decision Trees

Decision trees are graphical, easily understood representations of the dataset consisting of a root node, branch nodes, and terminal nodes, or leaves. Nodes are formed by maximizing *information gain*, a quantity calculated from class probabilities. New instances are classified by traversing the tree from the root node to the leaves, corresponding roughly to an "If (root node value=X) AND (branch node$_1$ value = Y) AND ... (branch node$_n$ = W) Then (leaf node class=Z)" production rule.

Tree algorithms make use of three rules to govern tree growth: a splitting, or node-forming, rule; a stopping rule to limit tree size; and a pruning rule to reduce tree error. This research uses See-5, a decision tree algorithm developed by Quinlan (Quinlan 1996). In See-5, the splitting rule is a variation of information theory, and is called the *gain ratio*. Information theory (Shannon 1948) states that if there are *n* equally probable messages, then the probability *p* of each message is $\frac{1}{n}$, and the optimal number of bits to encode that message is $-log_2(p)$.

In a probability distribution D=(p_1, p_2, ... , p_n) the entropy of D (the "impurity" of the information contained in the distribution) is given by Equation (1)

$$Entropy(D) = -\sum_{j=1}^{C} p(D, j) \times \log_2\left(p(D, j)\right) \qquad (1)$$

where C is the number of classes in the dataset and $p(D, j)$ is the number of records in a particular class divided by the total number of records in the database (Quinlan 1996). Therefore, a distribution that is uniform has a higher entropy.

At each node, the See-5 algorithm calculates the "split information" for each attribute value D_i over all classes in the dataset, defined as

$$Split(C) = -\sum_{i \in Values(C)} \frac{|D_i|}{D} Entropy(D_i) \qquad (2)$$

The information gain for a particular attribute is defined as

$$Gain(D, C) = Entropy(D) - Split(C) \qquad (3)$$

The *gain ratio* is then calculated as

$$\frac{Gain(D,C)}{Split(D,C)} \tag{4}$$

The test T that maximizes the gain ratio becomes the current node, and the algorithm begins to determine the next node. Stopping rules specify a minimum gain ratio; when this minimum value is reached, the algorithm stops growing the tree.

Pruning can be accomplished in many ways. See-5 examines the error rate of each subtree and compares it with an expected error rate if a leaf replaced the subtree. If the expected error rate for the leaf (Error(Leaf$_i$)) is lower, the subtree is deleted and the leaf substituted in its place. Error rates are calculated by the ratio of incorrect classifications to the total number of cases at the leaf. The error rate for the entire tree is defined as

$$Error(Tree) = \sum_{i=1}^{N} Error(Leaf_i) \times p\left(Leaf_i\right) \tag{5}$$

where N is the number of leaves and $p(Leaf_i)$ is the probability a record will belong to $Leaf_i$ (Berry 1997).

Decision trees have a number of strengths that make them an attractive choice for a data mining algorithm. They handle continuous, categorical, or missing values easily. Gain for attributes with missing values is calculated using only the records where that value is defined. Trees can generate easily understood rules, and show clearly the attributes that are most important for prediction or classification. Although trees are computationally expensive, training time (in CPU minutes) is short, especially when compared with neural networks.

As the mining algorithms are applied to the data, the data miner can adjust, or tune, certain parameters to improve the response. Tuning the See-5 algorithm involves two parameters: the pruning confidence factor and the minimum number of cases per node. Lowering the pruning confidence factor results in more simplistic trees; conversely, raising this factor results in larger, more complex trees. Raising the minimum number of cases per node simplifies the tree by requiring more support for the terminal node, while lowering the minimum number of cases results in more complex trees.

When reading a tree, coverage and error rates are important parameters to consider. Coverage shows how many records are contained in a leaf; when confronted with conflicting tree paths, the coverage and error rates of the different paths should be compared. Obviously, the path with the highest coverage and lowest error rate is more desirable.

Data Mining Methodology

Before actually mining data, we need to follow specific steps common to all data mining applications.

1. **Determine the type of learning**

   If we are performing a classification or prediction task, we choose a supervised learning algorithm. If we want to look for new or different patterns in the data, we choose an unsupervised learning algorithm such as clustering or k-nearest neighbor analysis.

2. **Choose the data mining algorithm**

   When choosing the data mining algorithm, we must consider the type of data available, the goal, the computing resources, and most importantly, the user of the induced knowledge. As shown in Table 1, some algorithms produce a more understandable output than others.

3. **Choose the target variable**

   Unsupervised learning methods do not require a target variable. However, in supervised learning, we must first identify what we are interested in learning. For preventive maintenance datasets, the target variable may be equipment faults, time intervals between events, or equipment availability.

4. **Pre-process the data**

   Once the target variable has been chosen, the data must be prepared to get the best result from the data mining algorithm. Pre-processing involves much more than formatting the data for the chosen algorithm. We must decide how to deal with missing or incorrect values and outliers; we may need to remove some variables from the dataset; we may decide to sample the data. Expert knowledge of both the data and the domain is invaluable in this step to ensure that the information in the data is exposed to the mining algorithm. Pyle (Pyle 1999) thoroughly covers critical issues in preparing data for a good mining outcome.

5. **Mine the data**

   Applying the data mining algorithm is the easiest and fastest step in the mining process. Usually, 70% of the data is used for training the algorithm; the remaining 30% is used to test the resulting model. We can tune or adjust algorithm parameters to improve the results and reapply the algorithm.

6. **Analyze the output**

   Here, as in the pre-processing step, expert knowledge is important. Experts can help sift through the results, identifying the important knowledge nuggets, deciding what is interesting and what is trivial in the output. Post-processing may be required to assist human users with output analysis.

7. **Refine the task (optional)**

   Data mining is an iterative process. Many times the output provides the miner with insight into ways to refine and improve the results, or suggest a different approach altogether.

Applications of Data Mining in Engineering and Production

Although data mining has been used extensively in retail, telecommunications, and scientific databases, engineering and manufacturing in particular are just beginning to apply these techniques. A few examples of these applications follow.

Sillitoe and Elomaa (Sillitoe and Elomaa 1994) used decision trees to classify and map surfaces for a mobile robot. An ultrasonic array mounted on the robot creates echoes that are transformed into 2D contours by the decision tree algorithm. Their experiment used the echo information directly, rather than relying on a model-based approach.

Riddle, Segal and Etzioni (Riddle 1994) used a brute-force induction classification technique to uncover flaws in a Boeing manufacturing process. Their algorithm, despite its exhaustive search method, performed a depth-bounded search within an acceptable CPU time and with excellent predictive results.

In another example, Moczulski (Moczulski 1998) used case-based reasoning to design anti-friction bearing systems. His knowledge base came from text sources, whose quantitative values were converted into qualitative ones by hand, and thus susceptible to the bias of the person doing the coding.

MINING MAINTENANCE DATA

The scheduled maintenance data used in this research covered 101 weeks of a packaging machine operation. The machine was composed of 16 repairable subsystems, each with several non-repairable components; a failure in any subsystem stopped the entire machine.

Planned maintenance tasks took place sporadically, from every 6 weeks to every 11-12 weeks. Average mean time between failure (MTBF) was 8.19 hours in the first 52 weeks of operation, and 3.8 hours in the last 49 weeks; average mean time to repair (MTTR) was 0.42 hours in the first 52 weeks and 0.61 hours in the last 49 weeks.

When analyzing this data, we made the following assumptions about the data and the preventive maintenance (PM) tasks:

- The cause of failure was accurately reported
- Maintenance personnel used the correct methods to repair the machine
- Repairs returned the machine to "good as new" status
- All PM tasks took approximately the same amount of time to complete.

Reliability Analysis of Repairable Systems

As a repairable system, this machine can be mathematically modeled using the Non-Homogenous Poisson Process (NHPP), with events (failures) that are neither identical nor independently distributed (Ascher 1984). The centroid, or Laplace test, can be applied to the event data to determine if a particular system meets these requirements. This test is calculated according to expression (6):

$$U = \frac{\dfrac{\sum\limits_{i=1}^{n-1} T_i}{n-1} - \dfrac{T_m}{2}}{T_m \sqrt{\dfrac{1}{12(n-1)}}} \tag{6}$$

where

T_m = total time the system is observed
T_i = time of the ith failure
n = total number of failures*
*When the end of the observation time, T_m, is not accompanied by a failure event, the $(n-1)$ terms are replaced by n.

The result, U, compares the centroid of the times of failure to the midpoint of the total time of observation. If U<0, the trend is decreasing and reliability is improving. If U>0, reliability is deteriorating because the interarrival times are becoming shorter. A U=0 shows a stationary process with no trend. The value of U approximates a z-score if $n \geq 4$, and can be used to test the null hypothesis that no trend in interarrival times is present (O'Connor 1991).

A centroid test on this dataset gave a U of 5.66, strongly rejecting the null hypothesis of no trend. Additionally, centroid tests were performed on each subsystem; all showed either an increasing or a decreasing trend, with some trends more marked than others.

The failure intensity function,

$$U(t) = \lambda \beta t^{(\beta-1)}, \ t>0 \tag{7}$$

where λ and β >0 and t is the age of the system, predicts the number of failures per hour (Ascher 1984). λ and β can be estimated by the following equations for a single system (Ascher 1984):

$$\hat{\lambda} = \frac{N}{T} \tag{8}$$

$$\hat{\beta} = \frac{N}{\sum\limits_{i=1}^{N} \ln\left(\dfrac{T}{X_i}\right)} \tag{9}$$

where

N = number of system failures

T = total time of observation

X_i = age of the system at the ith failure

Figure 2 shows the predicted failures/hour of the current system using Equation (7). The y-axis represents the number of failures per hour; the x-axis indicates hours of operation. The failure intensity increase indicates an increasing rate of occurrence of failure (ROCOF) in the system.

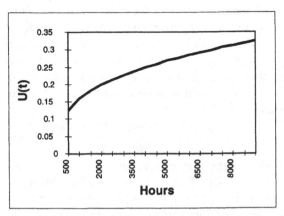

Figure 2: Failure Intensity Function

Mission reliability, or the probability that a system will successfully complete an interval of a fixed duration d>0 at given time t is defined as

$$R(t) = e^{-\left[\lambda(t+d)^\beta - \lambda(t)^\beta\right]}$$ (10)

R(t) is plotted in Figure 3 for d = 25 hours. The figure shows that the probability of this system operating without breakdowns for 25 consecutive hours is only 4% when the machine has been in service for 500 hours, and the probability quickly declines with running time. Clearly, the system is deteriorating and in the wear-out stage.

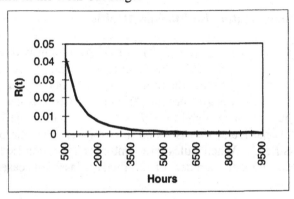

Figure 3: Reliability Function

Mining the Scheduled Maintenance Data

There are two main goals for this data mining task: first, to determine what subsystems or components are most responsible for downtime; and second, to predict when preventive maintenance tasks would be most effective in reducing failures. This information can then be used in setting maintenance policy guidelines such as planned maintenance schedules and alarm limits.

Because we were interested in classification and prediction, we chose to use a supervised learning method for this research. We decided to use decision trees for their robust performance and user-friendly output. The first mining task for the scheduled maintenance dataset is a classification of machine health; therefore, we chose equipment availability as the target variable, calculated using Equation (11). Equipment availability, because it takes into account the amount of time needed to repair a problem, is a good indicator of machine health.

$$\frac{ScheduledTime - DownTime}{ScheduledTime} * 100 \qquad (11)$$

The next step, data preparation, involved the removal of unneeded variables from the dataset. These variables tracked downtime resulting from production scheduling and operator issues, process inputs and outputs, and waste figures. The remaining variables track the number and type of machine faults and the associated downtime.

Classifiers such as decision trees work best when the number of classes is low. In this dataset, the output value of availability was rounded off to the nearest 10%, which gave five classes. The lowest class accounted for only one case, so it was combined with the next higher class, forming four classes (<70, 80, 90,and 100% availability).

Although wide ranges existed within some variables, we considered the outliers to be an important part of the overall picture, and therefore did not remove them. Also, since this dataset was relatively small, sampling was not necessary. The final dataset consisted of 35 variables and 101 instances.

The Data Mining Output for Machine Health

A decision tree can be read as an If-Then rule, beginning with the root node. In Figure 4, the root node is X4T, the amount of downtime for System 4. The right hand branch of the tree, therefore, reads "If X4T (the amount of time that System 4 is down) is greater than 90 minutes, AND System 2 is down at least once, the overall availability is 70%". The (2/1) at the terminal node means that this particular rule is found in two cases, and one of those cases was misclassified. When rules are ambiguous or conflicting, the rule containing the greater number of correctly classified cases should be followed.

Using 70% of the data for training and 30% for testing, we applied the decision tree algorithm to the dataset. Figure 5 shows the induced tree, which clearly shows that the emphasis on preventive maintenance should be on subsystems X2, X4, X12, and X16 since these subsystems are close to the root node and are associated with availability ratios at 80% or below. Subsystem X11 could also be included in the list, although the tree is somewhat ambiguous in that area. Of these 5 subsystems, X2, X11, and X16 had centroid test values that were strongly positive, indicating an increasing rate of occurrence of failure. X4 and X12 had the two largest total down times and correspondingly large standard deviations for reliability.

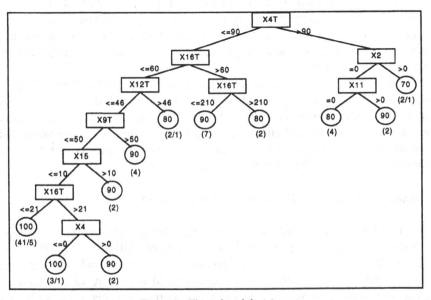

Figure 4: The induced decision tree

Second Data Mining Task- Determining the PM Interval

The machine used in this study is usually only scheduled for two shifts; only rarely is a third shift used. Thus, PM tasks can take place on the third shift without disrupting the production schedule. The target attribute was changed to a value indicating how many hours had passed since the last planned maintenance task; this value was binned into intervals of 200 actual operating hours (approximately 4 work weeks, a common PM interval in industrial practice), resulting in 4 classes for the PM attribute:

1. A week in which planned maintenance took place
2. 0 to 200 hours since the last planned maintenance task
3. 201 - 400 hours since the last planned maintenance task
4. > 400 hours since the last planned maintenance task.

Since the data was already pre-processed for the first mining task, the target variable changes were the only revisions to the dataset.

The Data Mining Output for PM Interval

The induced tree consisted of 23 nodes with an error rate of 28.7%; analysis of the output shows that Subsystems 2, 11 and 16 clearly experience an increase in both downtimes and breakdowns during the 200-400 hour period, while Subsystems 4 and 12 are more likely to fail during the 0-200 hour range. A reasonable PM schedule for these two subsystems (and others that tend to fail in the same interval) would be to perform PM tasks every 100 hours of operation (approximately every 2 weeks), until more data can be collected to refine the failure prediction. Maintenance tasks for subsystems that fail in the 200-400 hour range should be conducted after 200 hours of operation, or every month.

Since this machine is only operated on two shifts, more frequent maintenance can be performed without interfering with the production schedule. After implementation of the PM schedule, new data can be added to the dataset, the trees rebuilt using a smaller range for operational hours, and the interval adjusted if necessary. Note that once the data is in the proper form for mining, building the trees is a quick process.

Summary of Scheduled Maintenance Data Mining

The benefits of using decision trees in this domain are many. Scheduled maintenance datasets are, in many cases, scattered collections of written and electronic records. Errors and omissions are common, and often detailed information is unavailable. The dataset used in this study suffers from all these problems, and yet the decision tree output concurs with traditional statistical analysis as well as with the company's experience with this particular machine. In addition, the decision tree output is easily transformed into rules that facilitate understanding of the results by both technical and non-technical personnel.

CONDITION-BASED MAINTENANCE DATA

If a machine's normal vibration signature is known, comparisons can be made between the normal signal and the current signal to identify impending failures. In order to reduce the information overload, multi-sensor data fusion is often applied to reduce the amount of data to a signal composed of the most reliable information from each sensor. This task can be simplified by using data mining techniques to determine which sensors are most sensitive to particular defects at different torque levels.

As part of a United States Navy study on the use of artificial neural networks in condition-based maintenance, the Multi-Disciplinary University Research Initiative (MURI) for Integrated Predictive Diagnostics at Penn State performed a series of vibration tests on a CH-46 Westland helicopter transmission. Eight accelerometers were mounted on the gearbox using specially designed brackets. Two no-defect tests and seven faulted component tests were run, including three tests of crack propagation. A study using autoregressive modeling (Garga 1997) established that the problem space was separable even by high and low torque conditions, using reduced data from sensor 3 for all 9 test conditions. Several of the study's classification results produced error rates between 0% and 6%, outperforming neural networks trained with the same reduced data.

Many of the datasets from these studies are available for download at the MURI website, at http://wisdom.arl.psu.edu. The website also includes a report detailing the test conditions. For this research, we used data collected from one no-defect test, one crack propagation test (Helical Idler Gear Crack Propagation), and one faulted component test (High Speed Helical Input Pinion Tooth Chipping). The gear mesh frequency of the idler gear and pinion tooth is known to be 9088.80 Hz. Figure 5 shows a sample time series plot. Figure 6 shows a sample power spectrum plot of the same data.

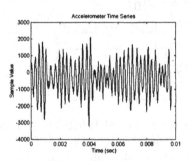

Figure 5: Time Series Plot

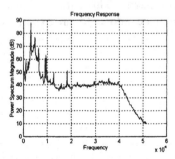

Figure 6: Frequency Response Plot

Mining Condition-Based Maintenance Data

For this task, we continued with the use of decision trees; however, the target variable for this dataset is the fault classification, as we are interested in finding the sensors, frequencies, and magnitudes that best predict the type of fault.

Data preparation consisted of transforming the 96 time series signal files (3 tests, 4 torque levels, 8 sensors) to a frequency response using Matlab. Variables in this dataset were the power spectrum magnitudes for the eight sensors, the torque level, and the frequency. Targeting the frequencies between 8590 Hz and 10500 Hz – an interval that included the gear mesh frequency +/- 1500 Hz – reduced the number of instances.

Separate training and test datasets in the target frequency range for each torque level were used to build the decision trees. The target range was extended to >7500 and <11500 Hz – to test the accuracy of the results with less accurate domain knowledge – and the trees regrown. Table 2 shows the error rates and tree sizes for each torque level.

Table 2: Error rates and tree sizes

Torque	8590 - 10500 Hz			75000 – 11500 Hz		
	Train Error	Test Error	Size	Train Error	Test Error	Size
100	0.5	2.7	10	0.5	4.0	6
80	1	3.5	6	0.7	1.7	6
75	0.0	0.5	4	0.2	0.6	7
70	0.5	1.5	6	0.0	3.0	4

The decision tree for the 8590-10500 Hz test (torque=100), shown in Figure 7, shows that four sensors (nos. 4, 6, 7 and 8) give the most information when classifying the three fault conditions in this frequency range. In addition, two frequency boundaries are identified that also contribute to the classification. This information can be used to tune data fusion algorithms or can be added to an expert system that monitors the machinery.

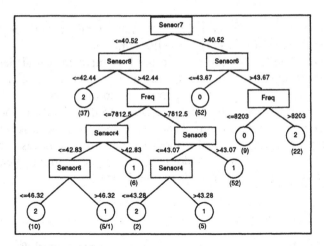

Figure 7: Decision tree for torque=100, 8950-10500 Hz
0: No Fault   1: Idler Gear Crack   2: Input Pinion Chip

Training and test sets for the entire frequency range, from 0 Hz to 40000 Hz, were used to build a series of trees, varying the pruning confidence factor and the minimum number of instances required to form a leaf. The low error rates show that, in this well-separated dataset, the decision tree algorithm can correctly identify relationships without using domain knowledge to narrow the frequency range. Error rates, pruning levels, and tree sizes are shown in Table 3.

Table 3: Error rates and tree sizes for 0-40000 Hz

MinCase	Prune Level	Train Error	Test Error	Size
2	25	2.0	8.2	638
10	25	6.1	10.2	288
25	25	9.5	12.9	179
2	15	2.5	8.0	571
10	15	5.9	9.7	287
25	15	10.0	12.3	158
2	5	3.5	7.9	454
10	5	5.9	9.4	257
25	5	10.0	12.0	156

Although the trees for these examples are very large, the trees can be readily transformed into rules for inclusion in an expert system. The error rates for these larger trees is comparable to the neural network error rates found by Garga et al. (Garga 1997), using the same data.

The decision tree output can also be used as input to neural networks. Using the full frequency range data, we built a backpropagation neural network and adjusted layer weights to correspond to the importance of sensor variables in the induced decision tree; the closer to the root node, the larger the weight given the variable. Figure 8 shows the decreased mean square error (MSE) and rapid convergence of this modified neural network, as compared to an unmodified network with random initial weights. Similar ideas have been used by researchers in other fields (Kubat 1998).

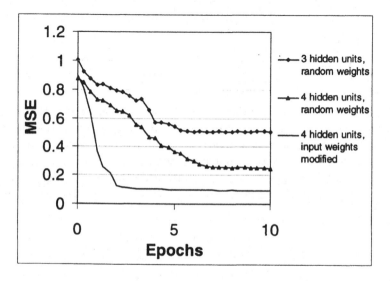

Figure 8: Decision tree initialization of neural networks

Clearly, decision tree learning of condition-based monitoring data can contribute greatly to the analysis of the data and the development of effective monitoring systems. Although decision trees, unlike neural networks, are not incremental learners and have to be regenerated with new data, their speed, robustness, and understandable output make trees an attractive and feasible choice for mining the large databases generated by monitoring equipment.

CONCLUSIONS AND FURTHER WORK

To be competitive in the current business climate, a manufacturing company must deliver high quality products in a responsive and cost effective manner. In this environment, machine maintenance requires renewed attention because improper machine maintenance schedules can lead to deteriorated product quality, high cost, and excessive downtime. Complex machine tools and advanced data acquisition systems generate "mountains" of sensor and system monitoring data that reside in gigabyte- and terabyte-sized databases. This dimensionality and the desire to uncover patterns automatically (if not unsupervised) has rendered traditional data analysis methods ineffective. In this chapter we demonstrate that data mining techniques are useful tools for understanding maintenance data and developing a plan for machine repair and monitoring. Decision trees in particular give the user a graphic representation of the information contained in the data, can be transformed into rules easily, and are quick to train even on large datasets.

The importance of accurate data cannot be overstated. Although the algorithms can still give good results with "dirty" data, having the correct variables in the correct form greatly improves results. The development of integrated data collection/data mining software will increase the usefulness of these methods.

Further work needed in scheduled maintenance management is the addition of other data mining methods such as rule induction and self-organizing maps, and genetic algorithms for an alternative method of feature selection. In the area of condition-based maintenance, areas of possible study include the use of decision trees and other mining algorithms in conjunction with data fusion, expert systems, and extensions into other areas of machine monitoring.

ACKNOWLEDGEMENTS

Carol Romanowski acknowledges the support of the Engineering Research Program of the Office of Basic Energy Sciences at the Department of Energy. Rakesh Nagi acknowledges the support of the National Science Foundation under career grant DMI-9624309.

REFERENCES

Ascher, H. and Feingold, H., Repairable Systems Reliability. New York: Marcel Dekker, Inc., 1984.

Berry, M. and Linoff, G., Data Mining Techniques: For Marketing, Sales and Customer Support. New York: John Wiley & Sons, 1997.

Cheung, K. and Hausmann, W. H., "Joint determination of preventive maintenance and safety stocks in an unreliable production environment," Naval Research Logistics, 44, 257-272, 1997.

Elder, J. F. and Pregibon, D., "A Statistical Perspective on Knowledge Discovery in Databases," in Advances in Knowledge Discovery and Data Mining, pp. 83-113, Menlo Park, CA: AAAI Press/The MIT Press, 1996.

Garga, A., Elverson, B. T., and Lang, D.C., "Fault classification in helicopter vibration signals," in American Helicopter Society 53rd Annual Forum, 1997.

Hsu, L.-F. and Kuo, S., "Design of optimal maintenance policies based on on-line sampling plans," European Journal of Operational Research, 86, 345-357, 1995.

Kelly, C. M., Mosier, C. T., and Mahmood, F., "Impact of maintenance policies on the performance of manufacturing cells," International Journal of Production Research, 35(3), 767-787, 1997.

Kennedy, R. L., Lee, Y., Van Roy, B., Reed, C., and Lippmann, R., Solving Data Mining Problems Through Pattern Recognition. New Jersey: Prentice Hall PTR, 1998.

Kubat, M., Kaprinska, I., and Pfurtscheller, G., "Learning to Classify Biomedical Signals," in Machine Learning and Data Mining, pp. 409-428, Chichester: John H. Wiley & Sons, Ltd., 1998.

Mobley, R. K., An introduction to predictive maintenance. New York, NY: Van Nostrand Reinhold, 1990.

Moczulski, W., "Inductive Learning in Design: A Method and Case Study Concerning Design of Antifriction Bearing Systems," in Machine Learning and Data Mining, pp. 203-219, Chichester: John H. Wiley & Sons, Ltd., 1998.

O'Connor, P. D. T., Practical Reliability Engineering. New York: Marcel Dekker, Inc, 1991.

Pyle, D., Data Preparation for Data Mining. San Francisco: Morgan Kaufmann Publishers, Inc., 1999.

Quinlan, J. R., "Improved use of continuous attributes in C.5," Journal of Artificial Intelligence Research, 4, 77-90, 1996.

Riddle, P., Segal, R., and Etzioni, O., "Representation Design and Brute Force Induction in a Boeing Manufacturing Domain," Applied Artificial Intelligence, 8, 125-147, 1994.

Shannon, C. E., "A mathematical theory of communication," Bell System Technical Journal, 27, 379-423 and 623-656, 1948.

Sillitoe, I. and Elomaa, T., "Learning decision trees for mapping the local environment in mobile robot navigation," in Proceedings of the MLC-COLT Workshop on Robot Learning, 1994.

Williams, J. H., Davies, A., and Drake, P. R., Condition-Based Maintenance and Machine Diagnostics. London: Chapman & Hall, 1994.

CHAPTER 11

Methodology of Mining Massive Data Sets for Improving Manufacturing Quality/Efficiency

Jye-Chyi (JC) Lu
JCLU@isye.gatech.edu
School of Industrial and Systems Engineering
Georgia Institute of Technology
Atlanta, GA 30332-0205

ABSTRACT

In this information era, many enterprises have begun exploring ways to utilize information stored in various databases for creating a competitive edge in managing their supply chain and networked manufacturing processes. This practice requires tools to automatically synthesize a large volume of data for getting needed knowledge. Although there are several existing data mining techniques, most of them are not effective in processing large amounts of data with possible nonstationary and dynamically changing trends. Our procedure first reduces the massive data sets into smaller size data by using data splitting and other data reduction techniques. Then, the traditionally used methods in data mining, signal/image processing and statistical analysis can be useful to handle the reduced-size data. Thus, decision rules for identifying and classifying process problems can be constructed based on these reduced-size data to improve manufacturing quality and efficiency. Finally, by using weighted averaging or voting procedures including artificial neural networks, the synthesized results obtained from the split-data can be integrated. Our real-life examples show a great potential of the proposed methods in mining knowledge from massive manufacturing data sets and in making significant impact in many fields including E-business operations.

D. Braha (ed.), Data Mining for Design and Manufacturing, 255–288.
© 2001 *Kluwer Academic Publishers. Printed in the Netherlands.*

INTRODUCTION

The advancement of automatic data acquisition instruments, networking systems and computers has facilitated the growth of data representing various operations at many distributed sites. Many companies begin to develop tools for extracting useful knowledge from their databases, including customer orders, material inventory levels, production schedules, quality/efficiency measures, service records and capital investments to support their decisions on production management. In a joint project with Nortel, we successfully developed an interface to collect data from several databases for mining information from numerous records. Timely synthesis of information is critical for product design, process troubleshooting and quality improvement decisions. A major obstacle in our project was that tools for processing a large volume of information coming from numerous stages of Nortel's operations were not available. Currently, there are many research studies and software packages in database building, data mining and factory operation simulation. However, there is very little research on tools for synthesizing massive data sets for supporting decision-making processes. The purpose of this article is to propose ideas and tools for synthesizing a large size of information from manufacturing processes, so that decisions can be made at different hierarchically organized resolution levels.

Due to the high cost-constraints typically associated with processes in the semiconductor and electronics industries, recent equipment/process quality/efficiency improvement techniques have relied on sophisticated process information synthesis tools to handle complicated data such as nonstationary and dynamically-shifting trends contributed from potential process faults. May and Spanos (1993) combined artificial neural networks (ANNs) and evidence reasoning in automated malfunction (fault) diagnosis; Gardner, et al. (1997) constructed spatial signatures for semiconductor process fault detection. However, these methods are not suitable for handling large volumes of data. The underlying theme of our methodology is to "reduce the size of data," and apply existing (and new) procedures developed for the smaller size data to the reduced-size data for decisions. For example, splitting the huge size data into several smaller pieces of data sets, one can process these data sets simultaneously using parallel computing techniques. Treating the coefficients from a wavelet model (see Section 3.1 for definitions) as the reduced-size data, one can apply pattern recognition methods or principal component analyses (PCA) to understand process behavior including fault patterns for improving manufacturing quality.

There are many challenges in the research addressed above. For examples: (1) Can the data be split intelligently for including crucial information such as temporal or spatial dependence? (2) Can the reduced-size data represent the original data well for accurate decision-making? (3) Is the data reduction approach effective and does it meet the requirement of computing speed? (4) How robust are the data-reduction procedures in noisy-data situations? (5) Can the integration of analyses applied to the "split-data" be effective? To answer these questions, Section 2 motivates our research with several real-life examples. Section 3 presents several data reduction and data splitting techniques. Section 4 develops a few decision rules based on the reduced-size data. Section 5 provides procedures of integrating results obtained from synthesizing the split-data. An extended wavelet neural network (EWNN) equipped with parallel data processing capability is proposed for data integration. Some of the ideas addressed are illustrated with examples from our partners' manufacturing systems. Section 6 concludes this article.

MOTIVATING EXAMPLES

Massive Data Sets from Production Processes   The following problem is encountered in many companies such as JDS Uniphase, IBM, Bayer, etc. This example uses one of our past projects to illustrate the situations faced in companies. In working with Nortel's (switching equipment) printed circuit board (PCB) assembly plant to develop their knowledge-based production systems, we collected many types of data across materials, manufacturing, service, repair and accounting departments. Because the type and cost of the electronic circuits and boards changed constantly, and because hundreds of different products went through the production line every week, Nortel needed an information system to update these product costs and identify areas of improvement continuously and timely. Our team thus developed a software tool which automatically collects data from many databases, synthesizes available data and finally, reports the information of production cost and process performance in an easy-to-read "9-up" chart on Nortel's intranet. The

9-up chart information was utilized in management meetings to allocate company resources for efficiently operating production system and planning future innovations of equipment and product. Although this project was successful for a time period, the software tool cannot respond to the dynamically changing nature of Nortel's operations. Many data collections involving direct-feed and the information synthesizing processes were implemented by fixed rules developed from manual data analyses. When the database system (or operator) or the data type (or information synthesis rule) is changed, our tools required constant maintenance.

Hierarchical Process-Fault Detection   Figure 1(a) presents three of seven sets of Ar^+ gas levels from a nominal process in a thin-film deposition experiment sponsored by Semiconductor Research Corporation (SRC). The four most important data features were the large bends (curve-changing points) exhibited in Figure 1(a). Figure 1(b) plots these features extracted from the modulus maximum (Mallat and Hwang, 1992; see Section 3.1 for its definition and Example 4 for illustration details) applied to these seven nominal data sets. The minima and maxima of these local features were connected to form the upper and lower "operation tolerance-bounds." This is a first "crude representation" of all the data including process variations. Any new process with data outside these bounds, especially at the four feature locations, was identified as a faulty process. Because reducing the entire data set to four data features may be too aggressive, omitting valuable details, we proceeded to construct a tolerance bound for a "second resolution" analysis. Figure 1(c) shows an additional data feature located in the middle of the data range. By using this new data feature, the tolerance bound was refined to include a small dip in the middle (see Figure 1(d) for a graphical presentation). Figure 1(e) illustrates that all of the data from the nominal processes were within these new bounds. Figure 1(f) shows a real example of a faulty process with data clearly outside the tolerance bounds. When processing a huge amount of data, the above analysis can be conducted by analytical methods without any graphical analysis. There might be more classes of these local data features creating a hierarchical multi-resolution analysis of data. See Example 5 for details.

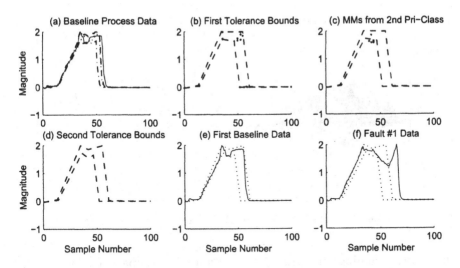

Figure 1: Multi-Resolution Process Fault Detection with Local Features

Complex Antenna Functionality-Testing Data   In a project with Nortel's wireless antenna manufacturing division, we collected 28 sets of data (with 32,671 data points in each set) similar to Figure 2(a). The goal was to locate representative features involving significantly less data for monitoring antenna quality in real-time and for different types of process trouble-shooting. The cusps and lobes of the azimuth cut of antenna data presented in Figure 2(b) are difficult to handle by standard techniques such as polynomial regression, nonparametric modeling or Fourier transform. Their locations and values are critical in process monitoring and trouble shooting. A recent experiment of using the two-dimensional (2D) wavelets to model and extract data features reinforced our confidence of wavelet's (see Section 3.2 for its definition) ability in handling this type of data. However, for building an intelligent system to synthesize these data in real-time, the 2D wavelets used too many coefficients and were not efficient. Recognizing the smooth pattern for the data at elevation cuts, our strategy was to use a combination of one-dimensional wavelets and polynomial regression for extracting important data features. See Examples 1 and 2 for details of implementing this idea.

Other Data Types   There are many other large size data sets in fields other than manufacturing. For example, large-scale consumer

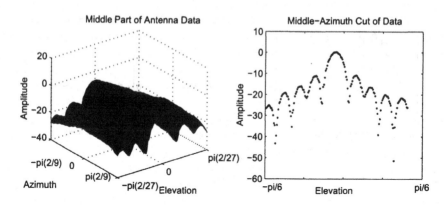

Figure 2: Data Signals from Testing a Wireless Antenna

databases including large transaction volumes from supermarkets, banks, insurance companies, telephone service, etc. These databases, with the use of our methodology, can provide improvement in customer service, marketing strategy, management and decision making. There are several databases containing various images, e.g., medical imagery or satellite imagery. For instance, the NASA Earth Observing System (EOS) generates 100+ gigabytes of remotely sensed image per hour when operational. These EOS data sets are heterogeneous with spatial and temporal structures, multi-formatted and geographically distributed. Moreover, there are many ways that decision makers, scientists, and the public need to interact with these data sources. Accessing to an entire data set in the examples given above is time-consuming and makes the standard computational methods infeasible. Thus, analysis of these massive data sets requires new approaches. In this article, we will focus on numerical data collected in various measurement tools including different kinds of sensors and product testing devices in manufacturing processes. This type of data can exhibit *nonstationary and dynamic trends*, but has distinct patterns contributed from process faults. Traditionally used data-mining, statistical and signal processing procedures such as association, clustering, decision tree, regression, time-series models and Fourier transform are not best suited to describe this type of data. Wavelet-based methods had been recommended by many researchers. See examples in Jin and Shi (1999) of tonnage signals used to detect faults of a sheet-metal stamping process and in Wang, et al. (1999) of different catalyst recycle rates for performing diagnostics of failures in a residual fluid catalytic cracking process. This article builds on wavelet's strength to develop data-mining procedures for massive data sets.

DATA REDUCTION METHODS

Sampling Approaches

The main idea of sampling approaches is to choose part of a data set to represent the original data. Several combinations of the following sampling methods can be used effectively to capture important data characteristics for reducing the size of data.

Systematic Sampling   Besides *randomly sampled* data points to represent the whole data set, if all the objects in a data set have similar forms, systematic sampling can be used to select every other data part for reducing the data size.

Stratified Sampling   When the data is less uniform, the stratified sampling method (Cochran, 1977) can be useful. This method divides a large data set into several subgroups with data of similar characteristics, and then selects a few representatives randomly from each subgroup to compose a reduced data set. When there exists prior knowledge about the importance of data parts in representing the whole data set, we can assign a mixture distribution probability according to the importance of the data parts in a decision-making process. The sizes of data in the stratified samples can also be used to determine the probability weights.

Segmentation Sampling   In real-life applications, very often the data set is divided into several segments according to the physics that generates the data. Then, a part of data from a single segment or several segments is sampled in further studies.

Feature Extraction   One special type of segmentation is the "feature extraction." Jin and Shi (1999) utilized the knowledge of a sheet-metal stamping process to identify physically motivated features from tonnage information. Example 4 shows ideas of using the following modulus maximum technique to automatically search key data features.

Modulus Maximum   In many applications, such as image processing and computer vision, local features such as curve change-points and picture edges carry the key information contained in a data set. Mallat and Hwang (1992) developed important measures based on *modulus maximum* to identify local singularities. The *modulus maximum* is a point (u_0, s_0) in scale-space such that $|Wf(u, s_0)|$ is a strict local maximum in either the right or left neighborhood of the location u_0, where $Wf(u, s) = s^{(-1/2)} \int_{-\infty}^{+\infty} f(t)\psi(t-u)/sdt$ is the continuous wavelet transform (CWT) for a signal $f(t)$. This transform measures the degree of similarity between $f(t)$ and the wavelets $\psi_{u,s}(t)$ as an inner product. Mallat (1998) defined the *maxima line* as a connected curve in the (u, s) plane along which the points are modulus maxima. In Figure 3, two signals (a1, a2) are shown with their scalogram representations (b1, b2), where the maxima lines (c1, c2) are plotted in scale-space with $\log_2 s$ as the scale unit. Mallat (1998) and many others used the maxima lines to identify the locations of local singularities, where a fine (small) scale of $s = 5$ was employed.

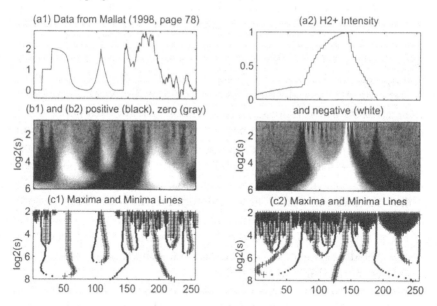

Figure 3: DerGauss CWT Plots and Modulus Maxima Lines

Modeling and Transformation Approaches

Summary of Distribution Characteristics   If data are from the same population, simple summary statistics such as mean, median, variance, skewness and percentiles can represent data well.

Regression Modeling   If there exists a certain structure (linear or nonlinear) of data variables, and this structure can be captured by regression models, the regression coefficients and the regression functional forms can "represent" the information in the data.

Principal Components Methods   Considering a data set made by several vectors in a multi-dimension linear space, the statistical principal components method (Johnson and Wichern, 1982) can "project" the vectors into a lower-dimension linear space, and treat the projection as a method of data reduction.

Wavelet Transforms   The idea of the regression method can be extended to other tools, such as Fourier and wavelet transforms, for handling more complicated data patterns. Wavelets are very popular in many engineering and computing fields for solving real-life problems including data compression for signal and image processing. Wavelet transforms can perform multi-resolution function approximation (Mallat, 1998) and describe irregular data patterns such as sharp "jumps" better than Fourier transforms and standard statistical procedures, e.g., spline and nonparametric regression. A wavelet is a function $\psi \in \mathbf{L}^2(\mathbf{R})$, which has zero average $\int_{-\infty}^{\infty} \psi(t)dt = 0$ and is normalized to one, where $\mathbf{R}$ denotes the real-line. Wavelets can be translated (s) and dilated (u) to create a family of time-frequency atoms, $\psi_{u,s} = \psi[(t - u)/s]$. An example of the ψ function is the Sombrero wavelet (see Mallat, 1998 page 77), which has a Mexican-hat like pattern. For small scale s, the "base function" $\psi(t)$ is scaled to be short in duration, thus providing good time-localization of high-frequency events, such as signal edges, which usually carries important data information. Conversely, for large scales s, $\psi(t)$ will have good frequency localization and poor-time localization properties, thereby extracting global, long-term low-frequency trends. The discrete wavelet transforms (DWT) is defined similarly to CWT as $Wf(2^{(-j)}k, 2^{(-j)}) = 2^{j/2} \int_{-\infty}^{\infty} f(t)\psi(2^j t - k)dt$, where j and k are integers.

The following presents an example of identifying important data patterns to reduce data size. Here, we apply a regression model on the wavelet coefficients for getting a 10 fold data reduction ratio compared to the regular two-dimensional wavelet transforms.

Structured Wavelet Modeling   By inspecting the data pattern in Figure 2, the model is structured as follows:

$$Y(x_i, y_j) = \sum_{k=1}^{k_{L,M}} s_{L,k}(x_i)\phi_{L,k}(y_j) + \sum_{l=1}^{L}\sum_{k=1}^{k_{l,M}} d_{l,k}(x_i)\psi_{l,k}(y_j) + \epsilon_{i,j}, \qquad (1)$$

where the wavelet coefficients $s_{L,k}$ and $d_{l,k}$ are modeled by polynomial functions, e.g., $s_{L,k}(x_i) = a_{0,L,k} + a_{1,L,k}\ x_i + a_{2,L,k}\ x_i^2 + ... + a_{p_m,L,k}\ x_i^{p_m}$, and $d_{l,k}(x_i)$ is defined similarly. The errors $\epsilon_{i,j}$ are independent, and have zero means and variances $\sigma^2(x_i, y_j)$. The variance has a similar functional structure as the mean function (1), but with different orders of wavelet and polynomial functions. With a set of wavelet and regression coefficients and their "bases," the original data can be reconstructed from the structured wavelet model. More coefficients will provide better reconstruction of the data. The model coefficients are the "reduced-size" data that can be useful in real-time process control. Examples 1 and 2 show a case of 200:1 data reduction with only limited errors.

Automatic Data Pattern Recognition Methods   Most of the above methods require human efforts to understand the background "physics" of the data and its structure for identifying an efficient way of data reduction. For synthesizing a huge size of data, it is infeasible to demand human operators to develop case-based data reduction methods. However, to support intelligent manufacturing systems, it is important to develop a method that can automatically recognize data patterns and select an appropriate data reduction method for the subsequent information synthesis such as process fault detection and classification. Although it is possible to derive a "rule-based" expert system to perform this task, it is important to note that the purpose of data reduction is not just "compressing" the data to a smaller size, but to facilitate an accurate decision-making process. Thus, we need to understand what types of decisions are required for selecting an appropriate data reduction methods. See Example 4 for more details.

Data Splitting Methods

As stated earlier, if the data size is too large, "data splitting" is needed for making the data size smaller. After reviewing many existing procedures of data splitting, we have experienced that these procedures are not "intelligent" enough to recognize the structure (or characteristic) for a very large size of data. For example, kd-trees (see Moore (1991) and other of his papers in http://www.cs.cmu.edu/ AUTON for details) have been used for indexing large databases involving spatial distributions. However, to answer certain statistical queries, it is assumed that the set of records found by the query is small. Another disadvantage of kd-trees is that when the new data are added to the database, the pre-processed tree may have to be rebuilt. Data reduction using fuzzy C-means clustering (Mascoli, 1995) is an example of how to apply "intelligence" to automate the large data reduction task. The advantage of this approach is that cluster-validity measures from unsupervised optimal fuzzy clustering methods have been incorporated such that no a priori assumptions about data structures (e.g., number of clusters, range of responses) are necessary. However, this approach may not be applicable for the large size of data. Hence, more research is needed in splitting large size data intelligently and effectively.

Illustrating Examples

Example 1 - Systematic Sampling and Structured Wavelet Modeling. In Model (1) the data points at each azimuth cut (181 of them; Figure 2(b) shows the irregularity of its data pattern) were modeled by the wavelet model selection method described in Section 3.5. The largest 40 DWT coefficients obtained by using the default wavelet base (S8) in the Splus (1996) Wavelets package gave a satisfactory fit with a minimized value of the objective function (2). Because the data pattern in each azimuth cut was similar to the data in the zero-azimuth cut as shown in Figure 2(b), the wavelet coefficients for each cut were all selected from the same wavelet base. Moreover, because these wavelet coefficient values had a polynomial-like structure across azimuth cuts, we systematically sampled every 10 azimuth locations (19 of them total) to further reduce the data size. The goodness-of-fit tests show that the fourth-order polynomial function gave a satisfactory fit to these wavelet coefficients each with 19 components. Thus, this combined wavelet-regression model can be used to predict data at all 32,671 locations. Overall, the model's reconstruction error is 8%. Considering all cusps, lobes and variations at the data boundaries, we concluded that an 8% error with only 160(40 (wavelet coefficients) $\times 4$ (regression coefficients)) parameters was reasonable. This procedure results in a 200:1 data reduction to the original data with a very limited data reconstruction error.

Example 2 - Process Fault Detection. In the Nortel project described in Section 2, there were 20 sets of antenna signals (for $r = 1, 2, \ldots, 20$) collected from the "nominal process," and eight sets of new data (for $r = 21, 22, \ldots, 28$) collected separately for validating our procedures. By using the procedures described above each data set resulted in 160 wavelet-regression coefficients, which served as the reduced-size data in this example. Due to the existence of process variations, the target antenna is created from the average of the coefficients in the 20 baseline antennae. Denote by D_β the sum of squares of the difference of the reduced-size data in the target and the new antenna. Denote by D the sum of the squares of differences of all predictions (32761 points) from the target and the new antenna model at the original data scale. The correlations between these two summary quality measures for the original data and the reduced-size data in 20 baseline antennae is 0.942. All faulty antennae were identified correctly with data in the reduced-size scale. This gives us confidence that the decision of process faults based on D_β statistics at the reduced-size scale is trustworthy. Next, we present a procedure for using sampling and segmentation methods to detect and classify process faults.

Example 3 - Process Fault Classification. By mapping the types of process faults to "signatures" of data patterns presented in the reduced-size data, it is possible to develop a "reverse-mapping" procedure to classify process faults. However, because there are only a few antenna data sets with faults in our study, engineers advised us to build process fault classification signature on the original data scale. This idea is further supported by the fact that engineers have more feelings about the departure of antenna signal patterns at several important locations. Due to the (potentially) large size of the data in the original scale, it is impractical to check the antenna quality on every location in real-time operations. Hence, segmentation and systematic sampling procedures were used to check antenna quality at selective locations.

To monitor the quality of new antennae and recognize the "fault" types, we developed simultaneous confidence intervals with the differ-

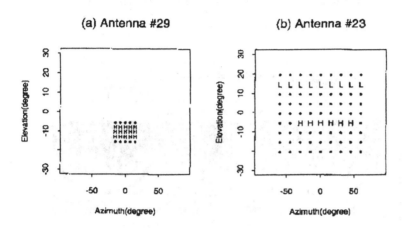

*: within the operation bounds; L: below the operation bounds; H: above the operation bounds.

Figure 4: Spatial Map for Classifying Process Faults

ences $d_{new,i} = Y_{new,i} - \hat{Y}_{target,i}$ at a set of positions i based on the large sample normal approximation method, where $Y_{new,i}$ is the new antenna data and $\hat{Y}_{target,i}$ is the target antenna data at the original data scale. By working out the variance $\hat{\sigma}_d(x_i, y_i)$ of the difference and the individual confidence interval from the Bonferroni inequality (Mendenhall, Wackerly and Scheaffer, 1990), we established a "spatial map" for summarizing the pattern of process signals outside the confidence intervals. The two plots shown in Figure 4 illustrate our idea. In Figure 4(a), we sampled only a region where engineers experienced many problems in the past. Eight out of the nine new antennae did not have problems. For the "faulty" antenna, its spatial map of the "out-of-confidence-interval" case shows that the antennae signals have higher values in a neighborhood near the center of the antenna signal map. In Figure 4(b), we systematically sampled a set of locations that cover a wide range of important areas. According to the clusters of the signals above and below the confidence intervals, engineers can match this pattern to a designated process problem (e,g., "mis-alignment" in several circuits at a few antenna positions).

Example 4 - Automatically Searching for Key Data Features   The

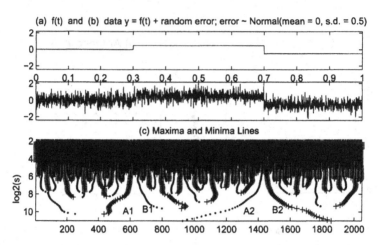

Figure 5: Automatically Identify Key Data Features from Data with Great Uncertainties

above examples required human efforts to identify the special data pattern that permitted reduction of the data. Consider on automating the data feature extraction process for sensor signals. If the data exhibit little noise, then by taking the first (or higher-order) difference of the data, one might be able to identify important local data features at those "jump-points" with larger differences. If the data are noiser, such as those plotted in Figure 5, where there are jumps, 0.3 and 0.5, at locations 0.3 and 0.7, respectively, in the mean data signal, this differencing idea does not work. Extending Mallat's (1998) idea of maxima lines, we construct both maxima (line $A2$ in Figure 5) and minima (line $B2$) lines of the original CWT values (not in absolute values as used by Mallat) at different scales. Ideally, when the scale becomes very fine (small s), the two lines $A2$ and $B2$ will converge to the same singularity location. However, when the data are noisy, the two neighboring maxima and minima lines usually will not converge to one point. Hence, a statistical "t-like test" based on the difference of neighboring maxima and minima CWT values in the finest scale might be developed to check if the two lines have converged "statistically." Then the converged point is the data feature. The variance of the "t-like test" can be estimated from pooling insignificant local maxima and minima CWT values together. However, because these maxima and minima CWT values might not be independent, the distribution of the test statistic can be quite complicated. Further theoretical research is needed to result this issue. Nevertheless, from Figure 5(c) we can visually identify the two data features close to 0.3 and 0.7 from the limits (s tends to zero) of lines (A1, B1) and (A2, B2).

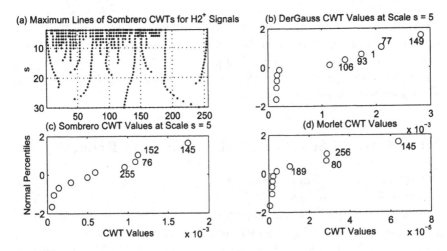

Figure 6: *Maxima Lines and QQ-Plots of MMs' CWT Values for $H2^+$ Signals*

Example 5 - Hierarchical Feature Extractions   To facilitate the hierarchical process-fault detection analysis similar to the study given in Section 2, the priority of the extracted local features needs to be identified. In general, a larger CWT value implies a stronger correlation (in the inner product sense) between the data and the wavelet bases. Thus, the features with larger CWT values should be set as higher priority. From our studies of CWT values (not shown here) of the local features identified from maxima-minima lines, we learned that there are usually only a few local features with larger (absolute) CWT values. A simple normal quantile-quantile (QQ) plot, such as in Figure 6 (b-d), can identify important local features with "significantly" larger CWT values. Smaller CWT values that were clustered on the left side of the QQ-plot can be ignored in further analyses. A more rigorous analysis based on the idea (Lenth, 1989) of pooling "insignificant effects" (CWT values) to estimate their variation in identifying significant effects can be used here. Finally, different wavelet bases can produce different maxima and minima lines. Figures 6(b) and (c) show the CWT values using two popular wavelet bases, DerGauss and Sombrero, respectively. Note that important local features were identified in a similar priority class in this case. For example, because location #145 (see Figure 3(a2) for the data curve) has a "distinct" CWT value, it was identified as a first priority local feature. The change-points around locations #77 and #145 were the most important local features for H_2^+ intensity. They were identified in all four kinds of wavelet bases available in our software. The CWT values around locations #1 (starting point), #189 (signal reaching zero) and #256 (ending point) and other "reflection points" at locations #93, #106 and #152 were large as well.

Wavelet Model Selection Methods for Data Reduction

Situation with A Single Data Set   The existing procedures in the literature focused on "de-noising" the data signal and did not address the "data reduction" needs. For the H_2^+ data shown in Figure 7(d)), the commonly used DWT model selection methods in the literature, such as SURE (Donoho and Johnstone, 1995) and AMDL (Antoniadis et al. 1997) both selected 77 out of 128 coefficients (see the upper-right plot of Figure 7 for the AMDL example). Thus, we modified the commonly used "mean square error (MSE)," $\sum_{i=1}^{N}[y_i - \hat{y}_i]^2$, in their objective function by adding a penalty term C/N for keeping the number C of coefficients small. The resulted objective function based on the "relative reconstruction error" is:

$$\text{RRE}(C) = \left[\sum_{i=1}^{n}(y_i - \hat{y}_{i,C})^2\right]^{1/2} \bigg/ \left(\sum_{i=1}^{n}y_i^2\right)^{1/2} + C/n. \tag{2}$$

The lower-left plot shows that the REE has a minimum value at $C = 10$. The lower-right plot shows that the approximation model with $C = 10$ captures almost all the details of the data pattern. Similar observations were found in modeling antenna data shown in Figure 1.

By matching the rate of decrease in the MSE and the rate of increase in number of coefficients used in the model, we further developed a modified model selection criterion (see Martell, 2000 for details). This criterion has several nice properties such as minimax property shown in

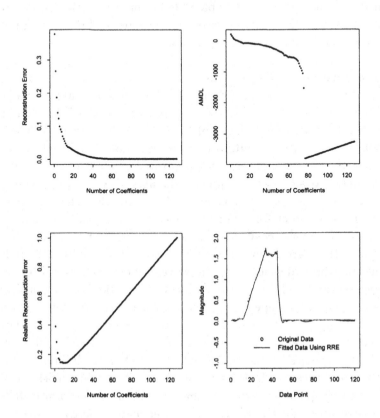

Figure 7: Model Selections for Single Data Set

classical wavelet shrinkage method (e.g., Donoho and Johnstone, 1995).

$$C(r, h) = \frac{\|\mathbf{w}_n - \hat{\mathbf{w}}(r)\|^2}{\|\mathbf{w}_n\|^2} + \frac{(r+2)ln(r+2) - (r+2)}{h\,n^h}, \tag{3}$$

where $h > 0$ is a parameter controlling the number of non-zero wavelet coefficients used in the $\hat{\mathbf{w}}(r)$. When h is closer to 0.0, only a few non-zero wavelet coefficients are selected for the estimator $\hat{\mathbf{w}}(r)$ of $\mathbf{w}_n$. The estimate $\hat{f} = \sum_{I \in M} w_I \psi_I$ is used to approximate the original data function, where M is the set of indices corresponding to the indices of the non-zero elements in $\hat{\mathbf{w}}(r)$. Applying this method to many simulated samples and test-data given in the literature (e.g., Donoho and Johnstone, 1995), we concluded that our method is compared favorably to existing procedures in reducing the size of coefficients used in the model and in limiting modeling errors. See Martell (2000) for details.

Situation with Noisy Replicates   Several sets of data streams from a process can be obtained as the "replicates" in decision-making analyses for the following reasons: [a] develop a "baseline process" for comparing with future process outcomes, [b] develop a representative model for all azimuth-cuts data (181 cuts totally) in Example 1, and [c] research the effect(s) of treatments, e.g., temperature/schedule changes, based on experimental design data for process quality/efficiency improvement. Regardless which model selection method is used, the selected wavelet coefficients are different for different replicates. Our challenge is to determine a set of wavelet coefficients which will represent the overall data model well. Here are a few strategies to solve this challenge: [i] Extend the procedures developed in the wavelet model selection literature (including our method with a new objective function) to a multivariate case. Then, the independent replicate data case becomes a special situation for applying the extended procedure. Woojing and Vidakovic (personal communication) gave an example of this research. [ii] Simply count the frequencies of the coefficients selected from different sets of replicates. Sort these coefficients according to their frequencies. Then, pick the most frequently used coefficients to minimize the sum of RREs from all replicates; [iii] Create an "across-replicate" energy measure by adding squares of all coefficients at the same wavelet atom position $\psi_{s,u}$. Sort these measures obtained from all N atom positions. Pick the largest C^* measures for minimizing the sum of RREs from all replicates; and [iv] Develop a *(atom position-) block-type* thresholding method to screen out the sets of wavelet coefficients with smaller energy defined in [iii].

DECISIONS BASED ON REDUCED-SIZE DATA

Various decision rules can be developed from the reduced-size data for solving manufacturing problems. This section starts with a few motivating examples followed by the proposed ideas.

Example 6 (Using Expert Information to Construct Test Statistics) Jin and Shi (1999) used an X-bar chart to monitor a wavelet coefficient selected from expert knowledge for detecting a specific type of process faults. Utilizing the physical background of the process, Koh, et al. (1999) split the stamping process data into disjoint sets of data signals. For each set of data, they developed a likelihood ratio test for checking if the mean (or variance) of a Haar wavelet coefficient differs from its target value significantly for detecting possible multiple faults.

Example 7 (Using the Linear Discriminant Analysis (LDA) to Distinguish Process Fault Types) In the experiment producing the Ar^+ signals given in Figure 1, we deliberately induced four different types of known faults, such as shutting down the key mass flow controller (MFC) for chemical materials, starting the MFC at the wrong setpoint, or a pressure controller failure, to develop process fault detection and classification methods. Rying, et al. (1997) applied a scale-dependent energy metric, E_s = sum of squares of all wavelet coefficients at atoms $\psi_{u,s}$ across all u positions at the same scale s, to the Ar^+ signals. Then, they used the LDA method to distinguish energy metrics calculated from the four types of process faults. From the results given in Figure 8(d), one can see that the LDA method did a better job of separating fault types #2 and #3 than #1 and #4 due to the reason that faults #2 and #3 have very different shapes of signals compared to the signals shown in Figure 1 from the nominal process. Because faults #1 and #4 have similar global trends as the nominal process data, the LDA procedure cannot detect and classify these types of faults very well. Tools detecting local changes such as the wavelet neural network presented in Section 5 can be more useful to distinguish these faults.

A Multivariate Two-Sample Comparison Method Assume that the new process data come from the same distribution as the nominal process data with a possible different mean but the same variance. To test if the new process produces the same outcomes as the nominal process, the following hypothesis testing procedure can be used: If $T_c^2 = (\hat{\beta}_N - \hat{\beta}_a^\circ)\hat{\Sigma}_o^{-1}(\hat{\beta}_N - \hat{\beta}_a^\circ)^t$ is greater than its distribution's $1 - (\alpha/2)$th percentile, the null hypothesis $H_0 : \beta_N = \beta^\circ$ is rejected, where β_N and $\hat{\beta}_N$ are the true and estimated p-dimensional coefficients selected from the new process outcomes, $\hat{\beta}_a^\circ$ is the average of wavelet coefficients obtained from all replicates in the nominal process, and $\hat{\Sigma}_o$ is an estimate of variance-covariance matrix Σ_o. The covariance estimate can be obtained from the multivariate samples $\hat{\beta}_i$, $i = 1, 2, \ldots, M$ (=#repli-

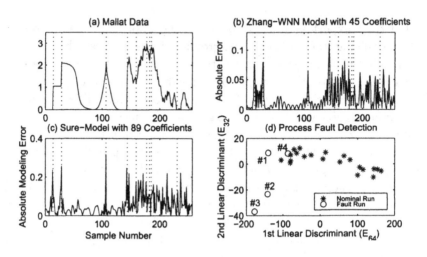

Figure 8: Process Fault Discrimination and Errors of Local Features

cates), by using the sample covariance. When the number of replicates is larger than the number of linearly independent vectors of combinations of wavelet coefficients, this covariance estimate can be inverted.

If the original data are normal random samples, because the DWT wavelet coefficient is a linear combination of data points, the joint distribution of a subset (or all) of the selected wavelet coefficients is a multivariate normal. Then, the distribution of T_c^2 can be related to the (non-central) F-distribution when multiplied by a constant depending on p and M. Thus, developing real-time decision rules for detecting process changes can be done in a straightforward manner.

If the original data is not normal, typical resampling ideas such as the bootstrap or jackknife (Efron, 1982), might be useful in generating the approximate distribution. However, there is a serious problem with the standard resampling procedures for this type of application. For example, when resampling M sets of new data with replacement from these M sets of the original data in the bootstrap sampling, at least p distinct data sets must be sampled; otherwise, the estimate of Σ_o will be singular. The chance of acquiring samples that provide a sensible estimate of Σ_o that is invertible is usually very small unless that M is much larger than p, which is not the case in typical applications where the number (M) of replicates is limited and p can be large depending on the process noise. We propose the following procedure to solve this problem. Step [i]: Take independent samples of size s without replacement from the M sets of p-dimensional wavelet coefficient vectors. Replicate this process K times. Because there are in general $\binom{M}{s}$ ways to choose s vectors from a set of M vectors, s must be chosen small enough to guarantee that there is a sufficient number of different samples of size s, and K must be chosen carefully to ensure the uniqueness of the selected K samples. The uniqueness of the samples becomes important when inverting the sample covariance matrix. For example, with $s = 5$ and $M = 21$, there are $\binom{21}{5} = 20{,}349$ ways to choose such a a sample. If K is set at 40, there is a high probability that the 40 samples of size five will be unique. Step [ii]: Randomly sample one out M sets of these wavelet coefficients and treat it as the new sample taken from the nominal process for calculating the T_c^2 statistic. Step [iii]: Repeat the above two steps N_s (e.g., 1000) times, to construct an empirical distribution of the test statistic. Use this distribution to identify its $1 - (\alpha/2)$th percentile for detecting process changes. This procedure has been applied to the process fault detection experiment with Ar^+ gas levels described in Section 2 with successful results (see Lada, 2000 for details).

Functional/Spatial Analysis of Variance (ANOVA)   In quality improvement activities, the ANOVA procedure is commonly used to investigate if the treatment effects (e.g., temperature or material type) contributed from process changes or production experiments are significant. When the data from each trial of a treatment is a large size data stream or sensor image, the functional or spatial ANOVAs are needed to discern, estimate and test the treatment effects. The commonly used "point-by-point" ANOVA procedure based on individual measurements is not appropriate for these types of data due to its inefficiency and the correlation between neighboring signal points. Traditionally, the estimation of treatment effects for multidimensional measurements is done by applying the standard statistical techniques (e.g., PCA) on a dimensional-reduced and decorrelated data set; see Ramsay and Silverman (1997) for details. However, the exhibiting of PCA becomes calculationally involved or even impossible when data size is huge. As indicated in Section 3.2, wavelet transforms are appropriate for data reduction. Moreover, the selected wavelet coefficients are less dependent than the original data (see Wornell (1996) for the discussion of wavelet's decorrelation property). Thus, it is natural to conduct ANOVAs based on the wavelet coefficients.

In a conference paper, Rosner and Vidakovic (2000) proposed to study the functional ANOVA problems in the wavelet domain. Their idea is to back-transform the treatment effects estimated by the wavelet coefficients to the original data scale for calculating treatment effects and for conducting functional ANOVAs. However, their procedures did not consider the selection of wavelet coefficients for "replicated" data and conducting ANOVAs based on the wavelet coefficients. The following describes the steps of our new method: Step [i]: Apply the DWT procedure to each set of data (stream or image); Step [ii]: Use the model selection procedures for replicates described in Section 3.5 to screen out non-important wavelet coefficients in a atom-position manner; Step [iii]: Apply regular ANOVA (or multivariate ANOVA) only to the selected wavelet coefficients from all replicates and treatments to estimate treatment effects in the wavelet space and test their significance. Since the treatment effects are functions or spatial surfaces, the definition of "significance" is open and potentially leads to different tests involving wavelet coefficients.

Other Applications   The above ANOVA idea can be extended to more versatile linear models. By treating the wavelet coefficients as the process performance or output variables, and treating the levels of controllable variables selected by design of experiment (DOE) plans as the input variables, one can employ multivariate regression modeling techniques to these input and output variables. With these regression models, one can obtain predictions at various input levels, for searching optimal process condition to improve production quality and efficiency. Instead of using the regression model, artificial neural networks can be useful in mapping these input and output variables for system characterization as addressed in Example 8, but applied to the wavelet coefficients instead of the original data. Finally, there is a potential of applying the traditional data mining techniques such as the association rules, classification tree and cluster analysis to the reduced-size data (e.g., wavelet coefficients) for finding hidden data patterns.

INTEGRATION OF SYNTHESIZED RESULTS

Review of Data Integration Procedures

Once the large size of data is split or reduced, the next challenging task is to develop techniques of integrating the synthesized results from the reduced-size data. Meta-data modeling technique can be helpful in this situation. We first review the "architectures" (Prothman, 2000) of constructing meda-data for managing information systems. Traditionally, three approaches (Thuraisingham, 1998) have been used to access data from multiple, external, heterogeneous data sources: multidatabases, federated databases and data warehouses. Our project with Nortel about the "9-up" charts described in Section 2 took the approach of multidatabases. Multidatabases provide a simple connection between systems, permitting the user to create queries across multiple databases at the same time. Unfortunately, because this approach does not provide a consistent view of the entire data, users are expected to formulate extremely complex queries. In particular, for each query, a user must understand the internal representation of each relevant source, manually resolve syntactic and semantic conflicts, and construct queries using the sources' native query language. Unless every potential user is intimately familiar with the detailed working of each connected data source, this is not a desirable approach.

When providing a resource to a broad community, it is critical to provide a consistent interface to the data. Federated databases define an integrated schema over the subset of available data that is interesting to the federation as a whole. This global schema represents a virtual database, combining data from each participating source to form a single, consistent representation. This is the approach taken by JDS Uniphase where the eMatrix is used to integrate product data, fiber optics design information and enterprise resource planning system for seamless information exchange. Queries posed over a federated database are sent to the applicable data sources, after being translated into the native query language, and the results are combined before being passed on to the user. There are three drawbacks to this approach. First, in order to process queries in an efficient way, sources participating in a federation may be required to contribute resources (e.g., query capabilities, storage facilities) to the federation. Second, because data is not represented locally, this approach is susceptible to long delays when answering queries. Third, misleading or incorrect results may be returned when a data source is unavailable, even temporarily.

To reduce the amount of network traffic and improve query results, data can be combined in a single database, resulting in the traditional data warehouse. The approach introduces two new problems. First is the tremendous amount of storage required to keep all of the data in a single database. To avoid this problem, warehouses typically contain only summary and aggregate data. The second problem is the difficulty keeping the data current and the warehouse fully functional.

DataFoundry (Critchlow et al. , 2000) combines many of the advantages of the systems described above to form a unique approach to database integration. It provides a consistent view of integrated data to the users. It uses a local data cache to reduce network traffic and improve performance and reliability. This approach provides a global schema containing a subset of the information in the data sources, similar to a federated database. However, the DataFoundry schema is expanded to include some summary and aggregation information. A local data store contains both this additional data and the most important and most frequently accessed source data. The result is a consistent view of the data and greatly improved query performance.

Based on our review of the literature concerning the "methodologies" used to integrate reduced-size data for global modeling, visualization and decision-making processes, we concluded that most of existing procedures relied on (weighted) averaging (for numerical predictions) or voting (for decisions). For example, Bayesian probability models (e.g., Schroder, et al. 1999) and neural networks for data integration are generalization of these approaches using more intelligent weighting schemes for integrating data from different sources/pieces. The significant advantage with these approaches is that it is capable of handling non-linear data functions and non-standard data types. The advantage of using the neural networks is that it does not require the prior information needed in the Bayesian approach. Both approaches require a long training time to build the model (or network), but are relatively fast in using the model for data integration (Benediktsson et al. 1998). In particular, using the popular backpropagation algorithm (e.g., Raghavan et al. 1997) in constructing ANNs can be computationally complicated and converge very slowly due to the use of iterative training procedures. This can be a serious drawback, especially when the data size is very large, or the data has nonstationary trends, or the dimensionality of the data is very high. Our extended wavelet neural network (EWNN) is proposed below to solve many of these problems.

Extended Wavelet Neural Networks

Neural networks are used in data mining, data integration, artificial intelligence, machine learning and many other applications. As noted above, there are many problems in applying the traditionally used ANNs to mine massive data sets. The following proposes an approach of utilizing parallel computing techniques and wavelet's ability of modeling "jumps" to increase network's construction speed and to enhance its learning ability of local features carrying key data information (Mallat and Hwang, 1992).

Wavelet neural networks (WNNs) (Zhang and Benveniste, 1992) replaced the global sigmoidal activation function of ANNs with wavelet atom functions. As illustrated in many recent applications such as Kunt et al. (1998), WNNs possess many better properties than the traditional ANNs commonly used in system identification and signal classification. Bakshi et al. (1994) concluded that the training and adaptation efficiency of his wave-nets are at least an order of magnitude better than ANNs. In this article, we proposed an extended wavelet neural network which performs better than ANNs and WNNs in capturing important local features, in reducing the data size and in utilizing the parallel computing capability for network training. We first use a few examples to motivate our ideas.

Example 8 - System Characterization Example In our research of developing a run-to-run process monitoring tool (see Example 10 for details) for a semiconductor ultra-thin film deposition process, we compared commonly used ANNs and Zhang's (1997) WNN - WaveARX, which consists of WNNs plus a linear regression function, in a system characterization experiment for understanding the "learning" ability of neural networks. Our study modeled the behavior of *in-situ* H_2^+ intensity based on information from process variables, e.g., wafer temperature and chamber pressure (see Figures 9(a-b) for examples). Figure

9(c) shows that WaveARX performed better than the ANNs in most locations. However, both WaveARX and the ANNs did not predict well at "curve-change" points such as the highest point at Location 145 of the H_2^+ signals. This will affect calculations of the area metric (see the shaded area in Figure 9(d)) needed to predict film thickness. Because the quality of the model of *in-situ* signals is critical in semiconductor ultra-thin film production due to the problem of obtaining *ex-situ* film thickness measurements, the improvement of network's ability of capturing the local features is very important in the future for improving the quality of advanced semiconductor manufacturing processes.

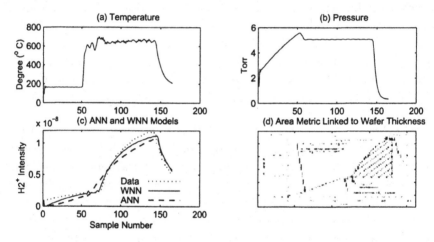

Figure 9: System Characterization with Neural Networks and An Area Metric

Example 9 - Modeling Local Features with EWNNs   WaveARX, wavelet approximation methods (e.g., SURE developed by Donoho and Johnstone, 1995) and all linear and nonlinear model selection methods (in Mallat, 1998 Chapter X) used the global MSE to guide the data modeling process. They usually do not perform well at local features such as curve change-points and 2D image edges. Figures 8 (a-c) show the errors in modeling Mallat's (1998) data with WaveARX and SURE methods. Although the models had used more coefficients than we preferred (45 or 89 number of coefficients with 256 data points), the errors at several "key" local features (marked with dots in Figures 8(a-c) plots) were larger than errors at other locations. In analyzing larger volumes of data, if the strategy is to capture these local features carrying key data information, then the quality of network modeling at these features should be emphasized.

Details of Extending Wavelet Neural Networks The first strategy of improving the above weakness of the WNNs is to use a simple "point-match" model. When the locations of key data features are given, the simplest method to model the data is to employ a single "jump-point" to match the data at those locations and use a global model (with coarse level wavelets, Fourier transforms or regression) to model the data between adjacent jump-points. The whole model is a piecewise smooth function at points other than the jump-point locations. If continuity of the global model and local "point-match" model is desired, the network needs to fit a more complicated global model with constraints to match the data at jump-points. This is a very difficult task and the network can be very complicated. Our second strategy is to use the following objective function for including the local feature modeling requirement in building neural networks.

A New Objective Function Assume that the data $Y(t_i)$ has a nonlinear structure given as $Y(t_i) = f(t_i) + \sigma_i \epsilon_i$, where $f(t_i)$ is an unknown function, t_i is a data point location, the errors ϵ_i's are random normals, and σ_i is unknown. The function $f(t)$ is smooth such that $|f(t) - f(t+\Delta)| \leq C\Delta^\alpha$, where C is a constant, Δ is a very small quantity which tends to zero, and $\alpha < 2$. Our function has three measures for quantifying the quality of [1] local feature modeling, [2] global trend modeling, and [3] data reduction.

$$
L_n(\delta, \mathbf{C}) = E_\epsilon \left\{ \sum_{j=1}^{J} \int_{I_j(\delta, C_j)} [f(t) - \hat{f}_{jL}(t)]^2 dt / norm(local) \right\} \quad (4)
$$
$$
+ E_\epsilon \left\{ \int_{U^c(\delta, \mathbf{C})} [f(t) - \hat{f}_G(t)]^2 dt / norm(global) \right\} + \lambda(C/n),
$$

where $I_j(\delta, C_j) = [u_{j\delta} - C_j/2, u_{j\delta} + C_j/2]$ is the C_j-size interval for the jth local feature centered at $u_{j\delta}$ selected by the curve change-point (or image edge) detection methods such as the modulus maximum, J is the number of local features (in a selected "priority-class" for the multi-resolution analysis), $norm(local)$ is the sum of the squares of all the local window lengths, $norm(global)$ is defined similarly for the supports of the global models. The local function, $\hat{f}_{jL}(t)$, can be a wavelet transform in a very fine scale for modeling detailed information. The global function, $\hat{f}_G(t)$, can be a Fourier or coarse level wavelet transform. They are evaluated at different regions as indicated in Eq. (4). The last term C/n is the data reduction measure, which serves as the Lagrange multiplier when C/n is bound above in a constraint equation.

Selecting the Thresholds δ's for Multi-Resolution Modeling   The parameter δ decides the priority-classes of local features. Let us see a simple example with the Ar^+ signals shown in Figure 1. The first priority class includes the four local features located at positions #18, #35, #43 and #46 extracted by the modulus maximum. The level-1 threshold δ_1 was set at the smallest CWT value (at the fine scale, e.g., $s = 5$) of data at these four locations. Any data feature with its CWT value smaller than this threshold was not selected. Notice that there is a dip of the data between positions #35 and #43. The second priority class includes an additional fifth feature at position #39. The level-2 threshold δ_2 was set at the CWT value of data at this position. Thus, this creates an opportunity of a two-level analysis of the data. For such simple data as these Ar^+ signals, only two priority-classes were needed. More complicated data, such as the 2D antenna signals shown in Figure 2, might require more priority classes. For the data set as seen in Figure 3(a), there are many smaller local features in the right side of the data. Those local features will not be included at higher priority levels of analyses. Hence, our network approximation function at those level will be smoother. However, when it goes down to lower priority levels, some of the more significant change-points in that area will show up. The challenge in our future research is to decide these threshold parameters δ's for defining local features included in those hierarchical priority-classes without "seeing" the large size data.

Selection of window size C_j's   The parameters C_j's govern the size of the feature windows. If the local model is a "point-match" function, the window size is one data point. If the local model is a linear combination of wavelets, the window size should be larger than the support of the smallest wavelet $\psi_{J=1,k}$, but should be smaller than the minimum of all half-distances between the locations of two adjacent local features. The smaller the window size, the larger the region of data that is modeled by the global model. In this case, the model will be simpler and use fewer coefficients. However, its fitting of data in the neighborhoods of local features can be poor. To balance these issues of global and local modeling and to minimize the cost in Eq. (4), we suggest to use a network with the following piecewise function integrated from nonoverlapping global and local regions. In this network, the window sizes C_j for each local feature may be different. See Rying, Bilbro and Lu (2000) for implementation details.

A Data-Modeling Network with Piecewise Functions   Based on the above idea, our EWNNs will have the following form as a sum of several piecewise functions: $\hat{f}(t) = \sum_{j=1}^{J} w_{1j}\psi[(t - u_{j\delta})/s_j] + \sum_{j=1}^{J+1} w_{2j} \phi[(t - u_{j*\delta})/(s^*)_j]$, where the function ψ models over the local windows and ϕ models over the global regions. Because there is no overlap between our local and global models, we can fit these models to "predetermined" data regions (from the initial windows) simultaneously with "parallel computing" techniques to expedite the network building process. In the case when continuity of models at the boundaries of the local and global regions is required, we propose the following "expansion" method for local modeling. First, fit the global models. Then, use predictions from the global models outside the local windows as data points in an "expanded" data set for local modeling. In applying the wavelet transform in local modeling, we must extend (see Ogden (1997) for details) the local window $I_j(\delta, C_j)$ into the real line. The global model predictions outside the local windows are the natural choice for expansion.

When the initial local and global models are selected based on the initial windows, we can update the window sizes and models by using optimization algorithms such as Newton's method (if the convexity condition holds) to efficiently change window sizes and re-calculate the weights or coefficients of wavelet transform bases for minimizing the objective function given in Eq. (4). The centers of local and global models are not changed in this updating process. Thus, the hierarchical levels of those key features are not changed. Because of the use of parallel computing techniques in the EWNNs, our method is more suitable than the traditional ANNs and WNNs in synthesizing larger quantities of data.

Application of the EWNNs to Solve Real-Life Problems

Example 10 - Characterization of Process Behavior   Many physical processes exhibit some form of process drift in system parameters. In our deposition process for the thin-film fabrication experiments, we observed a 15% process drift from 31 consecutive runs of film deposition thickness. This process drift is significant in semiconductor manufacturing, especially in the case of 12-inch wafers with smaller and more advanced devices. A poor-quality process can be very costly in an automated production environment. Thus, the problem has to be identified and solved in a short time period. The proposed EWNNs have the ability to provide a drift model for describing the link between the *in-situ* area metric defined in Figure 9(d) and the film thickness across different runs. The prediction of film thickness from the area metric model will indicate if there is a need for tuning the process conditions, and what is the amount of tuning in which direction (see Rying, et al., 2000 for an example). This can be further extended to a new "run-to-run controller". In the case that the *ex-situ* thickness measurement is not available (Rying et al., 1998), predictions from *in-situ* signals give the only "metrology" of *ex-situ* data. This is critical in monitoring process quality and in trouble-shooting process problems for ultra-thin film depositions.

Example 11 - Application to Stochastic Dynamic Programming (SDP)
Chen et al. (1999) used orthogonal arrays (OAs) to discretize the space of continuous state variables in SDP and eliminate the curse of dimensionality associated with SDP problems. The solution to a SDP problem optimizes a stochastic system over several discrete time periods. To approximate the optimal value function in each time period, they fit Friedman's (1991) multivariate adaptive regression splines (MARS) over the data collected with the OA design. The primary drawback to MARS is a suboptimal algorithm for selecting "knots" at which the approximation can bend. The process of selecting knots is analogous to the process of identifying "local features." Thus, we can employ our feature extraction technique to locate the knots for better capturing surface curvatures. The multi-resolution approach can be applied to identify features that are common to the optimal value functions in all periods versus detail features that are specific to one period. Finally, the combined parallelization of the computing algorithms for MARS knot selection will permit solutions to a larger size of optimization problems.

CONCLUSION

Our world is moving from the industrial age to the information age. Production automation, E-business and many information technology supported operations are growing rapidly. There are tremendous opportunities to utilize synthesized knowledge for improving efficiency of these operations. The proposed procedures will make significant contributions to data reduction, decision-making and neural networking methods for synthesizing large volumes of dynamic and nonstationary data in one- , two- and multiple-sample situations. Applications of the developed procedures include establishing process standards, monitoring potential process changes, identifying and classifying process faults, building regression and neural network models for predicting manufacturing system behavior in process quality/efficiency improvement and control. Our ideas can be useful in a wide range of applications including signal and image processing, machine and artificial intelligence, satellite image analysis and bioinformatics.

ACKNOWLEDGMENT

The author is grateful to Drs. Vidakovic, B. and Martell, L. for their insides in wavelet model selection methods, to Dr. Huo, X. for his helps in computations, and to Miss Kongthon, A. for literature review. This research was supported by the NSF in the DMS-0072960 award, Nortel and JDS Uniphase.

REFERENCE

Antoniadis, A., Gijbels, I. and Gregoire, G., "Model Selection Using Wavelet Decomposition and Applications," Biometrika, 84(4), 751-763, 1997.

Bakshi, B., Koulouris, A. and Stephanopoulos, G., "Wave-Nets: Novel Learning Techniques, and the Induction of Physically Interpretable Models," in Wavelet Applications, 2242, 637-648, 1994.

Benediktsson, J., Sveinsson, J., Arnason, K., "Classification and Integration of Multitype Data," in Proceedings of the IEEE 1998 International Geoscience and Remote Sensing Symposium, 177-179, 1998.

Castillo, E. D. and Hurwitz, A. M., "Run-to-Run Process Control: Literature Review and Extensions," Journal of Quality Technology, 29(2), 184-196, 1997.

Chen, V. C. P., Ruppert, D. and Shoemaker, C. A., "Applying Experimental Design and Regression Splines to High-Dimensional Continuous-State Stochastic Dynamic Programming," Operations Research, 47(1), 38-53, 1999.

Cochran, W. G., Sampling Techniques. New York: Academic Press, 1977.

Critchlow T. et al., "DataFoundry: Information Management for Scientific Data," IEEE Transactions on Information Technology in Biomedicine, 4(1), 2000.

Donoho, D. and Johnstone, I., "Adapting to Unknown Smoothness via Wavelet Shrinkage," Journal of the American Statistical Society, 90, 1200-1224, 1995.

Efron, B., The Jackknife, the Bootstrap, and Other Resampling Plans, in CBMS-NSF Regional Conference Series in Applied Mathematics, Number 38. New York: SIAM, 1982.

Friedman, J. H., "Multivariate Adaptive Regression Splines (with discussion)," Annals of Statistics, 19, 1-141, 1991.

Gardner, M. M., Lu, J. C., Gyurcsik, R. S., Wortman, J. J., Horning, B. E., Heinish, H. H., Rying, E. A., Rao, S., Davis, J. C., Mozumder, P. K., "Equipment Fault Detection Using Spatial Signatures," IEEE Transactions on Components, Hybrids and Manufacturing Technology, Part C: Manufacturing, 20(4), 295-304, 1997.

Jin, J. Shi, J., "Feature-Preserving Data Compression of Stamping Tonnage Information Using Wavelets," Technometrics, 41(4), 327-339, 1999.

Johnson, R. A. Wichern, D. W., Applied Multivariate Statistical Analysis. New Jersey: Prentice Hall, 1982.

Koh, C. K. H., Shi, J., Williams, W. J., Ni, J., "Multiple Fault Detection and Isolation Using the Haar Transform," Transactions of the ASME, Part 1: Theory, 290-294; Part 2: Application to the Stamping Process, 295-299, 1999.

Kunt, T. A., McAvoy, R. E., Cavicchi, R. E., Semancik, S., "Optimization of Temperature Programmed Sensing for Gas Identification Using Micro-Hotplate Sensors," Sensors and Actuators, 53, 24-43, 1998.

Lada, E, Process Fault Detection with Wavelet-based Data Reduction Methods, unpublished M.S. thesis (under Dr. J.-C. Lu's supervision), Operations Research Program, North Carolina State University, Raleigh, 2000.

Lenth, R. V., "Quick and Easy Analysis of Unreplicated Factorials," Technometrics, 31(4), 469-473, 1989.

Mallat, C. G. Hwang, W., "Singularity detection and processing with wavelets," IEEE Transactions on Information Theory, 38(2), 617-643, 1992.

Mallat, C. G., A Wavelet Tour of Signal Processing. Boston: Academic Press, 1998.

May, G. S. Spanos, C. J., "Automated Malfunction Diagnosis of Semiconductor Fabrication Equipment: A Plasma Etch Application," IEEE Transactions on Semiconductor Manufacturing, 6(1), 28-40, 1993.

Martell, L, Wavelet Model Selection and Data Reduction, unpublished Ph.D. thesis (under Dr. J.-C. Lu's supervision), Department of Statistics, North Carolina State University, Raleigh, 2000.

Mascoli G. J., "Automated Dynamic Strain Gage Data Reduction Using Fuzzy C-Means Clustering," in Proceedings of the IEEE International Conference on Fuzzy Systems, 2207-2214, 1995.

Mendenhall, W., Wackerly, D. D., Scheaffer, R. L., Mathematical Statistics with Applications (fourth edition). Boston: PWS-KENT Publishing Company, 1990.

Moore A. W., "An Introductory Tutorial on Kd-trees," Computer Laboratory, University of Cambridge, Technical Report No. 209, 1991.

Ogden, R. T.. Essential Wavelets for Statistical Applications and Data Analysis. Boston: Birkhauser, 1997.

Prothman B., "Meta data: Managing Needles in the Proverbial Haystacks," IEEE Potentials, 19(1), 20-23, 2000.

Raghavan S., Cromp, R., Srinivasan, S., Poovendran, R., Campbell, W. and Kanal L., "Extracting an Image Similarity Index using Meta Data Content for Image Mining Applications," in Proceedings of the SPIE, The International Society for Optical Engineering, 2962, 78-91, 1997.

Ramsay, J.O. and Silverman, B. W., Functional Data Analysis. New York: Springer-Verlag, 1997

Rosner, G. Vidakovic, B., "Wavelet Functional ANOVA, Bayesian False Discovery Rate, and Longitudinal Measurements of Oxygen Pressure in Rats," Technical Report, Statistics Group at ISyE, Georgia Institute of Technology, Atlanta, 2000.

Rying, E. A., Gyurcsik, R. S., Lu, J. C., Bilbro, G., Parsons, G., and Sorrell,

F. Y., "Wavelet Analysis of Mass Spectrometry Signals for Transient Event Detection and Run-To-Run Process Control," in Proceedings of the Second International Symposium on Process Control, Diagnostics, and Modeling in Semiconductor Manufacturing, editors: Meyyappan, M., Economou D., J., Bulter, S. W., 37-44, 1997.

Rying E. A., Hodge, D. W., Oberhofer, A., Young, K. M., Fenner, J. S., Miller, K., Lu, J. C., Maher, D. M., Kuehn, D., "Continuous Quality Improvement in the NC State University Research Environment," in Proceedings of SRC TECHCON'98, 1-14, 1998.

Rying E. A., Bilbro, G., Lu, J. C., "Focused Local Learning with Wavelet Neural Networks," manuscript submitted for publication, 2001.

Rying E. A., Bilbro, G., Ozturk, M. C., Lu, J. C., "In-Situ Fault Detection and Thickness Metrology Using Quadrupole Mass Spectrometry," manuscript submitted for publication, 2001.

Sachs E., Hu, A., Ingolfsson, A., "Run by Run Process Control: Combining SPC and Feedback Control," IEEE Transactions on Semiconductor Manufacturing, 8(1), 26-43, 1995.

Schroder M., Seidel K., Datcu M., "Bayesian Modeling of Remote Sensing Image Content," IEEE 1999 International Geoscience and Remote Sensing Symposium, 3, 1810-1812, 1999.

Splus WAVELETS, A Statistical Software Module for Wavelets sold by Math-Soft, Inc., Seattle, WA, 1996.

Thuraisingham B., Data Mining: Technologies, Techniques, Tools, and Trends. New York: CRC Press, 1998.

Wornell, G. W., Signal Processing with Fractals: A Wavelet Based Approach. Englewood Cliffs, NJ: Prentice Hall, 1996.

Wang, X. Z., Chen, B. H., Yang, S. H., McGreavy, C., "Application of Wavelets and Neural Networks to Diagnostic System Development, 2, An Integrated Framework and its Application," Computers and Chemical Engineering, 23, 945-954, 1999.

Zhang, Q., Benveniste, A., "Wavelet Networks," IEEE Transactions on Neural Networks, 3(6), 889-898, 1992.

Zhang, Q., "Using Wavelet Network in Nonparametric Estimation," in IEEE Transactions on Neural Networks, 8(2), 227-236, 1997.

CHAPTER 12

Intelligent Process Control System for Quality Improvement by Data Mining in the Process Industry

Sewon Oh
witness@postech.ac.kr
Division of Mechanical & Industrial Engineering, Pohang University of
Science & Technology, San 31 Hyoja, Pohang 790-784, Republic of Korea

Jooyung Han
luke@postech.ac.kr
Division of Mechanical & Industrial Engineering, Pohang University of
Science & Technology, San 31 Hyoja, Pohang 790-784, Republic of Korea

Hyunbo Cho
hcho@postech.ac.kr
Division of Mechanical & Industrial Engineering, Pohang University of
Science & Technology, San 31 Hyoja, Pohang 790-784, Republic of Korea

ABSTRACT

The large amount of bulky and noisy shop floor data is one of the characteristics of the process industry. These data should be effectively processed to extract working knowledge needed for the enhancement of productivity and the optimization of quality. The objective of the chapter is to present an intelligent process control system integrated with data mining architecture in order to improve quality. The proposed system is composed of three data mining modules performed in the shop floor in real time: preprocessing, modeling, and knowledge identification. To consider the relationship between multiple process variables and multiple quality variables, the Neural-Network/Partial Least Squares (NNPLS) modeling method is employed. For our case study, the proposed system is configured as three control applications: *feedback control, feed-forward control*, and *in-process control*, and then applied to the shadow mask manufacturing process. The experimental results show that the system identifies the main causes of quality faults and provides the optimized parameter adjustments.

D. Braha (ed.), Data Mining for Design and Manufacturing, 289–309.
© 2001 *Kluwer Academic Publishers. Printed in the Netherlands.*

INTRODUCTION

A shop floor in the process industry is associated with many sources of processing data. Their careful analysis would help produce products of high quality. With the support of the analyzed results, an intelligent processcontrol system plans, schedules, monitors, and executes various control tasks. For these analyses to be more fruitful, useful control knowledge must be extracted from the collected data sets. The term data mining is usually taken as a synonym for the knowledge extraction activity.

Data mining techniques have their origins in various methods such as statistics, pattern recognition, artificial neural networks, machine learning, high performance and parallel computing, and visualization (Fayyad and Stolorz, 1997; Koonce et al., 1997; Lavingston et al., 1999). However, most of them are not suitable for data mining of the data sets collected from a shop floor in the process industry for the following reasons. First, the shop floor generates a wide variety of data kinds, such as categorical data (e.g., model, raw material, product properties, and dimension), controllable data (e.g., device speed, device pressure, internal temperature, chemical liquid attributes), uncontrollable data (e.g., outer temperature, humidity), and quality data (e.g., dimensions, surface finish, fault percentage). Second, the shop floor data have nonlinear and complex correlations. Third, control knowledge must be extracted from a large number of observations in real-time for real-time control.

Thus, several known techniques lack some capabilities when applied to the shop floor data. The statistical approaches make impracticable assumptions, such as uncorrelated inputs, small-scale data, and only one response variable. To make matters worse, they take too much time in modeling. Multivariate projection methods, such as PCA (Principal Components Analysis) and PLS (Partial Least Squares), are used to handle a large data set having only the linear relationship between process and quality data (Geladi and Kowalski, 1986; Qin and McAvoy, 1992). Although learning methods and artificial neural networks can handle the nonlinear relationship (Cybenko, 1989; Hornik et al., 1989), the learning time and over-fitting problems need to be solved. In summary, a new approach to data mining for the shop floor data has yet to appear.

The objective of the chapter is to propose an intelligent process control system framework combined with data mining architecture. The detailed objectives are 1) to construct data mining architecture in the context of the shop floor and 2) to apply data mining modules to an industrial process control system. In data mining architecture, a Neural Network/Partial Least Squares (NNPLS) algorithm is adopted as a key methodology for modeling the relationships of process and quality variables. The proposed framework and methodologies will help improve quality and further increase productivity in the process industry.

The remainder of the paper is organized as follows: Section 2 provides a literature survey. Section 3 proposes the system framework integrated with data mining modules and control applications. Section 4 describes the data mining modules consisting of data preprocessing, relationship modeling, and knowledge identification. Section 5 discusses their applications to the industrial process control. Section 6 shows the experimental results of the case study performed in the shadow mask manufacturing process. Finally, Section 7 provides the conclusion.

RELATED WORK

A process control system, also referred to as "shop floor control system (SFCS)", is the central part of a CIM (Computer Integrated Manufacturing) system needed in controlling the progress of production in a shop floor. The process control system is concerned with monitoring the progress of an order, making decisions, and execution of the scheduled tasks required to fill the orders received from the factory-level control system (Cho and Wysk, 1995). Several process control system architectures and prototypes have been reported at the National Institute of Standards and Technology (NIST), European Strategic Program for Research and Development in Information Technology, and Computer Aided Manufacturing International (CAM-i) (Jones and Saleh, 1990; Boulet et al., 1991; Davis et al., 1992). However, issues and problems associated with on-line quality optimization integrated with the existing shop floor control system have not been reported. The functional and informational structure for control of inspection tasks has been proposed to consider the quality control tasks at higher levels of the CIM reference model (Arentsen et al., 1996). Nevertheless, it gives no comprehensive description and methodologies of the extraction and management of control knowledge for quality improvement.

Many approaches have been introduced to optimize the machining parameters in discrete part manufacturing. The optimal cutting parameters have been determined mainly through three methodologies: mathematical modeling approaches (Okushima and Hitomi, 1964), AI-based approaches (Lingarkar et al., 1990), and neural network-based approaches (Rangwala and Dornfel, 1989). Although those methodologies can be used for data mining techniques, they are not appropriate for on-line quality assurance and real-time optimization in the context of industrial process control.

Recently, several quality engineering techniques such as control chart, design of experiments, response surfaces, Taguchi methods, and control of variation became the driving forces in achieving a significant quality improvement (Taguchi, 1986; Mitra, 1993; Myers and Montgomery, 1995;

Montgomery, 1997; Escalante, 1999). However, their off-line properties prevent the process control system from rapidly coping with unexpected quality faults. In addition, the statistical process control (SPC) technique, the only on-line quality improvement tool, ignores decision-making. For instance, in the event of instability on an SPC chart, an operator's intervention is normally required to search for and remove its root causes. Thus, a new technique appropriate for intelligent process control in real-time is required for improving and assuring quality.

The term "intelligence" makes it possible for a process control system to act appropriately in an uncertain environment in order to increase the probability of success of the system's ultimate goal in view of the given criteria of success (Albus, 1991). The core of an intelligent system is the existence of the knowledge repository, which is also called "a world model or knowledge base," which can be developed with architecture of knowledge extraction and identification. Hence, the integration of a process control system and data mining architecture is essential to overcome the limitations known for on-line quality improvement.

SYSTEM FRAMEWORK AND ARCHITECTURE

The proposed process control system is constructed with various components in the data mining architecture. As shown in Figure 1, the data mining architecture consists of three core modules: data preprocessing, relationship modeling, and knowledge identification. Data preprocessing collects and refines the process data and domain expert's experienced knowledge. Relationship modeling analyzes the preprocessed data and then builds the relationship between process and quality variables (Hereafter we call it as "prediction model"). From the prediction model, each variable can be weighed as it is compared to those of other variables. This knowledge is useful for weighing variable, since only a small subset among many possible factors in the shop floor is anticipated to be "real" or to exhibit the so-called *effect sparsity* phenomenon (Hocking, 1976; Lin, 1998).

In order to find the optimized process adjustments, a control model is then formulated as the optimization problem consisting of non-linear objective functions and constraints. The objective functions are to minimize the deviations of quality variables from the desired levels as well as the change-numbers of process variables from the current settings. The constraints represent the lower and upper bounds of each controllable process variable. They also include the prediction model as the relationship between process and quality variables. Knowledge identification plays an important role in identifying and representing various knowledge contents such as process diagnosis, quality prediction, and optimization.

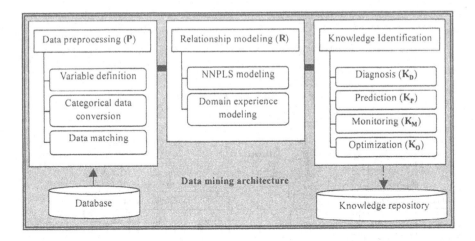

Figure 1. Data mining architecture for process control system

The process control system, as shown in Figure 2, consists of the three control applications: *feedback control, in-process control*, and *feed-forward control*, each of which is constructed by mixing several modules in the data mining architecture. The *feedback control application* built with data preprocessing (P), relationship modeling (R), process diagnosis (K_D), and prediction (K_P) modules creates and updates the prediction and control models by analyzing the raw process data in the database. The *in-process control application* keeps monitoring the process in real-time to detect the occurrence of assignable causes or process shifts. The *feed-forward control application* is invoked to regulate the process parameters just before a particular type of product is loaded. These are combined with monitoring (K_M), prediction (K_P) and optimization (K_O) modules.

For description of the prediction and control models, there exist process variables ($\mathbf{x} = [x_1, x_2, ..., x_m]$) and quality variables ($\mathbf{y} = [y_1, y_2, ..., y_n]$). The process variables can be rearranged into three types: initial conditions ($\mathbf{x}_I = [x_1, x_2, ..., x_{m'}]$), uncontrollable factors ($\mathbf{x}_U = [x_1, x_2, ..., x_{m''}]$), and controllable factors ($\mathbf{x}_C = [x_1, x_2, ..., x_{m'''}]$). The paired data are represented as matrices $\mathbf{X}$ and $\mathbf{Y}$, where elements, x_{ik} and y_{jk}, are the kth observation of ith process variable and j^{th} quality variable. Row vectors, $\tilde{\mathbf{x}}$ and $\tilde{\mathbf{y}}$, are the observed process and quality data to be monitored, diagnosed and optimized. $\tilde{\mathbf{x}}$ is further decomposed into $\tilde{\mathbf{x}}_I$ (observed initial condition data), $\tilde{\mathbf{x}}_U$ (observed uncontrollable factors), and $\tilde{\mathbf{x}}_C$ (observed controllable factors). Figure 3 shows the description of the two data matrices.

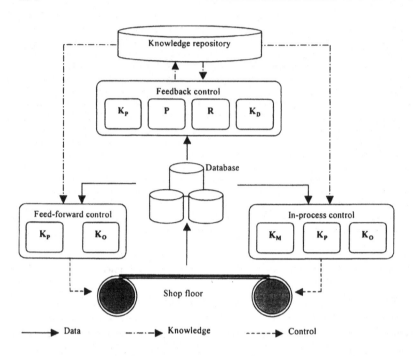

Figure 2. The proposed framework of intelligent process control system

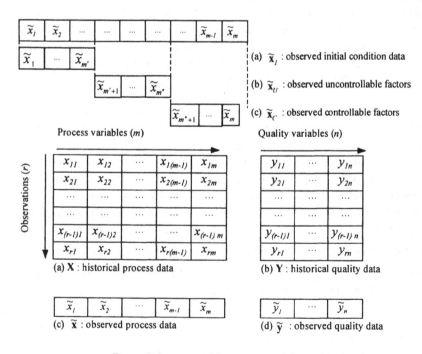

Figure 3. Structure of the process and the quality data

In general, the domain experts experience the correlations between process and quality variables in a descriptive manner. A key issue is how to convert those descriptive correlations into quantitative ones. There have been many researches on evaluating the subjective preference to the multiple attributes, such as weighted sum, ordinary ranking, cardinal ranking. Kknowledge extracted through several data mining modules can be classified for efficient utilization in decision-making of process control, as shown in Table 1

Table 1. Classification of the extracted knowledge

Knowledge	Description
Prediction model	Relationship models between process and quality variables
Diagnosis measure	Relative importance of each process variable
Control model	Model suggesting optimal setting value for process control

CORE MODULES IN THE DATA MINING ARCHITECTURE

Data Preprocessing

Variable Definition

The first step for data mining is to identify and arrange process and quality variables, as shown in Figure 4.

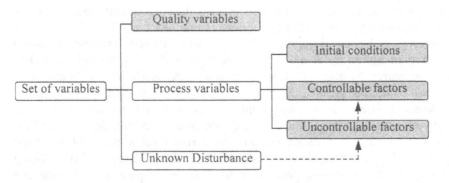

Figure 4. Taxonomy of variable for the shop floor data

A process variable is one of three elementary kinds: initial conditions of the manufacturing environment, controllable factors, and uncontrollable factors. The initial conditions include the properties of raw materials and process lines, such as supplier, color, weight, shape, mechanical structure, thickness, melting point, and queuing time to be processed. The controllable factors are what can be adjusted either manually or automatically, while the uncontrollable factors may not be changed. Some unknown disturbances can be transformed into either controllable or uncontrollable factors. A quality variable, which is also called a response variable, takes the measured value representing the product quality.

Conversion of Categorical Data

The shop floor data have two data types: numerical and categorical. The categorical data must be transformed for full utilization of the existing NNPLS algorithm. The transformation technique applied in this research is to expand data columns as the number of possible categorical value and insert either binary value 0 or 1. The process of creating these dichotomous variables from categorical variables is called *dummy coding* (Stockburger, 1997). For instance, assume that an initial condition variable, **Material**, is of categorical type, whose possible values are A, B, and C. In this case, the concerned data column should be expanded to three numerical data columns having binary values (for example, '100' for A) in accordance with the value of **Material**.

Data Matching

Process data (**X**) and quality data (**Y**) should be matched to find out under which process conditions a particular product has been processed. If the unit of the final product is the same as that of raw material, it is easy to match process and quality data. Otherwise, process data and quality data should be matched carefully. For instance, if a raw material is a type of coil and the final product is a sheared piece, the **X** is measured in a fixed time interval, while the **Y** is measured for every piece of the final product. In this case, **X** and **Y** can be paired by the time-based matching rule as shown in Figure 5. A process variable is measured by its sensing clock and the remaining processing time of its sensing point. The latter implies the time difference between estimated inspection time and process data sensing time. To find the measured value of a particular process variable for a piece of the final product, the inspection clock is compared with the summation of the sensing clock and the remaining processing time. If the inspection clock and the summation of the sensing clock and remaining processing time are identical within some tolerance, both process and quality data are matched. For instance, if $q_{jl} \approx p_{i2} + f_{i2}$, y_{jl} is matched with x_{i2}. This process is repeated for every piece of the final product. The past data collected based on the above framework will be utilized for prediction modeling.

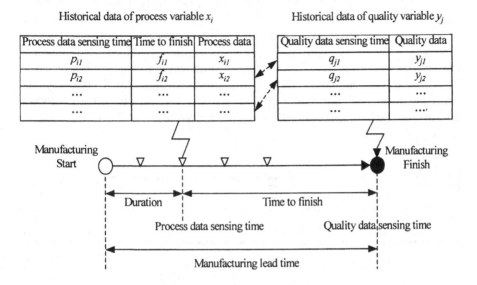

Figure 5. Conceptual procedure of data matching

Relationship Modeling

NNPLS Modeling

To handle the practical "short and fat" and the large-scale Multi-Input and Multi-Output (MIMO) data, the multivariate projection approaches that reduce the original large-scale data to the lower dimensional data have been developed (Geladi and Kowalski, 1986). In particular, an NNPLS method becomes popular for dealing with highly nonlinear correlated information between process variables and quality variables (Qin and McAvoy, 1992; Malthouse *et al.*, 1997). In the NNPLS algorithm, $\mathbf{X}$ and $\mathbf{Y}$ are decomposed as a sum of series of lower dimensional matrices as follows:

$$\mathbf{X} = \mathbf{t}_1\mathbf{p}_1^T + \mathbf{t}_2\mathbf{p}_2^T + \cdots + \mathbf{t}_A\mathbf{p}_A^T + \mathbf{E}_A = \mathbf{T}\mathbf{P}^T + \mathbf{E}_A$$

$$\mathbf{Y} = \mathbf{u}_1\mathbf{q}_1^T + \mathbf{u}_2\mathbf{q}_2^T + \cdots + \mathbf{u}_A\mathbf{q}_A^T + \mathbf{F}_A = \mathbf{U}\mathbf{Q}^T + \mathbf{F}_A$$

In this representation, $\mathbf{T}(n \times A)$ and $\mathbf{U}(n \times A)$ represent the score matrix, while $\mathbf{P}(m \times A)$ and $\mathbf{Q}(p \times A)$ represent the loading matrix for $\mathbf{X}(n \times m)$ and $\mathbf{Y}(n \times p)$. In constructing the relationship model, a weight matrix corresponding to $\mathbf{P}$, $\mathbf{W}(m \times A)$ ($=[w_{ia}]$, i=1,...,m, a=1,...,A) is also generated. The first set of loading matrix, $\mathbf{p}_1$ and $\mathbf{q}_1$, is obtained by maximizing the covariance between

X and **Y**. The projection of the **X** and **Y**, respectively, onto p_1 and q_1 gives the first set of score matrix, t_1 and u_1 Then, **X** and **Y** are indirectly related through their scores by "inner relation," which is a nonlinear functional relationship from t_1 and u_1, $u_1 = N(t_1) + r_1$, where $N(\cdot)$ stands for the function constructed by neural networks. The same procedure of determining the scores and loading matrix is iteratively continued until the required number of NNPLS dimensions (A), also called the number of principal components, is extracted. In practice, the number of NNPLS dimensions is determined either by the percentage of variance explained or by the use of statistically sound approaches such as cross validation. The overview of NNPLS algorithm is illustrated in Figure 6.

Process factors

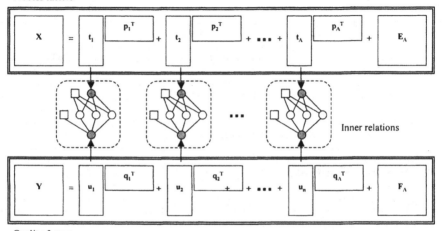

Quality factors

Figure 6. Overview of NNPLS algorithm

Modeling Domain Operator's Experience

A correlation matrix among process data and quality data can be obtained from the domain expert's experience as shown in Figure 7. The relation matrix consists of the importance indicator of each quality variable (QI) and correlation values (R). The latter may have four weight values: strong (5), medium (3), weak (1), and no relationship (0). The value of zero implies that the related process variable has no effect on the given quality variable. The domain operator's experienced knowledge can be transformed into the importance indicator of each process variable (WV). This can be further normalized (NWV).

$$WV_i = \sum_{j=1}^{n} (QI_j \times R_{ij}) \qquad i = 1, \cdots, m$$

				Process variables		
			x_1	x_2	...	x_m
Quality variables	y_1	QI_1	R_{11}	R_{21}	...	R_{m1}
	y_2	QI_2	R_{12}	R_{22}	...	R_{m2}
	...	...	...	...	...	...
	...	...	...	...	...	...
	y_n	QI_n	R_{1n}	R_{2n}	...	R_{mn}

x_i = ith process variable
y_j = jth quality variable
R_{ij}= Correlation between x_i and y_j
QI_j=jth quality importance indicator

Figure 7. Process-quality relation matrix from the domain expert's experience

Knowledge Identification

Diagnosis

The two indices inferred from the NNPLS prediction model – squared prediction error (*SPE*) and variable importance (*VI*) – are used for diagnosing quality faults. The *SPE* index implies how far each process variable is away from its initially set value, while the *VI* index means how much contribution the each process variable gives for describing the quality variables. The quality fault diagnosis identifies the causes of the observed quality faults and then presents the contribution of the process variables to the observed quality fault. The diagnosis procedure is performed by the feedback control application.

First, an *SPE* index of each process variable is obtained from the predicted values of process variables ($\hat{x} = [\hat{x}_1, \hat{x}_2, \cdots, \hat{x}_m]$) as follows:

$$\hat{x}_i = \sum_{a=1}^{A} \hat{t}_a p_{ai} \ , \ i = 1, \cdots, m$$

$$SPE_i = \sqrt{(\tilde{x}_i - \hat{x}_i)^2} \ , \ i = 1, \cdots, m$$

Second, a *VI* index of each process variable is conceptually represented as follows:

$$VI_i = \sum_{a=1}^{A} (w_{ia}^2 \times \frac{\sum_{z=1}^{a} Var(x_i)}{\sum_{i=1}^{m} \sum_{z=1}^{a} Var(x_i)}) \ , \ i = 1, \cdots, m$$

which means the proportion of the effect of x_i against the total effect of whole variables (i.e. $x_1, \ldots, x_m$) after z principal components being extracted. The *VI* is normalized into the *NVI*. Consequently, the diagnosis index (*DI*) of how much each process variable has an effect on the target quality variable is computed as follows:

$$DI_i = [\lambda \cdot NWV_i + (1-\lambda) \cdot NVI_i] \times SPE_i, \ 0 \le \lambda \le 1, \ i = 1, \cdots, m$$

Prediction

The task of prediction is to forecast the quality of products. The prediction result can be gained when the present dataset goes into the relationship model.

Monitoring

Process monitoring is performed by the in-process control application in real-time. Based on the t-score generated from the prediction model, monitoring is performed to check whether the current process variables are stable or not. For each principal component, newly generated t-score ($\tilde{\mathbf{t}} = [\tilde{t}_1, \tilde{t}_2, \cdots, \tilde{t}_A]$) is obtained as follows:

$$\tilde{t}_a = \sum_{i=1}^{m} w_{ai} \tilde{x}_i^{a-1} \qquad a = 1, \cdots, A$$

$$\tilde{x}_i^0 = \tilde{x}_i$$

$$\tilde{x}_i^a = \tilde{x}_i^{a-1} - \tilde{t}_a p_{ai} \qquad i = 1, \cdots, m$$

The current process is stable if $|\tilde{t}_a - \bar{t}_a|$ for all as is less than the critical value of student-t with r-1 degree of freedom. $\bar{t}_a$ is the estimated mean of the t-scores of the ath principal component obtained from the historical data.

Optimization

To effectively accommodate the large-scale manufacturing process, the two modifications are required in the traditional control model (Camacho, 1995). First, the change-numbers of process variables in the objective function should be minimized instead of their change-magnitudes. This would reduce the number of changes in process variables. Second, some process variables that the operators do not want to change must remain unchanged, which should be included in the constraints. Sometimes, the disturbances should remain constant (for example, external temperature) during the optimization. This modification makes only control factors vary, and therefore it can exclude impractical solutions.

With these modifications, a control model is formulated as a bounded non-linear programming problem as follows.

$$\min_{x_C} \quad \beta_0 \sum_{j=1}^{n} \left\| y_j^{target} - y_j \right\|^2 + \beta_1 \sum_{i=m^*+1}^{m} \left\| z_i \right\|^2$$

$$z_i = \begin{cases} 1, if\ x_i \neq \tilde{x}_i \\ 0, \text{otherwise} \end{cases} \quad \text{for } m'' + 1 \leq i \leq m$$

s.t.

$$l_i \leq x_i \leq u_i \quad \text{for } m'' + 1 \leq i \leq m$$

In formulating the optimization problem, the differences between predicted values and target quality levels are minimized. The predicted quality can be computed by the PLS prediction model. The right-hand side of the objective function makes it possible to minimize the number of changes in the controllable process variables. The coefficients β_0 and β_1 are chosen depending on the situations of the process industry. Sequential quadratic programming is used to solve the optimization problem (Biegler, 1998). The control model can be also solved by using response surface optimization techniques such as desirability function approach, generalized distance approach, and fuzzy modeling approach (Myers and Montgomery, 1995).

CONTROL APPLICATIONS FOR QUALITY IMPROVEMENT

Feedback Control

Each control application utilizes several combinations of the core data mining components to achieve its own goal as a control system. When the process control system is initialized, the *feedback control application* builds the prediction model, the control model, and the diagnosis measures using the historical paired data (X-Y). Its detailed structure is illustrated in Figure 8. It also updates the models when some defects are found at the inspection stage at the end of the process. If the models could correct, the *feed-forward control* and *in-process control applications* would predict and optimize quality appropriately, and therefore quality faults would not occur. The constructed models and diagnosis measures are used for process diagnosis and quality optimization in the control applications.

Feed-forward Control

The *feed-forward control application* determines the optimal values of controllable process factors ($x^*_{m''+1} \cdots x^*_m$), when the values of initial conditions

$(\widetilde{x}_1 \cdots \widetilde{x}_{m'})$ and uncontrollable factors $(\widetilde{x}_{m'+1} \cdots \widetilde{x}_{m^*})$ are given. As shown in Figure 9, based on current process variables $(\widetilde{x}_1 \cdots \widetilde{x}_m)$, the quality predictor conjectures the quality of product by using the existing prediction model. If the current process variables are appropriate for producing the products of good quality, they remain unchanged. However, if the prediction component forecasts the products of low quality due to the changes in $\widetilde{x}_I$ and $\widetilde{x}_U$, the optimization component prioritizes the controllable process variables by using the knowledge from K_D and then solves an optimization problem by using the existing control model to suggest a set of optimal control parameters $(x^*_{m'+1} \cdots x^*_m)$.

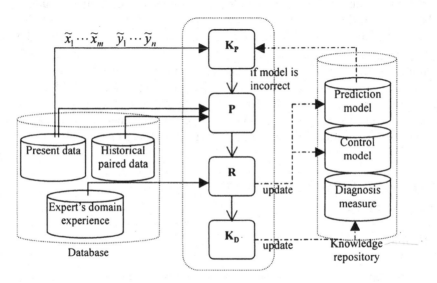

Figure 8. Detailed feedback control application

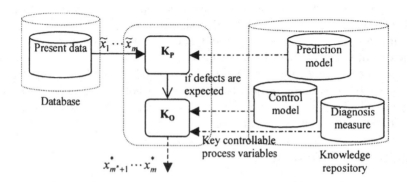

Figure 9. Detailed feed-forward control application

In-process Control

The *in-process control application* plays an important role in coping with the sudden changes of process variables because a typical process shop floor is operated in a complex manner. Although the *feed-forward application* suggests a set of optimal process parameters, both controllable and uncontrollable factors may become unstable during production. The *in-process control application* checks continuously to see if the manufacturing process is running in a desired stable state. As shown in Figure 10, the monitoring component receives $\tilde{x}_1 \cdots \tilde{x}_m$ and checks whether they are stable or not by using the t-score generated from the prediction model. If the process is unstable, the prediction component speculates whether the products are of good quality or not. Quality prediction, sensitivity analysis, and quality optimization are performed in the same way as those of the *feed-forward control application*.

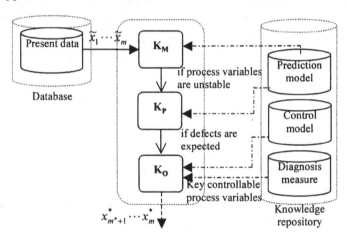

Figure 10. Detailed in-process control application

CASE STUDY

Prototype Shop Floor

The proposed architecture has been implemented in the process manufacturing that produces the shadow mask, which prevents the outer edges of the electron beams from hitting the wrong phosphor dots in CRT or CDT. The shadow mask is made of metal rolls called INVAR, which has an extremely low coefficient of thermal expansion and good thermal fatigue resistance. Since it should be very precisely manufactured, photochemical blanking, also called photo etching, is employed rather than the use of "hard tooling." Because of the characteristics of shadow masks, such as sensitivity to the temperature and pressure, various defects frequently affect their quality. Since many types of defects occur whose causes are not disclosed explicitly, they must be identified at the beginning. However, too many process variables prevent even the domain experts from investigating and identifying the causes of defects. A close look into the manufacturing facility reveals the number of the process variables and quality variables to be 763 (10 initial conditions, 643 controllable process variables and 119 uncontrollable process variables), and 46, respectively. The optimal values of controllable process variables are even harder to obtain.

The detailed system architecture applied to the case study is shown in Figure 11. The *feed-forward control* and *in-process control applications* have been installed at the each phase (coating, exposing, developing, and etching, respectively). The *feedback control application* has been implemented in the inspection stage and invoked when some defects were detected. All the models are built in the MATLAB™ environment: the function *nnpls* is used to identify the nonlinear PLS model, and the function *constr* is invoked to solve the nonlinear optimization problem.

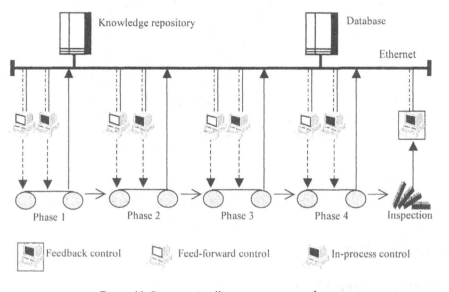

Figure 11. Prototype intelligent process control system

Experimental Design and Results

One hundred ten (110) process variables and two (2) quality variables (*Second type small hole* and *First type blocking*) were selected for a numerical illustration. The sixty-four (64) instances from the historical paired data were used to construct the prediction and control models through NNPLS modeling, and the four (4) separate observations were tested for the validation of the *feed-forward control* and *in-process control applications*. Using domain operator's experienced knowledge, the two process variables (*Hardening liquid temperature* and *A/K temperature*) were identified as the most important controllable process variables. First, a prediction model has been constructed. The percentage of variation explained by the model is cumulatively computed until ten principal components are extracted, as shown in Table 2. In other words, only ten new variables can explain the all the variables by 85.89%. Using the prediction model with ten new variables are the feed-forward control and in-process control applications tested and validated.

Table 2. Explained variation captured by NNPLS modeling

Number of principal components	X		Y	
	Explained variation (%)	Accumulated explained variation (%)	Explained variation (%)	Accumulated explained variation (%)
1	25.25	25.25	44.92	44.92
2	7.07	32.32	27.64	72.56
3	11.47	43.79	6.92	79.48
4	11.02	54.82	1.24	80.72
5	12.01	66.82	0.27	81.00
6	4.20	71.03	0.74	81.73
7	8.75	79.78	0.73	82.46
8	3.16	82.93	1.78	84.24
9	2.96	85.90	0.63	84.87
10	0.86	86.76	1.02	85.89

Before the raw materials are put into the manufacturing process, the *feed-forward control application* is executed to check if the current process parameters are optimally set for the particular materials. The application identifies the causal variables and then suggests the optimal values of the controllable process variables for quality improvement, as shown in Table 3. The pre-etching spray pressure remains unchanged, while other four variables should be adjusted to the suggested values.

Table 3. Results of feed-forward control application

	Process variables					Quality variables	
	Driving tensile force	Pre-etching electric potential	Driving speed	Outer moisture of EPT roller	Pre-etching spray pressure	Second type small hole	First type blocking
Current	39.5	232.0	1.8	90.0	0.5	10.54	4.3
Suggested	41.5	236.5	2.3	87.3	0.5	0	0

The key process variables suggested by the *feed-forward control application* are reset in the manufacturing facility. After production begins, the *in-process control application* checks continuously if the process variables are stable. It is noted that not all the variables but reduced variables are monitored. Among 10 principal components, the first and second ones are used to monitor the process, as shown in Table 4. For each observation, the application evaluates process status. The results indicate that observations 3 and 4 are out of control, since one of the 1st and the 2nd principal components indicates an unstable situation. In the case where the application hits observation 4, the prediction module is executed to see how the quality of products shall become in the future. If the quality faults are expected, several correction actions should be taken.

Table 4. Quality monitoring results of in-process control application ($\alpha=0.05$)

	For the 1st principal component			For the 2nd principal component		
	$\left\|\tilde{t_1} - \bar{t_1}\right\|$	Student's t	Control	$\left\|\tilde{t_2} - \bar{t_2}\right\|$	Student's t	Control
Obs. 1	0.54	1.17	In	0.93	1.14	In
Obs. 2	0.28	1.17	In	0.64	1.14	In
Obs. 3	2.94	1.17	Out	0.35	1.14	In
Obs. 4	0.32	1.17	In	2.82	1.14	Out

The *in-process control application* finds that the two quality variables shall be degraded in the future, as shown in Table 5. To identify the causes of the defects, several diagnosis measures such as *SPE, NVI* and *DI* can be utilized at this point. The *DI* index indicates that the **Driving tensile force** is identified to be the most influential variable, because the *SPE* index is high even for the low *NWV*. This implies that the process variable is extremely unstable with regard to its set value. To the contrary, the two critical process variables inferred from the domain expert's experience modeling, **Hardening liquid temperature** and **A/K temperature**, are disregarded due to the low *SPE* index. This means that the operators would control the two process variables tightly. The optimized values of the five controllable process variables have been suggested. The optimized values of process variables suggested by the *feed-forward control application* must be optimized again by the *in-process control application*, since the unstable process variables need to be compensated for.

Table5. Results of in-process control application

	Process variables					Predicted quality	
	Driving tensile force	Pre-etching electric potential	Driving speed	Outer moisture of EPT roller	Pre-etching spray pressure	Second type small hole	First type blocking
Current	41.5	236.5	2.3	87.3	0.5	4.21	3.99
NWV	0.072	0.072	0.217	0.072	0.072	-	-
SPE	3.108	0.822	0.572	0.936	1.023	-	-
NVI	0.089	0.092	0.080	0.046	0.052	-	-
DI ($\lambda = 0.5$)	0.250	0.067	0.085	0.055	0.063	-	-
Suggested	38.3	243.5	2.2	94.2	0.5	0	0

CONCLUSION

Although quality became considered a success factor for the manufacturing companies, but it was isolated from the shop floor process control activities. Quality improvement should be integrated into the conventional process control system to improve both quality and productivity. In this chapter, a functional and implementation framework for intelligent process control has been proposed based on the data mining architecture that *integrates* modeling and identification of knowledge about prediction, monitoring, diagnosis, and optimization of quality. The proposed framework includes the systematic procedure for data mining, which consists of data preprocessing, modeling, and knowledge identification. It also contains the three primary control applications for quality improvement, *feedback control, feed-forward control,* and *in-process control application* to assure quality by way of process monitoring, process diagnosis, quality prediction, and quality optimization. Consequently, the data mining components have successfully been integrated into the process control system.

The real world application of the proposed framework to the shadow mask manufacturing process has confirmed that it plays a crucial role in on-line quality assurance and improvement. The cost and time for handling quality faults have been greatly reduced due to the use of the efficient prediction model. Moreover, the operators could cope with the quality faults consistently. The quality improvement tasks no longer need to depend on the availability of highly skilled technicians.

ACKNOWLEDGEMENTS

This work was supported by the Brain Korea 21 Project.

REFERENCES

Albus, J. S., "Outline for a theory of intelligence," *IEEE Transactions on Systems, Man, and Cybernetics*, 21(3), 473-509, 1991.

Arentsen, L., Tiemersma, J. J., and Kals, H. J. J., "The integration of quality control and shop floor control," *International Journal of Computer Integrated Manufacturing*, 9(2), 113-130, 1996.

Biegler, L., "Advances in nonlinear programming concepts for process control," *Journal of Production Control*, 8(5), 301-311, 1998.

Boulet, B., Chhabra, B., Harhalakis, G., Minis, I., and Proth, J. M., "Cell controllers: Analysis and comparison of three major projects," *Computers in Industry*, 16(3), 239-254, 1991.

Camacho, E., *Model Predictive Control in the Process Industry*, Springer Verlag: New York, 1995.

Cho, H. and Wysk, R. A., "Intelligent workstation controller for computer integrated manufacturing: Problems and models," *Journal of Manufacturing Systems*, 14(4), 252-263, 1995.

Cybenko, G., "Approximation by superposition of a sigmoidal function," *Mathematics of Control, Signals, and Systems*, 2(4), 303-314, 1989.

Davis, W. J., Jones, A. T. and Saleh, A., "Generic architecture for intelligent control systems," *Computer Integrated Manufacturing Systems*, 5(2), 105-113, 1992.

Escalante, E. J., "Quality and productivity improvement: A study of variation and defects in manufacturing," *Quality Engineering*, 11(3), 427-442, 1999.

Fayyad, U. and Stolorz, P., "Data mining and KDD: Promise and challenges," *Future Generation Computer Systems*, 13, 99-115, 1997.

Geladi P. and Kowalski, B. R., "Partial least-squares regression: A tutorial," *Anlytica Chimica Acta*, 185, 1-17, 1986.

Hocking, R. R., "The analysis and selection of variables in linear regression," *Biometrics*, 32, 1-51, 1976.

Hornik, K., Stinchcombe, M., and White, H., "Multilayer feedforward neural networks are universal approximators," *Neural Networks*, 2, 359-366, 1989.

Jones, A. T. and Saleh, A., "A multi-level/multi-layer architecture for intelligent shop floor control," *International Journal of Computer Integrated Manufacturing*, 3, 60-70, 1990.

Koonce, D.A., Fang, C., and Tsai, S., "A data mining tool for learning from manufacturing systems," *Computers and Industrial Engineering*, 33(1-2), 27-30, 1997.

Lavington, S., Dewhurst, N., Wilkins, E., and Freitas, A., "Interfacing knowledge discovery algorithms to large database management systems," *Information and Software Technology*, 41, 605-617, 1999.

Lin, D., "Spotlight interaction effects in main-effect plans: A supersaturated design approach," *Quality Engineering*, 11(1), 133-139, 1998.

Lingarkar, R., Liu, L., Elbestawi, M. A., and Sinha, N. K., "Knowledge-based adaptive computer control in manufacturing systems: A case study," *IEEE Transactions on Systems, Man, and Cybernetics*, 20(3), 606-618, 1990.

Malthouse, E., Tamhane, A., and Mah, R., "Nonlinear partial least squares," *Computers and Chemical Engineering*, 21(8), 875-890, 1997.

Mitra, A., *Fundamentals of Quality Control and Improvement*, Macmillan: New York, 1993.

Montgomery, D. C., *Design and Analysis of Experiments*, 4th ed., Wiley: New York, 1997.

Myers, R. H. and Montgomery, D. C., *Response Surface methodology: Process and Product Optimization Using Designed Experiments*, Wiley: New York, 1995.

Okushima, I. and Hitomi, K., "A study of economy machining: An analysis of maximum-profit cutting speed," *International Journal of Production Research*, 3, 73-84, 1964.

Qin S. J. and McAvoy, T. J., "Nonlinear PLS modeling using neural networks," *Computers and Chemical Engineering*, 16(4), 379-391, 1992.

Rangwala, S. S. and Dornfeld, D. A., "Learning and optimization of machining operations using computing abilities of neural networks," *IEEE Transactions on Systems, Man, and Cybernetics*, 19(2), 299-314, 1989.

Stockburger, D. W., Multivariate Statistics: Concepts, Models, and Applications, *http://www.psychstat.smsu.edu/*, 1997.

Taguchi, G., *Introduction to Quality Engineering*, Asian Productivity Organization: Tokyo, 1986.

CHAPTER 13

Data Mining by Attribute Decomposition with Semiconductor Manufacturing Case Study

Oded Maimon
maimon@eng.tau.ac.il
Department of Industrial Engineering, Tel-Aviv University

Lior S. Rokach
liorr@il.kamoon.com
Department of Industrial Engineering, Tel-Aviv University

ABSTRACT

This chapter examines the Attribute Decomposition Approach with simple Bayesian combination for dealing with classification problems in the semiconductor industry. Often classification problems in this industry contain high number of attributes (due to the complexity of the manufacturing process) and moderate numbers of records. According to the Attribute Decomposition Approach, the set of input attributes is automatically decomposed into several subsets. A classification model is built for each subset, then all the models are combined using simple Bayesian combination. This chapter presents theoretical and practical foundation for the Attribute Decomposition Approach. A greedy procedure, called D-IFN, is developed to decompose the input attributes set into subsets and build a classification model for each subset separately. The algorithm has been applied to problems in the semiconductor industry and on variety of databases from other application domains. The results achieved in the empirical comparison testing with well-known classification methods (like C4.5) indicate the superiority of the decomposition approach.

D. Braha (ed.), Data Mining for Design and Manufacturing, 311–336.
© 2001 *Kluwer Academic Publishers. Printed in the Netherlands.*

INTRODUCTION

One of the main techniques for data mining is creating a classification model (classifier). In classification problem the induction algorithm is given a set of training instances and the corresponding class labels and outputs a classification model. The classification model takes an unlabeled instance and predicts its class. The classification techniques can be implemented on variety of problems, for example it can be useful to predict the quality of production batch given its Work-in-Process (WIP) information.

Fayyad et al. (1996) claim that the explicit challenges for the KDD research community is to develop methods that facilitate the use of data mining algorithms for real-world databases. One of the characteristics of a real world databases is high volume. The difficulties in implementing classification algorithms as-is on high volume databases derives from the increase in the number of records in the database and from the increase in the number of fields/attributes in each record (high dimensionality). High numbers of records primarily create difficulties in storage and computing complexity. Approaches for dealing with high number of records include sampling methods; massively parallel processing and efficient storage methods. However high dimensionality increases the size of the search space in an exponential manner, and thus increases the chance that the algorithm will find spurious models that are not valid in general. Elder and Pregibon (1996) define this phenomenon as the "curse of dimensionality". Techniques that are efficient in low dimensions (e.g., nearest neighbors) fail to provide meaningful results when the number of dimensions goes beyond a 'modest' size of 10 attributes. Furthermore smaller data mining models, involving less attributes (probably less than 10), are much more understandable by humans. Smaller models are also more appropriate for user-driven data mining, based on visualization techniques.

Most of the methods for dealing with high dimensionality focus mainly on feature selection techniques, i.e. selecting some subset of attributes upon which the induction algorithm will run, while ignoring the rest. The selection of the subset can be done manually by using prior knowledge to identify irrelevant variables or by using proper algorithms.

In the last decade, Feature Selection has enjoyed increased interest by the data mining community. Consequently many feature selection algorithms have been proposed, some of which have reported remarkable accuracy improvement. The literature on this subject is too wide to survey here, however, we recommend Langley (1994), Liu and Motoda (1998) and Maimon and Last (2000) on this topic.

Despite its popularity, using feature selection methodology for overcoming the high dimensionality obstacles has several shortcomings:

1. The assumption that a large set of input attributes can be reduced to a small subset of relevant attributes is not always true; in some cases the target attribute is actually affected by most of the input attributes, and removing attributes will cause a significant loss of important information.

2. The outcome (i.e. the subset) of many algorithms for features selection (for instance almost any of the algorithms that are based upon the wrapper model methodology) is strongly dependent on the training set size. That is, if the training set is small the size of the reduced subset will be small as well due to the elimination of relevant attributes.

3. In some cases, even after eliminating a set of irrelevant attributes, the researcher is left with relatively large numbers of relevant attributes.

4. The backward elimination strategy, used by some methods, is extremely inefficient for working with large-scale databases, where the number of original attributes are more than 100.

A number of linear dimension reducers have been developed over the years. The linear methods of dimensionality reduction include projection pursuit (see Friedman and Tukey, 1974), factor analysis (see Kim and Mueller, 1978), and principal components analysis (see Dunteman, 1989). These methods are not aimed directly at eliminating irrelevant and redundant attributes, but are rather concerned with transforming the observed variables into a small number of "projections" or "dimensions." The underlying assumptions are that the variables are numeric and the dimensions can be expressed as linear combinations of the observed variables (and vice versa). Each discovered dimension is assumed to represent an unobserved factor and thus provide a new way of understanding the data (similar to the curve equation in the regression models). However, the linear methods are not able to reduce the number of original features as long as all the variables have non-zero weights in the linear combination.

The linear dimension reducers have been enhanced by constructive induction systems that use a set of existing attributes and a set of predefined constructive operators to derive new attributes (Pfahringer 1994, Regavan and Rendell 1993).

One way to deal with the aforementioned disadvantages is to use a very large training set (which should increase in an exponential manner as the number of input attributes increase). However, the researcher rarely enjoys this privilege, and even if it does happen, the researcher will probably encounter the aforementioned difficulties derived from high number of records.

The rest of the chapter is organized into three parts. In the first part we introduce the Attribute Decomposition Approach literally and theoretically. In the second part we develop a heuristic algorithm for implementing the decomposition approach. In the third part we examine the algorithm on semiconductor manufacturing data and on other applications.

ATTRIBUTE DECOMPOSITION APPROACH

The purpose of decomposition methodology is to break down a complex problem into several manageable problems. Problem decomposition's benefits include: conceptual simplification of the problem, making the problem more feasible by reducing its dimensionality, achieving clearer results (more understandable), reducing run time by solving smaller problems and by using parallel/distributed computation, and allowing different solution techniques for individual sub problems.

Three types of decomposition are commonly found in the literature: object, aspect, and sequential decomposition. Object decomposition divides the problem into physical components. Aspect decomposition divides a problem according to the different specialties involved in its solution. Sequential decomposition is applied to problems involving the flow of elements or information.

The decomposition approach is frequently used for operation research and engineering design, however, as Buntine (1996) states, it has not attracted as much attention in KDD and machine learning. The decomposition approach to machine learning was first used by Samuel (1967). He proposed a method based on a signature table system as an evaluation mechanism for his checkers playing programs. This approach was later improved by Biermann (1982).

Function decomposition was originally developed in 50's and 60's to be used in the design of switching circuits. Recently it has been adopted by the machine learning community. Goldman (1995) introduces FLASH, a Boolean function decomposer, on a set of eight-attribute binary functions and show its robustness in comparison with C4.5 decision tree inducer. Zupa et al. (1998, 1999) presented a general-purpose function decomposition approach for machine learning. According to this approach, attributes are transformed into new concepts in an iterative manner and create a hierarchy of concepts.

Buntine (1996) used a sequential decomposition approach to classify free text documents into predefined topics. Buntine suggests breaking the topics up into groups (co-topics), and, instead of predicting the document's topic directly, first classifying the document to one of the co-topics and then using another model to predict the actual topic in that co-topic.

Within machine learning, there are other approaches that are based on problem decomposition, but where the expert decomposes the problem and it is not discovered by a machine. A well-known example is structured induction applied by Shapiro (1987) and Michie (1995). Their approach is based on a manual decomposition of the problem and an expert-assisted selection of examples to construct rules for the concepts in the hierarchy. In comparison with standard decision tree induction techniques, structured induction exhibits about the same classification accuracy with the increased transparency and lower complexity of the developed models.

In this chapter we present an object-type decomposition where the attribute set is decomposed. This method facilitates the creation of a classification model for high dimensionality databases. The idea we propose can be summarized as follows:

- The original input attribute set is decomposed to mutually exclusive subsets.
- An inducer algorithm is run upon the training data for each subset independently.
- The generated models are combined in order to make classification for new instances.

Notation

Throughout this chapter we will use the following notation:

- $A = \{a_1,...,a_n,...,a_N\}$ - a set of N input attributes, where a_n is an attribute No. n.

- a_{target} - Represents the target attribute, obviously $a_{target} \notin A$.

- $S_m = \{a_{m,j(n)} | j = 1,...,l_m\}$ - indicates a subset m of the input attribute where n is the original attribute index in the set A.

- $R_m = \{n / a_{m,j(n)} \in S_m, j = 1,...,l_m\}$ denotes the correspondence indexes of subset m in the complete attribute set A.

○ $Z = \{S_1, \ldots S_m \ldots, S_M\}$ A decomposition of the attribute set A into M mutually exclusive subsets

$$S_m = \{a_{m,j(n)} \mid j = 1, \ldots, l_m\} \ m = 1, \ldots, M \text{ where } (\bigcup_{m=1}^{M} S_m) \subseteq A$$

○ V_n - Domain of an attribute a_n. We assume that each domain is a set of k_n discrete values. $\forall n:\ k_n \geq 2$, finite. For numeric attributes having continuous domains, each value represents an interval between two continuous values. In similar way V_{target} represents the domain of the target attribute.

○ $v_{n,j}$ a value No. j of domain V_n, i.e.

$$V_n = \{v_{n,j} \mid j = 1, \ldots, k_n\} \ \forall n = 1, \ldots, N.$$

○ The *universe set* U is defined as Cartesian product of all the domains, i.e. : $U = V_1 \times V_2 \times \ldots \times V_N \times V_{target}$. We assume that the universe set has static distribution D.

○ $T = (t_1, \ldots, t_Q)$. – indicate the training set, containing Q observations (sometime called "instances," "records," "rows" or "tuples") where t_q is a observation No. q. Obviously $t_q \in U$.

○ $a_n(t_q)$ - value of attribute No. n in a observation No. q. $\forall q, n$: $a_n(t_q) \in V_n$. For discretized attributes, this means that any continuous value can be related to one of the intervals. Null (missing) values are not allowed (unless they are encoded as a special value.

○ We use the notation I to represent a Bayesian learning method, and by $I(T, S_m)$ to represent a classifier which was induced by activating the learning method I on a training set T using the input attributes in the subset S_m. Using $I(T, S_m)$ one can estimate the probability

$$\hat{P}_I(a_{target} = v_{target,j} / a_n = a_n(t) \ n \in R_m) \text{ of an observation } t.$$

THE PROBLEM DEFINITION

The problem of decomposing the input attribute set is that of finding the best decomposition, such that if a specific induction algorithm is run on each attribute subset data, then the combination of the generated classifiers will have the highest possible accuracy.

In this chapter we focus on simple Bayesian combination, which is an extension of the Simple Bayesian classifier. The Simple Bayesian classifier (Domingos and Pazzani, 1997) uses Bayes rule to compute the probability of each possible value of the target attribute given the instance, assuming the input attributes are conditionally independent given the target attribute. The predicted value of the target attribute is the one which maximize the calculated probability, i.e.:

$$v_{MAP} = \underset{\substack{v_{target,j} \in V_{target} \\ \hat{P}_I(a_{target}=v_{target,j})>0}}{\arg\max} \quad \hat{P}_I(a_{target} = v_{target,j}) \cdot \prod_{n=1}^{N} \hat{P}_I(a_n = a_n(t) / a_{target} = v_{target,j})$$

Due to the last assumption, this method is also known as "naïve" Bayesian. However, a variety of empirical research shows surprisingly that the simple Bayesian classifier can perform quite well compared to other methods even in domains where clear attribute dependencies exist. Furthermore, simple Bayesian classifiers are also very simple and easy to understand (Kononenko, 1990).

In the attribute decomposition approach with simple Bayesian combination we use a similar idea, namely the prediction of a new instance t is based on the product of the conditionally probability of the target attribute, given the values of the input attributes in each subset. Mathematically it can be formulated as follows:

$$v_{MAP} = \underset{\substack{v_{target,j} \in V_{target} \\ \hat{P}_I(a_{target}=v_{target,j})>0}}{\arg\max} \quad \frac{\prod_{m=1}^{M} \hat{P}_I(a_{target} = v_{target,j} / a_n = a_n(t) \ n \in R_m)}{\hat{P}_I(a_{target} = v_{target,j})^{M-1}}$$

or it can be also written using log function as follows:

$$v_{MAP} = \underset{v_{target,j} \in V_{target}}{\arg\max} \{\log(\hat{P}_I(a_{target} = v_{target,j})) +$$

$$\sum_{m=1}^{M} [\log(\hat{P}_I(a_{target} = v_{target,j} / a_n = a_n(t) \ n \in R_m)) - \log(\hat{P}_I(a_{target} = v_{target,j}))]\}$$

Where $a_n(t)$ indicates a specific value of the attribute n. M is the number of used subsets and R_m is a set of index of attributes belongs to subset m.

In fact extending the simple Bayesian classifier by joining attributes is not new. Kononenko (1991) used a conditional independence test to join attributes. Domingos and Pazzani (1997) used estimated accuracy (as determined by leave-one-out cross validation on the training set). In both cases, the suggested algorithm finds the single best pair of attributes to join by considering all possible joins. However these methods have not noticeably improved accuracy. The reasons for the moderate results are two-fold. First both algorithms used a limited criterion for joining attributes. Second and more importantly, attributes are joined by creating a new attribute, whose values are the Cartesian product of its ingredients' values, specifically the number of attributes that can be joined together is restricted to a small number. Furthermore the problem have not been formally defined and explored.

Duda and Hart (1973) showed that Bayesian classifier has highest possible accuracy (i.e. Bayesian classifier predicts the most probable class of a given instance based on the full distribution). However in practical learning scenarios, where the training set is very small compared to the whole space, the complete distribution can hardly be estimated directly.

According to the decomposition concept the complete distribution is estimated by combining several partial distributions. Bear in mind that it is easier to build a simpler model from limited number of training instances because it has less parameters. This makes classification problems with hyper input attributes more feasible.

This problem can be related to the bias-variance tradeoff (see for example Friedman, 1997). The bias of a learning algorithm for a given learning problem is the persistent or systematic error that the learning algorithm is expected to make. A concept closely related to bias is variance. The variance captures random variation in the algorithm from one training set to another. This variation can result from variation in the training sample, from random noise in the training data, or from random behavior in the learning algorithm itself. The smaller each subset is, the less probabilities we have to estimate and potentially less variance in the estimation of each one of them. On the other hand, when there are more subsets, we expect that the approximation of the full distribution using the partial distributions is less accurate (i.e. higher bias error). Formally the problem can be phrased as follows:

Given an Learning method I, and a training set T with input attribute set $A = \{a_1, a_2, ..., a_N\}$ and target attribute $a_{t \arg et}$ from a distribution U over the labeled instance space, the goal is to find an optimal decomposition Z_{opt} $= \{S_1, ..., S_m, ..., S_M\}$, such that the average accuracy of the simple Bayesian *combination of the induced classifiers* $I(T, S_m), m = 1, ..., M$ *will be maximal over the distribution U.*

It should be noted that the optimal is not necessarily unique. Furthermore the problem can be treated as an extension of the feature/attribute subset selection problem, i.e. finding the optimal decomposition of the form $Z_{opt} = \{S_1\}$, as the non-relevant features is in fact NR=A/S$_1$. Moreover, the Naïve Bayes method can be treated as specific decomposition: $Z = \{S_1, S_2, ..., S_N\}$, where $S_i = \{A_i\}$.

Definition: Complete Equivalence

The decomposition $Z = \{S_1, ...S_m..., S_M\}$ is said to be *completely equivalent* if for each combination of the input attributes in the universe set ($\forall v_{n,j*} \in V_n \quad \{v_{1,j*}, ...v_{n,j*}..., v_{N,j*}\}$), with positive probability, the following is satisfied:

$$\underset{\substack{v_{t\,arget,j} \in V_{t\,arget} \\ P(a_{t\,arget} = v_{t\,arget,j}) > 0}}{\arg\max} \quad \prod_{m=1}^{M_1} \frac{P(a_{target} = v_{target,j} / a_n = v_{n,j*} \quad n \in R_m^{(1)})}{P(a_{t\,arget} = v_{t\,arget,j})^{M_1 - 1}} =$$

$$\underset{\substack{v_{t\,arget,j} \in l'_{t\,arget} \\ P(a_{t\,arget} = v_{t\,arget,j}) > 0}}{\arg\max} \quad P(a_{t\,arget} = v_{t\,arget,j} / a_n = v_{n,j*} \quad n = 1, ..., N))$$

Since Duda and Hart showed that the right term of the equation is optimal, it follows that a completely equivalent decomposition is optimal as well. The importance of finding a completely equivalent decomposition is derived from the fact that in real problems with limited training sets, we can only estimate the probability and it is easier to approximate probabilities with lower dimensions.

Theorem 1

Let Z be a decomposition which satisfies the following conditions:

- The subsets $m = 1...M$ S_m and the $NR = A - \bigcup_{m=1}^{M} Z_m$ are conditionally independent given the target attribute;

- The NR set and the target attribute are independent.

Then Z is *completely equivalent*.

The proof of the theorem is straightforward, and it is left for the reader. The above theorem represents a sufficient condition for complete equivalence. However it is important to note that it does not represent a necessary condition, for instance consider a Boolean target attribute, described by four Boolean attributes as follows: $A_{t \arg et} = (A_1 \bigcup A_2) \bigcap (A_3 \bigcup A_4)$. In this case it is simple to show that the decomposition $Z = \{S_1 = \{A_1, A_2\}, S_2 = \{A_3, A_4\}\}$ is *completely equivalent*, although S_1 and S_2 are not conditionally independent given A_{target}.

Search Space

The number of possible decompositions (i.e. the size of the search space) that can be considered increases in a strong exponential manner as the number of input attributes increases. The number of combinations that n input attributes may be decomposed into exactly M relevant subsets is:

$$P(n, M) = \frac{1}{M!} \sum_{j=0}^{M} \binom{M}{j} (-1)^j (M - j)^n$$

Dieterich and Michalski (1983) show that when n is big enough the above expression can be approximated to:

$$P(n, M) \approx \frac{M^n}{M!} \approx M^{n-M} e^M \sqrt{2 \cdot \pi \cdot M}$$

Evidently the number combinations that n input attributes may be decomposed into, up to n subsets, is

$$C(n) = \sum_{M=1}^{n} P(n, M) = \sum_{M=1}^{n} \frac{1}{M!} \sum_{j=0}^{M} \binom{M}{j} (-1)^j (M - j)^n$$

Due to the fact that in the attribute decomposition problem defined above it is possible that part of the input attribute will not be used by the inducers (the irrelevant set), the total search space is:

$$T(N) = \sum_{n=0}^{N} \binom{N}{n} C(n) = \sum_{n=0}^{N} \binom{N}{n} \sum_{M=1}^{n} \frac{1}{M!} \sum_{j=0}^{M} \binom{M}{j} (-1)^j (M-j)^n$$

The conclusion is that an exhaustive search is practical only for small number input attributes.

INFORMATION THEORY FOR DATA MINING

Different heuristic algorithms can be developed to search the state space described above; for instance genetic algorithm using the wrapper approach. In this chapter we choose to use the theoretic connectionist approach.

Uncertainty is an inherent part of our lives. Most real-world phenomena cannot be predicted with perfect accuracy. The quality of product manufactured can not predefined, the customer's demand for a product is not constant, and neither is the processing time of a product. The reasons for inaccurate predictions include both the lack of understanding about the true causes of a phenomenon (e.g., what affect customer's demand) and missing or erroneous data about those causes (e.g., the price of the competing product in the future).

For a small subset of natural phenomena, the cause-effect relationships are already known, either from scientific research or from our common knowledge. In many other cases, the relationship between the input attributes and the target attribute is extremely hard to comprehend. For example, semiconductor companies collect data on each batch manufactured in their plan but they are still faced with some percentage of failures.

Many classification algorithms are aimed at reducing the amount of uncertainty, or gaining information, about the target attribute. More information means higher prediction accuracy for future outputs. Information theory (see Cover, 1991) suggests a distribution-free modeling of conditional dependency between random variables. If nothing is known on the causes of variable X, its degree of uncertainty can be measured by its unconditional entropy as $H(X) = -\Sigma\ p(x)\ \log_2\ p(x)$. In communication engineering, entropy represents the minimum expected number of bits required to transmit values of X. The entropy reaches its maximum value of log [domain size of X], when X is uniformly distributed in its domain. Entropy is different from variance by its metric-free nature: it is dependent only on the probability distribution of a random variable and not on its values. Thus, in learning algorithms where the metric of attributes is unimportant (e.g., decision tree learning), the entropy can be used for choosing the best classification hypothesis (e.g., see the Minimum Description Length principle in Mitchell, 1997).

The entropy of a random variable can be decreased by adding information on its direct and indirect causes (moreover, it can be shown that additional information never increases the entropy). The entropy of a random variable Y, given another random variable X, (the *conditional entropy*) is given by $H(Y/X) = -\Sigma\ p(x,y)\ \log\ p(y/x)$. The association between two random variables X and Y (the mutual information) is defined as a decrease in entropy of Y as a result of knowing X (and vice versa), namely:

$$I(X;Y) = \sum_{x,y} p(x,y) \bullet \log \frac{p(x/y)}{p(x)} = H(Y) - H(Y/X) = H(X) - H(X/Y) = I(Y;X)$$

The decrease in entropy of Y as a result of knowing n variables (X_1,..., X_n) can be calculated by using the following chain rule (Cover, 1991):

$$I(X_1,..., X_n; Y) = \Sigma\ I(X_i; Y / X_{i-1},..., X_1)$$

The difference between mutual information (known also as *Information Gain*) and correlation coefficient resembles the difference between entropy and variance; mutual information is a metric-free measure, while a correlation coefficient measures a degree of functional (e.g., linear) dependency between values of random variables.

The information theoretic connectionist approach to learning suggests a clear advantage for discovering information patterns in large sets of imperfect data, since it utilizes a meaningful network structure that is based on the information theory. Maimon and Last (2000) present a *Multi-Layer Info-Fuzzy Network (IFN)* aimed at finding the minimum number of input attributes required for predicting a target attribute. Each vertical layer is uniquely associated with an input attribute by representing the interaction of that attribute and the input attributes of the previous layers. The first layer (layer No. 0) includes only the root node and is not associated with any input attribute. The multi-layer network can also be used to predict values of target attributes in a disjunctive manner, similar to the decision-tree. The principal difference between the structure of a multi-layer network and a decision-tree structure (Quinlan, 1986 and 1993) is the *constant ordering* of input attributes at every predicting node of the multi-layer network, the property which is necessary for minimizing the overall subset of input attributes (resulting in dimensionality reduction). Figure 1 represents a typical IFN with three input layers: the slicing machine model used in the manufacturing process, the rotation speed of the slicing machine and the shift (i.e. when the item was manufactured), and the Boolean target attribute representing whether that item passed the quality assurance test. The arcs that connect the terminal hidden nodes and the nodes of the target layer are labeled with the number of records that fit this path. For instance there are twelve items in the training set which were produced using the old slicing machine that was setup to rotate in a speed less than 1000 RPM and that were classified as "good" items (i.e. passed the QA test). The profound description of the IFN Methodology can be found in a new book by Maimon and Last (2000).

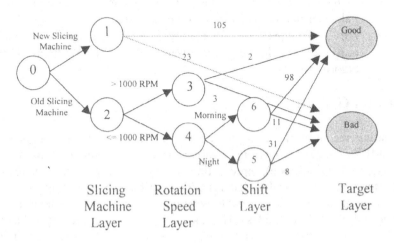

Figure 1. Illustration of Information-Theoretic Network

ATTRIBUTE DECOMPOSITON USING IFN

As the IFN method was found to be effective in discovering the relevant attributes and their relations to the target attribute, we further extended it aiming to approach the optimal decomposition (Z_{opt}) with our problem. For our purpose each subset is represented by a different network, while each attribute in this subset is located in a different layer. Attributes in the same network should be dependent, namely independent attributes should be in different networks as Theorem 1 suggests. However it does not mean that attributes in different networks are necessarily independent. In some cases assigning dependent attributes into different groups contributes to the overall prediction accuracy.

Obviously D-IFN can construct up to N networks (in the extreme case where each network may represent one input attribute) and up to N attributes in each network (in the case where we have one network that include all input attributes).

Each valid observation (observation that belongs to the universal set), is associated in each network with exactly one path between the root node and the target layer. We make the following two assumptions:

- There are no missing values. Either they are replaced with valid values or the observation should be ignored.
- All attributes have discretely valued domains. Continuous attributes should be made discreet before using them in the algorithm.

Algorithm Overview

For creating the multiple networks we use a greedy depth-first algorithm, called D-IFN. The D-IFN learning algorithm starts with a single network with a single node (the root node), representing an empty subset of input attributes. In each iteration the algorithm decides which attribute should be added to the current network as a new layer, and to what nodes on the previous layer it will be connected (the splitted nodes). The nodes of a new layer are defined as all Cartesian product combinations of the splitted nodes with the values of the new input attribute, which have at least one observation associated with it.

The selection of the a new input attribute is made according the following criteria:

1. The selected attribute should maximize the total significant decrease in the conditional entropy, as a result of adding it as a new layer. In order to calculate the total significant decrease in the conditional entropy, we estimate for each node in the last layer, the decrease in the conditional entropy as a result of splitting that node according the candidate attribute values. Furthermore the decrease in the conditional entropy is tested for significance by a likelihood-ratio test (Attneave, 1959). The zero hypothesis (H_0) is that a candidate input attribute and a target attribute are conditionally independent, given the node (implying that their conditional mutual information is zero). If H_0 holds, the test statistic is distributed as χ^2 with degrees of freedom equal to the number of independent proportions estimated from the observations associated with this node. Finally all significant decreases of specific nodes are summed up to achieve the total significant decrease.

2. The attribute is conditionally dependent on the splitted nodes (nodes with significant decrease in the conditional entropy) given the target attribute. For testing the conditional independency we use a-χ^2 as described in Walpole and Myers (1986). The zero hypothesis (H_0) is that a candidate input attribute and all splitted nodes are conditionally independent, given the target attribute.

3. The attribute should increase the value of Bayesian Information Criteria (see Schwarz, 1978) of the combined networks so far. McMenamin and Monforte (1998) used the BIC for constructing a neural network, while Heckerman (1997) used the BIC for building Bayesian networks. In this case there are two purposes for using it: A. As adding a new attribute to the current network increases this network's complexity. Under the Occam's-razor assumption that simplicity is a virtue in conflict with training accuracy, we verify whether the decrease in the entropy is worth the addition in complexity. B. The addition of the attribute to the current network, contributes to the combined networks structure. The Bayesian Information Criteria is formulated as follows:

$$BIC = \log L(T) - \frac{k\log(|T|)}{2},$$ where L(T) is the likelihood of

getting the training set (T) using the classifier and k representing the complexity of the classifier. In the current case

$$k = (\sum_{m=1}^{M} l_m + 1)$$ where lm is the number of arcs in the network

number m, while L(T) is calculated as follows:

$$\log(L(T)) = \sum_{q=1}^{Q} \log \left(\frac{\dfrac{\prod_{m=1}^{M} P(a_{t\,\arg et} = a_{t\,\arg et}(t_q)/a_n = a_n(t_q)\ n \in R_m)}{P(a_{t\,\arg et} = a_{t\,\arg et}(t_q))^{M-1}}}{\sum_{\substack{v_{t\arg et.j} \cdot v_{t\arg et} \\ P(a_{t\arg et} \cdot v_{t\arg et.j}) \cdot 0}} \dfrac{\prod_{m=1}^{M} P(a_{t\,\arg et} = v_{t\,\arg et.j}/a_n = a_n(t_q)\ n \in R_m)}{P(a_{t\,\arg et} = v_{t\,\arg et.j})^{M-1}}} \right)$$

If no input attribute was selected an attempt is made to build a new network (a new subset) with the input attributes that were not used by previous networks. This procedure continues until there is no more unused attributes or until the last network is left empty. Figure 2 illustrates the main flow of the algorithm.

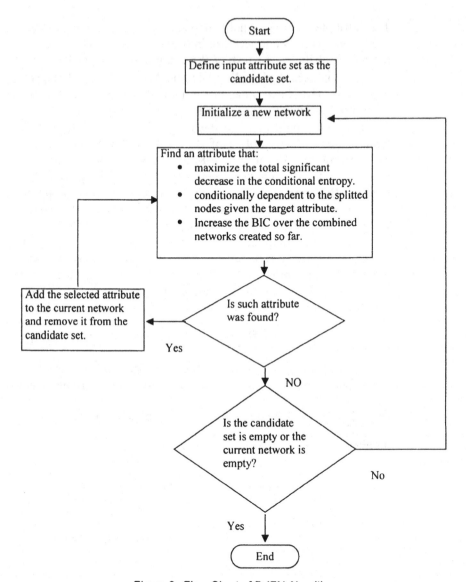

Figure 2. Flow Chart of D-IFN Algorithm

Computational Complexity

To calculate the computational complexity of constructing the networks, we are using the following notation:

N   Total number of candidate input attributes

$\max\limits_{i=1,\ldots,N}(k_i)$ Maximum domain size of a candidate input attribute.

$k_{\arg et}$ The domain size of a target attribute.

Q   The total number of observation in the training set.

The total number of layers in all networks can exceed N. In each layer we examine no more than N candidate attributes. Examination of a candidate attribute requires the calculation of the estimated conditional mutual information of the candidate input attribute and the target attribute, given a node.   We restrict defining a new node by the requirement that there is at least one observation associated with it. Thus, the total number of nodes at any hidden layer cannot exceed the total number of observations (Q). In most cases the number of nodes will be much smaller than Q. The calculation is performed by enumerating all members of the Cartesian product of the target attribute values ($k_{\arg et}$) with the candidate input attribute values (which is bounded by $\max_{i=1,...,N}(k_i)$) for each of the nodes in the last layer (which is bounded by Q). A similar enumeration is required for calculating the statistic for verifying that the candidate attribute is conditionally dependent on the splitted nodes given the target attribute. The calculation of the BIC requires an enumeration on all observation (Q) over all values of the target attribute. (*Comment*: when reviewing the BIC formula, it seems that we should enumerate over all networks too, however this can be avoided in practice as the part of the formula corresponding to the networks that were constructed before the current network can be kept in memory as constant and not recalculated again and again). This implies that the total number of calculations is bounded by: $O(N^2 \cdot Q \cdot k_{\arg et} \cdot \max_i(k_i))$

Prediction

Predicting the target attribute of an unseen instance is made by the following steps:

For each network:

1. Locate the relevant node (final or unsplitted) that satisfies the unseen instance.
2. Extract the frequency vector (how many instances relate to each possible value of the target attribute.)
3. Transform the frequency vector to probability vector. Using the frequency as is will typically over-estimate the probability so we use the Laplace correction to avoid this phenomena (See

Domingos and Pazzani, 1997). According to Laplace, the probability of the event $ai=v_{i,j}$ which has been observed fi times of F instances is $(f_i+1)/(F+ki)$ where ki denotes the domain size of attribute a_i.

4. Combine the probability vectors using naïve Bayesian combination.
5. Select the target value maximizing the naïve Bayesian combination.

CLASSIFICATION PROBLEMS IN THE SEMICONDUCTOR INDUSTRY

Integrated Circuit Manufacturing Overview

An integrated circuit (IC) is a miniature electric circuit containing large numbers of electronic devices packaged on a single chip made of semiconductor material.

The manufacturing of an integrated circuit begins with the production of a semiconductor wafer. While the number and variety of process steps in a wafer's manufacture may change from manufacturer to manufacturer, usually it contains more than 100 process steps (Van Zant, 1997) like Crystal Growth, Wafer Slicing, Thickness Sorting, Lapping, Etching, Polishing and Deposition. The wafer manufacturing process is largely mechanical. Measurements (for instance Flatness, Surface Quality Verification, Visual Inspection) are taken at various stages of the process to identify defects induced by the manufacturing process, to eliminate unsatisfactory wafer materials and to sort the wafers into batches of uniform thickness to increase productivity.

The wafer's manufacture is followed by the fabrication of integrated circuits on the surface of the wafer. According to Fountain et al (2000) several integrated circuits are produced all together on a single wafer. The wafers are then cut into chips using a high-precision diamond saw and mounted into packages.

Mining IC Manufacturing Data

IC manufacturing lines provide a lot of opportunities for applying data mining. In this case, data mining may have tremendous economic impact in increasing throughput, reducing costs, and consequently raising profitability. (Fountain et al, 2000).

In this chapter we will focus on classification problems. In many classification problems in this industry the target attribute represents a quality measure: in some cases it is formatted in binary values that correspond to the success or failure of the production of a specific product or lot and in other cases it is formatted in a continuously numeric value that represents the yield.

Classification problems in IC manufacturing lines are also characterized by an enormous number of input attributes of the following types:

- Various parameters of the process itself, for instance the parameters setting of each machine in the process, or the temperature of the liquid that a material is immersed in.
- Parameters related to the raw material that is used in the process, for instance the manufacturer of the raw material.
- Parameters that relate to the environment, such as moistness, temperature, etc.
- Attributes related to the human force that operate the production line, for example the parameters related to the worker assigned to a specific machine such as the experience level of that worker, the hour of the day, the day in the week, etc.

The Yield Problem

Background and Objectives

The main goal of data mining in IC manufacturing databases is to understand how different parameters in the process affect the line throughput. The throughput of IC manufacturing processes is measured by "yield," which is the number of good products (chips) obtained from silicon wafer. Since the number of wafers processed per time unit is usually limited by the capability of very expensive microelectronics equipment, the yield is the most important criterion for the effectiveness of a IC process.

Training Set & Problem Dimensionality

The training data set includes 1,000 records. Each record has 247 input attributes labeled p1,...,p247 that represent the setting of various parameters (for instance the speed of a certain machine). Due to the high commercial confidentiality of the process data, we will not explain here the specific meaning of the measured parameters.

Data Pre-processing

The databases passed a simple preprocessing stage. In this stage missing values are replaced by a distinctive value, and numeric attributes are made discreet by dividing their original range to ten equal-size intervals (or one per observed value, whichever was least). We could improve the results by using more sophisticated discretization methods (see for example Dougherty et al, 1995) or a more robust way to treat missing values (see for example Quinlan, 1993). However, since the main purpose of this research is to compare the attribute decomposition concept to other state-of-the-art data mining algorithms, we decided that any non-relevant differences between the algorithms should be neutralized.

Using D-IFN

Figure 3 presents textually two IFNs created by running D-IFN algorithm on the yield problem. The first IFN presents how p140 and p108 affect the target attribute, while the second IFN present how p99 affect the target attribute. For each terminal node a numeric vector is attached. Item i in the vector stands for the number of instances in the training set that correspond to that node and the value $v_{t \arg et, i}$.

```
Group 1
-------
p140 = 0: (0,0,0,0,0,0,0,0,21,0)
p140 = 1:
|    p108 = 0: (0,2,0,0,1,3,15,4,2,0)
|    p108 = 1: (0,0,0,0,1,1,2,1,4,32)
|    p108 = 2: (0,0,0,0,0,0,1,0,0,16)
|    p108 = 3: (1,0,0,0,1,1,2,4,1,72)
|    p108 = 4: (0,0,0,0,0,0,0,0,0,43)
|    p108 = 5: (0,0,0,0,0,0,0,0,0,12)
|    p108 = 6: (0,0,0,0,0,0,0,0,0,17)
|    p108 = 7: (0,0,0,0,0,0,0,0,0,93)
|    p108 = 8: (0,0,0,0,0,0,0,0,0,54)
|    p108 = 9: (0,0,0,0,0,0,0,0,0,5)
p140 = 2: (0,0,0,0,0,0,1,0,19,1)
p140 = 3:
|    p108 = 0: (0,0,0,0,0,0,0,0,9,0)
|    p108 = 1: (0,0,0,0,0,0,0,0,5,0)
|    p108 = 2: (0,0,0,0,0,0,0,0,4,0)
|    p108 = 3: (0,0,0,0,0,0,0,2,6,0)
|    p108 = 4: (0,0,0,0,0,2,0,1,0,0)
|    p108 = 5: (0,0,0,0,0,0,2,0,8,0)
|    p108 = 6: (0,0,0,0,0,0,0,0,34,0)
|    p108 = 7: (0,0,0,0,0,0,0,0,21,0)
|    p108 = 8: (0,0,0,0,0,0,0,0,12,0)
|    p108 = 9: (0,0,0,0,0,0,0,0,14,0)
p140 = 4:
|    p108 = 0: (0,0,0,0,0,13,0,0,0,0)
|    p108 = 1: (0,0,0,0,0,53,0,0,0,0)
|    p108 = 2: (0,0,0,0,0,25,0,0,0,0)
|    p108 = 3: (0,0,0,0,0,73,0,0,0,0)
|    p108 = 4: (0,0,0,0,0,0,1,2,6,0,0)
|    p108 = 5: (0,0,0,0,32,0,0,0,0,0)
|    p108 = 6: (0,0,0,1,0,0,14,0,0,0,0)
|    p108 = 7: (0,0,0,0,0,84,0,0,0,0)
|    p108 = 8: (0,0,0,0,0,9,0,0,0,0)
|    p108 = 9: (0,0,0,0,0,72,0,0,0,0)
p140 = 5: (0,0,0,0,3,0,0,18,1,2)
p140 = 6: (0,0,0,0,0,0,11,0,1,0)
p140 = 7: (0,0,0,0,6,0,0,0,0,0)
p140 = 8: (0,0,10,0,1,0,0,1,0,0)
p140 = 9: (9,0,0,0,0,0,0,0,0,0)


Group 2
-------
p99 = 0: (0,0,0,0,4,42,12,0,0,98)
p99 = 1: (1,0,0,0,4,28,0,0,0,120)
p99 = 2: (1,0,6,0,1,13,5,3,0,17)
p99 = 3: (4,0,0,0,2,41,0,0,0,40)
p99 = 4: (3,0,0,0,0,39,2,0,0,15)
p99 = 5: (0,1,0,0,9,23,1,2,0,16)
p99 = 6: (0,0,3,0,1,13,4,10,0,14)
p99 = 7: (1,0,0,0,13,0,0,15,0,12)
p99 = 8: (0,0,0,0,0,2,0,3,0,14)
p99 = 9: (0,1,2,0,11,150,12,4,0,0)
```

Figure 3: D-IFN output - Yield Problem

Prediction Power

We are using *10-fold cross-validation* (see Mitchell, 1997) for estimating prediction accuracy. Under this approach, the training set is randomly partitioned into 10 disjointed records subsets. Each subset is used once in a test set and nine times in a training set. Since the average prediction accuracy on the validation instances is a random variable, we are calculating its confidence interval by using the normal approximation of the Binomial distribution. Table 1 shows the average accuracy and sample standard deviation obtained by using 10-fold-cross-validation. The first row stands for the accuracy achieved using a simple average.

Table 1: Predication Accuracy - Yield Problem

Average	Simple Bayes	C4.5	D-IFN
84.11±2.3	88.29±2.1	89.46±3.2	92.21±1.9

The IC Test Problem

The fabricated ICs undergo two series of exhaustive electric tests that measure the operational quality. The first series, which is used for reducing costs by avoiding packaging defective chips, is performed while ICs are still in the wafer form. The second series, which is used for quality assurance of the final chip, is carried out immediately after the wafers are cut into chips and mounted into packages.

The electric test is performed by feeding various combinations of input signals into the IC. The output signal is measured in each case and it is compared to the expected behavior. As the second testing series is not essential for quality assurance, we would like to reduce it as much as possible.

Training Set and Problem Dimensionality

The training data set includes 251 records. Each record has 199 input attributes labeled p1,...,p199. The target attribute is Binary representing good and bad devices.

Data Pre-processing

Similar to the yield problem missing values are replaced by a distinctive value, and numeric attributes are discretized by dividing their original range to ten equal-size intervals (or one per observed value, whichever was least). Figure 4 shows a typical output of the pre-processing process of the input attribute p25.

```
Parameter 25
Discretized Value              Lower Bound      Upper Bound
0                              -Inf             2.84e-006
1                              2.84e-006        3.08e-006
2                              3.08e-006        3.39e-006
3                              3.39e-006        4.1e-006
4                              4.1e-006         5.21e-006
5                              5.21e-006        6.25e-006
6                              6.25e-006        7.56e-006
7                              7.56e-006        8.94e-006
8                              8.94e-006        1.21e-005
9                              1.21e-005        +Inf
10                             Missing Value
```

Figure 4: Output of pre-processing – Wafer Test Problem

Using D-IFN

Running D-IFN on has created 7 Networks containing only 19 electronic tests. The results showed that the manufacture can rely on these tests with very high confidence (more than 99%).

EXPERIMENTAL RESULTS FOR OTHER APPLICATIONS

In order to illustrate the strength of the decomposition approach in the general case we present here a comparative study. The D-IFN approach has been applied to 15 representative public domain data sets from the UCI Repository (Merz and Murphy, 1998). Table 2 shows the average accuracy and sample standard deviation obtained by using 10-fold-cross-validation. One tailed paired t-test with confidence level of 95% was used in order to verify whether the differences in accuracy are statistically significant. The results of our experimental study are very encouraging. In fact there was no significant case where Naïve Bayes was more accurate than D-IFN, while D-IFN was more accurate than Naïve Bayes in 8 databases. Furthermore D-IFN was significantly more accurate than C4.5 in 8 databases, and less accurate in only 2 databases.

TABLE 2: Summary of experimental results. The superscript "+" indicates that the accuracy rate of D-IFN was significantly higher than the corresponding algorithm at the level 5%. A superscript "−" indicates the accuracy was significantly lower.

Database	Simple Bayes	C4.5	D-IFN
Aust	84.93±2.7	85.36±5.1	84.49±2.9
BCAN	97.29±1.6	+92.43±3.5	97.29±1.6
LED17	+63.18±8.7	+59.09±6.9	73.64±5.5
LETTER	+73.29±1	+74.96±0.8	79.07±0.9
Monks1	+73.39±6.7	+75.81±8.2	92.74±11
Monks2	+56.21±6.1	+52.07±8.6	62.13±6.4
Monks3	93.44±3.7	93.44±3.7	92.62±3.3
MUSH	+95.48±0.9	100±0	100±0
Nurse	+65.39±24	−97.65±0.4	92.67±0.6
OPTIC	91.73±1.3	+62.42±2	91.73±1.4
Sonar	75.48±7.3	69.71±5.4	75±8.7
SPI	+94.2±0.9	+91.2±1.9	95.8±0.9
TTT	+69.27±3.2	−85.31±2.7	73.33±4
Wine	96.63±3.9	+85.96±6.9	96.63±3.9
Zoo	89.11±7	93.07±5.8	92.71±7.3
Average	81.83	80.5	86.84

CONLUSION AND FUTURE WORK

In this chapter a new concept of attribute decomposition for classification problems has been proposed and defined. In order to illustrate the potential of this approach, the original IFN algorithm has been extended. The algorithm has been implemented on real-life datasets from the semiconductor industry, which typically have a high number of attributes and may benefit from using the suggested approach. Furthermore we have applied the D-IFN to a variety of datasets, to demonstrate it is a generic methodology.

Finally, the issues to be further studied include: considering other search methods for the problem defined, examining how the attribute decomposition concept can be implemented using other classification methods like Neural Networks, examining other techniques to combine the generated classifiers, and exploring different decomposition paradigms other than attribute decomposition.

REFERENCES

Attneave, F., Applications of Information Theory to Psychology. Holt, Rinehart and Winston, 1959.

Biermann, A. W., Fierfield, J., and Beres, T., "Signature table systems and learning," IEEE Transactions on Systems, Man, and Cybernetics, 12(5): 635-648, 1982.

Buntine, W., "Graphical Models for Discovering Knowledge", in U. Fayyad, G. Piatetsky-Shapiro, P. Smyth, and R. Uthurusamy, editors, Advances in Knowledge Discovery and Data Mining, pp 59-82. AAAI/MIT Press, 1996.

Cover T. M., Elements of Information Theory. Wiley, 1991.

Dietterich, T. G., and Michalski, R. S., "A comparative review of selected methods for learning from examples," Machine Learning, an Artificial Intelligence approach, 1: 41-81, 1983.

Domingos, P., and Pazzani, M., "On the Optimality of the Simple Bayesian Classifier under Zero-One Loss," Machine Learning, 29: 103-130, 1997.

Dougherty, J., Kohavi, R., and Sahami, M., "Supervised and unsupervised discretization of continuous features," in Proceedings of the Twelfth International Conference on Machine Learning, pp. 194-202, 1995.

Duda, R., and Hart, P., Pattern Classification and Scene Analysis, New-York, NY: Wiley, 1973.

Dunteman, G.H., Principal Components Analysis, Sage Publications, 1989.

Elder IV, J.F. and Pregibon, D., "A Statistical Perspective on Knowledge Discovery in Databases," in U. Fayyad, G. Piatetsky-Shapiro, P. Smyth, and R. Uthurusamy, editors, Advances in Knowledge Discovery and Data Mining, pp 83-113. AAAI/MIT Press, 1996.

Fayyad, U., Piatesky-Shapiro, G., and Smyth P., "From Data Minig to Knowledge Discovery: An Overview," in U. Fayyad, G. Piatetsky-Shapiro, P. Smyth, and R. Uthurusamy, editors, Advances in Knowledge Discovery and Data Mining, pp 1-30, MIT Press, 1996.

Friedman, J.H., and Tukey, J.W., "A Projection Pursuit Algorithm for Exploratory Data Analysis," IEEE Transactions on Computers, 23 (9): 881-889, 1974.

Friedman, J.H., "On bias, variance, 0/1 - loss and the curse of dimensionality," Data Mining and Knowledge Discovery, 1 (1) : 55-77, 1997.

Heckerman, D., "Bayesian Networks for Data Mining," Data Mining and Knowledge Discovery, 1:1, pp. 79-119, 1997.

Kim J.O., and C.W. Mueller, Factor Analysis: Statistical Methods and Practical Issues. Sage Publications, 1978.

Kononenko, I., "Comparison of inductive and naive Bayesian learning approaches to automatic knowledge acquisition". In Current Trends in Reply to: Knowledge Acquisition, IOS Press, 1990.

Kononenko, I., "Semi-naive Bayesian classifier," in Proceedings of the Sixth European Working Session on Learning, Springer-Verlag, pp. 206-219, 1991.

Langley, P., "Selection of relevant features in machine learning," in Proceedings of the AAAI Fall Symposium on Relevance. AAAI Press, 1994.

Liu and H. Motoda, Feature Selection for Knowledge Discovery and Data Mining, Kluwer Academic Publishers, 1998.

Maimon, O., and M. Last, Knowledge Discovery and Data Mining: The Info-Fuzzy network (IFN) methodology, Kluwer Academic Publishers, 2000.

McMenamin, S., and Monforte, F., "Short Term Energy Forecasting with Neural Networks," The energy journal, 19 (4): 43-61, 1998.

Merz, C.J, and Murphy. P.M., UCI Repository of machine learning databases. Irvine, CA: University of California, Department of Information and Computer Science, 1998.

Michie, D., "Problem decomposition and the learning of skills," in Proceedings of the European Conference on Machine Learning, Springer-Verlag, PP. 17-31, 1995.

Mitchell T.M., Machine Learning, McGraw-Hill, 1997.

Pfahringer, B., "Controlling constructive induction in CiPF," in Proceedings of the European Conference on Machine Learning, Springer-Verlag, pp. 242-256. 1994.

Quinlan J.R., "Induction of Decision Trees," Machine Learning, 1(1): 81-106, 1986.

Quinlan J. R., C4.5: Programs for Machine Learning, Morgan Kaufmann, 1993.

Ragavan, H., and Rendell, L., "Look ahead feature construction for learning hard concepts," in Proceedings of the Tenth International Machine Learning Conference, Morgan Kaufman, pp. 252-259, 1993.

Samuel, A., "Some studied in machine learning using the game of checkers II: Recent progress," IBM Journal of Research and Development, 11: 601-617, 1967.

Shapiro, A. D., Structured induction in expert systems, Turing Institute Press in association with Addison-Wesley Publishing Company, 1987.

Schwarz G., "Estimation Dimension of a Model," Ann., Stat., 6: 461-464, 1978.Van Zant, P., Microchip fabrication: a practical guide to semiconductor processing, third edition, New York: McGraw-Hill, 1997.

Walpole, R. E., and Myers, R. H., Probability and Statistics for Engineers and Scientists, pp. 268-272, 1986.

Zupan, B., Bohanec, M., Demsar, J., and Bratko, I., "Learning by discovering concept hierarchies," Artificial Intelligence, 109: 211-242, 1999.

Zupan, B., Bohanec, M., Demsar, J., and Bratko, I., "Feature transformation by function decomposition," IEEE intelligent systems & their applications, 13: 38-43, 1998.

CHAPTER 14

Derivation of Decision Rules for the Evaluation of Product Performance Using Genetic Algorithms and Rough Set Theory

Zhai Lian-Yin
mlyzhai@ntu.edu.sg
School of Mechanical and Production Engineering, Nanyang Technological University, Nanyang Avenue, Singapore 639798

Khoo Li-Pheng
mlpkhoo@ntu.edu.sg
School of Mechanical and Production Engineering, Nanyang Technological University, Nanyang Avenue, Singapore 639798

Fok Sai-Cheong
mscfok@ntu.edu.sg
School of Mechanical and Production Engineering, Nanyang Technological University, Nanyang Avenue, Singapore 639798

ABSTRACT

In the manufacturing of critical components of a product, it is important to ascertain the performance and behaviour of those components being produced before assembly. Frequently, these part components are subject to stringent acceptance tests in order to confirm their conformance to the required specifications. Such acceptance tests are normally monotonous and tedious. At times, they may be costly to carry out and may affect the cycle time of production. This work proposes an approach that is based on genetic algorithms and rough set theory to uncover the characteristics of the part components in relation to their performance using past acceptance test data, that is, the historical data. Such characteristics are described in terms of decision rules. By examining the characteristics exhibited, it may be possible to relax the rigour of acceptance tests. A case study was used to illustrate the proposed approach. It was found that the cost in conducting the acceptance tests and the production cycle time could be reduced remarkably without compromising the overall specifications of the acceptance tests.

D. Braha (ed.), Data Mining for Design and Manufacturing, 337–353.
© 2001 Kluwer Academic Publishers. Printed in the Netherlands.

INTRODUCTION

Background

In the manufacturing of critical components of a product, it is important to ascertain the performance and behaviour of the part components being produced before assembly. Frequently, these part components are subject to stringent acceptance tests in order to confirm their conformance to the required specifications. Such acceptance tests are normally monotonous and tedious. At times, they may be costly to carry out and may affect the cycle time of production. This chapter describes an approach that is based on genetic algorithms and rough set theory to uncover the characteristics of the part components in relation to their performance using past acceptance test data, that is, the historical data. As pointed out by Mitchell (1981), historical information is of great importance to diagnostic tasks. It is often used as a basis for monitoring and diagnosis. Basically, when a part component or a product has a defect, it will exhibit some kind of symptoms. As a result, diagnosis, to some extent, can be viewed as a process to classify the operating status of a product or a part component into different concepts or categories such as operational and defective according to the symptom detected. The diagnostic process is fairly similar to the bespoke acceptance tests carried out to ascertain the quality of a product or a part component during production. However, the work described in this chapter goes beyond diagnosis. It is aimed at uncovering the characteristics of the most significant attributes of a part component that best describe the symptoms observed, in terms of decision rules extracted from the historical data available, so as to rationalize the procedure for acceptance tests.

The ability of acquiring decision rules from empirical data is an important requirement for both natural and artificial organisms. In artificial intelligence (AI), decision rules can be acquired by performing inductive learning (Wong et. al., 1986). Many techniques such as decision tree learning (Quinlan, 1986), neural network learning (Fausett, 1994) and genetic algorithm-based learning (Goldberg, 1989) have been developed to carry out such a task. In reality, a machine learning system constantly encounters raw data containing uncertainty such as imprecise or incomplete data, which tends to complicate the task of machine learning. Imprecise data, in this work, refers mainly to raw data that is fuzzy, conflicting or contradicting. For example, the opinion about the performance of a boiler that is perceived by two engineers could be different. This would introduce inconsistency to the raw data describing the performance of the boiler. On the other hand, incomplete information refers mainly to missing data in the raw data obtained. It may be caused by the unavailability of equipment or the oversight of operators during testing. This imprecise and incomplete nature of raw data is the greatest obstacle to automated decision rule extraction.

Over the past decades, many theories and techniques have been developed to deal with uncertainty in information analysis. Among them, fuzzy set theory (Zadeh, 1965) and the Dempster-Shafer theory of belief functions (Shafer, 1976; 1982) are two of the most popular techniques. Rough set theory, which was introduced by Pawlak (1982) in the early 80's, provides a novel way of dealing with vagueness and uncertainty. It focuses on the discovery of patterns in incomplete data and can be used as a basis to perform formal reasoning, machine learning and rule discovery. One of the most important applications of rough sets is in the area of classification and concept formation (Pawlak, 1992; 1994). The main advantages of using rough set theory can be found in the work of Pawlak (1996; 1997). In particular, rough set theory is also applicable to situations in which the set of experimental data is too small to employ standard statistical method (Pawlak, 1991). However, the basic notions of rough set theory are limited to dealing with discrete-valued attributes. Unfortunately, in reality, continuous-valued attributes are more common. For example, the speed of a vehicle or the temperature of boiler water cannot be characterized by just three linguistic descriptors such as '*low*', '*normal*', and '*high*'. It may take real values such as 51 km/h, 60 km/h and 70 km/h or 75.5°C, 97.3°C and so on. In order to resolve such a limitation, it is necessary to find an approach to handle continuous-valued attributes prior to the treatment of data using rough set theory. In the following sections, the scenario of a case study on the manufacture of an electronic device is described. An approach to the discretization of continuous-valued attributes as well as the framework of a prototype system to discover the characteristics of the part components in relation to their performance using past performance test data is also included. A detailed analysis on the results obtained is summarized.

The Manufacture of An Electronic Device

A manufacturing company producing a kind of electronic device for industrial applications is the subject of this study. It has been established that all the devices manufactured by the company must undergo 100% inspection in order to ascertain its performance. The most critical output parameter (O) of the device is influenced by the environment where the surrounding physical characteristics, P and Q, vary within a wide range, that is, O is associated with P and Q settings under which it works. The criterion for acceptance is based on the measurement of the output parameter of the device. If the value of O at a particular P-Q setting exceeds the corresponding limit, the device is considered failing the acceptance test and is rejected. Currently, the acceptance limit is set in accordance to the domain knowledge of engineers about the device. For simplicity, a set of linear limit threshold curves are normally used as the criterion for acceptance test (Figure 1). The device is tested at 12 P-Q settings (Table 1) defined by the set of linear threshold curves. More specifically, three Q and four P settings, that is, a total combination of 12 acceptance tests, are used to ascertain the performance of each device. Each of these acceptance tests is carried out on a production equipment under controlled environment, and is very costly and time consuming. As a result, the productivity is significantly affected. In order to improve the productivity, it is highly desirable to simplify the procedure of acceptance tests. One of the ways is to have a better understanding of the characteristics of the device through extracting the pattern of decision making in terms of decision rules from past acceptance test data. Once this is achieved, simplification of test procedure may be possible.

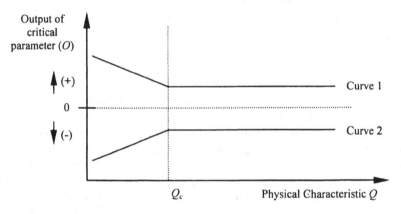

Note: Q_c is the critical value of Q.

Figure 1: The Linear Threshold Limit Curves

Figure 1 depicts the relationship between the value O and Q setting at a given P setting. It is obvious that the device is more sensitive to the lower range of Q, which is characterized by its critical value, Q_c. In the current acceptance test procedure, any device whose output falls within the boundary between Curves 1 and 2 is considered as good; otherwise, it is viewed as a bad unit and is rejected. A sample of acceptance test result for one device is shown in Table 1. As shown in the table, every device will be examined 12 times in total under different combinations of P-Q settings.

Table 1: P-Q Settings for Acceptance Test

Physcial Characteristic P	Physical Characteristic Q		
	Q_1	Q_2	Q_3
P_1	20.7	5.85	9.82
P_2	16.0	11.6	13.3
P_3	30.1	2.07	5.04
P_4	59.8	19.4	10.4

Notes: 1. Entries in the table are the output values of the critical parameter under corresponding P-Q settings.

2. $Q_1 < Q_2 < Q_3$; $P_1 < P_2 < P_3$.

As mentioned earlier, the objective of this study is to discover hidden patterns in the historical data that can correlate the quality of the device and the output values under different test conditions. In order to analyse the acceptance test data, it is necessary to re-format the data into *attribute-value* pairs, which is prevailing in machine learning tasks. One simple method for doing this is to treat every P-Q setting as an attribute (A_i, $1 \le i \le 12$), and the output values as its corresponding attribute values. This results in 12 condition attributes. The decision attribute ($QLTY$) is naturally the quality of the device, that is, *pass* or *fail*. The transformation of the attributes is shown as follows:

Condition Attributes:

$(Q_1, P_1) \Rightarrow A_1;$	$(Q_1, P_2) \Rightarrow A_2;$	$(Q_1, P_3) \Rightarrow A_3;$
$(Q_1, P_4) \Rightarrow A_4;$	$(Q_2, P_1) \Rightarrow A_5;$	$(Q_2, P_2) \Rightarrow A_6;$
$(Q_2, P_3) \Rightarrow A_7;$	$(Q_2, P_4) \Rightarrow A_8;$	$(Q_3, P_1) \Rightarrow A_9;$
$(Q_3, P_2) \Rightarrow A_{10};$	$(Q_3, P_3) \Rightarrow A_{11};$	$(Q_3, P_4) \Rightarrow A_{12};$

Decision Attribute:

Quality of Product $\Rightarrow$ $QLTY$

After the transformation, the information table for the device comprising 170 sets of data is depicted in Table 2. As shown in the table, the entries for the condition attributes are real numbers representing the output values of the electronics parameter of the device. The decision attribute $QLTY$ has two discrete values: '0' stands for a *good unit*, and '1' stands for a *bad unit*. As already mentioned, the continuous-valued condition attributes need to be discretized prior to the treatment by RClass* (Khoo and Zhai, 2000), which was developed based on genetic algorithms and rough set theory. The details of the discretization process are presented in the following section.

Table 2: Information Table for the Acceptance Test of 170 Electronic Devices

Unit	Condition Attributes												QLTY
	A_1	A_2	A_3	A_4	A_5	A_6	A_7	A_8	A_9	A_{10}	A_{11}	A_{12}	
1	20.7	16.0	30.1	59.8	5.85	11.6	2.07	19.4	9.82	13.3	5.04	10.4	0
2	14.5	7.71	27.1	48.4	20.2	18.5	3.46	30.3	16.9	17.4	6.02	7.08	0
3	63.2	58.0	66.4	85.1	12.9	8.94	16.6	31.9	13.8	16.0	9.98	0.38	1
4	100.	70.3	28.0	5.50	41.8	23.5	1.47	6.98	7.06	9.32	3.32	6.11	1
5	0.47	4.66	0.22	9.59	22.6	27.4	22.9	13.6	13.7	14.9	8.37	1.80	0
6	0.77	0.82	21.3	52.3	1.34	6.16	1.15	19.0	4.28	2.95	4.25	1.92	0
⋮	⋮	⋮	⋮	⋮	⋮	⋮	⋮	⋮	⋮	⋮	⋮	⋮	⋮
170	11.4	3.23	5.06	17.6	1.33	6.65	7.19	0.50	15.3	17.7	11.8	3.10	0

DISCRETISATION OF CONTINUOUS-VALUED ATTRIBUTES

When dealing with attributes in concept classification, it is obvious that they may have varying importance in the problem being considered. Their importance can be pre-assumed using the auxiliary knowledge about the problem and expressed by properly chosen '*weights*'. However, in the case of using the rough set approach to concept classification, it avoids any additional information aside from what is included in an information table. Basically, the rough set approach attempts to determine from the data available in an information table whether all the attributes are of the same strength and, if not, how they differ in respect of the classificatory power. For example, it is well known that when describing the condition of a machine in terms of the symptoms observed, some of these symptoms may have greater significance in the assessment of the machine's condition status than others do.

Learning algorithms frequently use heuristics to guide their search through the large space of possible relations among attribute values and classes. Minimising the *information entropy* of the classes in a data set is one of the popular heuristics used by many researchers (Quinlan, 1986, 1992). These learning algorithms assume all the attributes are categorical, that is, discrete. As a result, continuous attributes, which are more common in reality, must be discretized prior to the search. Thus, concept learning, pattern analysis, and decision tree generation algorithms should have the ability in handling continuous-valued attributes. In general, a continuous-valued attribute takes on numerical values (integer or real number) and has a linearly ordered range of values. Typically, the range of a continuous-valued attribute can be partitioned into sub-ranges. Such a process is known as discretization. However, discretization should be done in such a way to provide useful classification information with respect to the classes within an attribute's range.

In a decision tree algorithm, a continuous-valued attribute can be discretized during the generation of a decision tree by partitioning its range into two intervals (binary-interval). A threshold value, T, for a continuous-valued attribute, A, is first determined. The set $A{\le}T$ (the value of Attribute A that is less than or equal to T) is assigned to the left branch, while $A{>}T$ (the value of Attribute A that is greater than T) is assigned to the right branch. Such a threshold value, T, is called a *cut-point*. This method of selecting a cut-point was used in the algorithm of ID3 and its extensions (Quinlan, 1986; 1992). More specifically, given a set, S, which is composed of N training examples, the discretization of a continuous attribute, A, can be based on the entropy minimization for the selection of a cut-point to partition the domain range of A into two sub-domains as follows.

Step 1. The training examples, which are randomly selected from the historical data, are first sorted in accordance to increasing value of Attribute A;

Step 2. The midpoint between each successive pair of training examples in the sorted sequence is evaluated as a candidate cut-point. Thus, for each continuous-valued attribute, $N-1$ evaluations will take place;

Step 3. At each candidate cut-point, the set of training examples, S, is partitioned into two sub-sets (S_1 and S_2) and the class entropy of the resulting partition is computed. Mathematically, if S consists of k classes, namely C_1, C_2, ..., C_k, and let $P(C_i, S_j)$ ($j = 1$ or 2, the same below) denote the proportion of training examples in subset S_j having Class C_i for $i=1, ..k$, the class entropy of a subset S_j is given by

$$ENP(S_j) = -\sum_{i=1}^{k} P(C_i, S_j)\ln(P(C_i S_j)) \qquad (1)$$

The class entropy, $ENP(S_j)$, measures the amount of information needed (in terms of bits). This measure shows the degree of *randomness* exhibited by a class. The smaller the value of $ENP(S_j)$, the less even is the distribution;

Step 4. The resulting class entropy of Attribute A, $INENP(A,T;S)$, is then evaluated using the weighted average of $ENP(S_1)$ and $ENP(S_2)$ after partitioning into S_1 and S_2 as follows.

$$INENP(A,T;S) = \frac{|S_1|}{N} ENP(S_1) + \frac{|S_2|}{N} ENP(S_2) \qquad (2)$$

where T is the cut-point value, and $|S_1|$, $|S_2|$ and N are the number of training examples in S_1, S_2 and S respectively.

Thus, $INENP(A, T; S)$ is also known as the *class information entropy of partition* introduced by T.

Step 5. Based on Eq. 2, the cut-point T_A, for which *INENP(A, T; S)* is minimum among all the candidate cut-points is taken as the best cut-point. Using T_A, the continuous-valued attribute, A, is then discretized. The same procedure can be applied to other continuous-valued attributes.

As mentioned above, the midpoint value between each successive pair of training examples in a sorted sequence needs to be evaluated in order to identify the potential cut-points. However, such a process may not be necessary. Analysis shows that the potential cut-point for partitioning the continuous-valued attribute will not lie between a pair of training examples with the same concept in the sorted sequence. In other words, the best cut-point always lies between a pair of training examples belonging to different concepts or classes in the sorted sequence (Figure 2). Figure 2 shows an example with 3 concepts (classes) to be classified. In this case, only two points, namely V_{T1} and V_{T2} need to be evaluated, instead of all the *n-1* points. Thus, Step 2 of the search procedure for the best cut-point as outlined above can be simplified. In this case, the evaluation of potential cut-point will only be carried out when a pair of training examples with different concepts or classes is spotted.

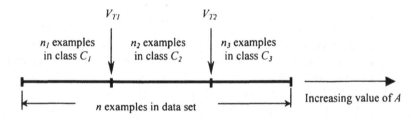

Figure 2: An Example of Potential Cut-points

Based on the above discussion, the efficiency of the search algorithm for the determination of best cut-point can be remarkably improved. In the worst case scenario where the training examples in all the possible pairs formed have different concepts, an exhaustive search involving all possible pairs of training examples is then in order. In practice, such a scenario is extremely unlikely. Under the worst-case scenario, the quality of discretization will be affected, as the information entropy of the attribute under discussion will fluctuate within a narrow band (Figure 3a). As a result, the attribute will contribute little to concept recognition. Such an attribute can therefore be viewed as a redundant attribute. How it is partitioned will not influence the final results significantly. From a mathematical perspective, the information entropy computed at every possible cut-point for this attribute will not change so remarkably compared to a typical attribute as shown in Figure 3b.

More specifically, Figure 3a shows a typical distribution of information entropy computed for a worst-case scenario at every possible cut-point. It is easy to find that the entropy values calculated fluctuate around a relatively higher level (around 0.75 in this case), and there is no apparent trend in the distribution. Thus, the minimal entropy value found does not have any useful implications. This also shows that such an attribute is less informative for classification problems. Figure 3b shows a typical informative attribute. There is an obvious minimal entropy value and the corresponding cut-point is assumed as the best cut-point that can separate different concepts successfully. This implies that the entropy information not only provides the criterion for discretization, but also, to a certain extent, measures how informative an attribute is.

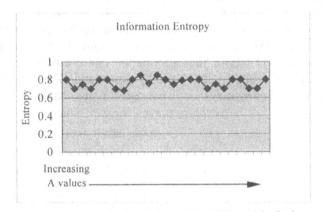

(a) Entropy trend of a redundant attribute without obvious minimal value

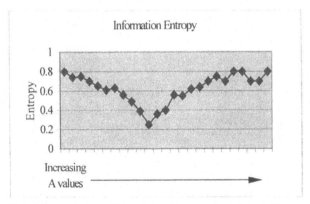

(b) Entropy trend of a typical attribute with an obvious minimal value

Figure 3: Entropy Distribution for Different Attributes

However, with the values attained by the decision attributes remain unchanged, the discretization operation imposed on the continuous-valued attributes may introduce another problem in rule extraction, that is, inconsistency. The reason is that in the training data set, different attribute values in the original training data set may have the same value as a result of the discretization. Such a problem can be easily overcome using rough set theory as outlined in the work of Khoo et al. (1999).

FRAMEWORK OF RClass*

The prototype system described here, RClass*, was designed for extracting diagnostic knowledge from raw information in the form of discrete attribute values (Khoo and Zhai, 2000). RClass* is a hybrid system, which integrates the strength of rough set theory, the unique searching engine of genetic algorithms and Boolean algebraic operations (Figure 4). It aims at dealing with inconsistencies in the training information. Since a detailed description of RClass* can be found in the work by Khoo and Zhai (2000), a brief introduction to the prototype system is presented in this chapter.

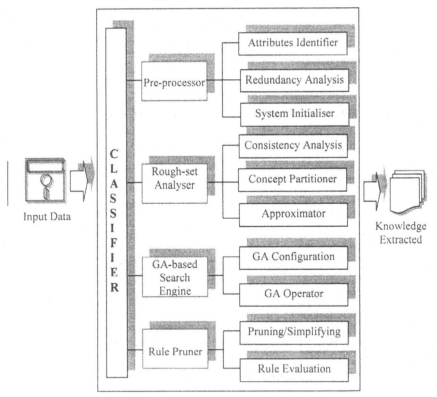

Figure 4: Framework of RClass*

Briefly, RClass* consists of four modules, namely a pre-processor, a rough-set analyser, a GA-based search engine and a rule pruner. The pre-processor is designed to identify attributes and their values; perform redundancy check and re-organise new data set with no superfluous observations for subsequent use; and initialise all the necessary parameters for the GA-based searching engine, such as the length of chromosome, population size, number of generation and the probabilities of crossover and mutation.

The rough-set analyser carries out three sub-tasks namely consistency check, concept forming and approximation. Once an inconsistency is spotted, it activates the concept partitioner and the approximation operator to carry out an analysis based on rough set theory, and subsequently the data are forwarded to the GA-based searching engine for rule extraction.

The GA-based search engine, once invoked, performs genetic operations such as crossover, mutation, and reproduction to gather certain rules and possible rules from the data sets obtained from the previous rough set analysis. Essentially, certain rules (or certainly valid rules) are defined as rules that can definitely classify some observations into a certain concept (class), and possible rules (or possibly valid rules) are rules that can only classify some observations into a certain concept (class) at some degree of possibility.

The rule pruner performs two tasks namely pruning (or simplifying) and rule evaluation. It examines all the rules extracted by the GA-based search engine and employs Boolean algebraic operators such as union and intersection, to prune and simplify the rules. Redundant rules are removed and related rules are clustered and generalised during the simplification process. Following that, the reliability of possible rules is assessed by calculating the probability of a possible rule to classify observations correctly. On the other hand, for each certain rule extracted, a so-called completeness index is computed, which indicates the number of observations correctly classified by the said rule. The completeness index represents the usefulness or the effectiveness of a certain rule. Basically, the usefulness of a certain rule or the reliability of a possible rule is measured by how well the rule covers the observations.

ANALYSIS AND RESULTS

The 170 sets of data (Table 2) is first analyzed using the bespoke discretization procedure. The analysis is aimed at filtering off attributes that contribute little to concept recognition, that is, redundant attributes. The problem of redundant attributes arises in many practical applications. The process of attribute reduction is referred here as knowledge reduction and has a great impact on knowledge acquisition tasks. Intuitively, a reduct of knowledge is its essential part, which suffices to define all basic concepts occurring in the knowledge under consideration. In other words, it is, in a certain sense, the most important part of the available knowledge. Thus, it can be interpreted as the part that characterizes the knowledge and cannot be eliminated further when pruning the knowledge.

In the case study on the electronics device, it is desirable to remove superfluous attributes in the 170 sets of data in such a way that the elementary categories in the data set are preserved. Through the bespoke discretization procedure, it is possible to eliminate all the unnecessary attributes from the data set, preserving only the part of knowledge, which is useful. The ranking of the optimal entropy values (starting from the smallest to the largest) calculated for the 12 attributes is shown in Table 3.

Table 3: Ranking of Entropy Values Computed for Attributes

Attributes	A_5	A_1	A_2	A_6	A_4	A_8
Cut-points	28.7	82.05	58.25	22.8	82.35	52.0
Entropy Values	0.194070	0.214936	0.223128	0.256952	0.266078	0.268422
Ranking	1	2	3	4	5	6
Attributes	A_7	A_9	A_{12}	A_{10}	A_3	A_{11}
Cut-points	16.2	29.65	131.8	28.20	60.6	24.8
Entropy Values	0.288351	0.292926	0.292926	0.303680	0.315982	0.315982
Ranking	7	8	8	9	10	10

It has been established that out of the 170 sets of data randomly sampled from the manufacturing system, there are 158 good units and 12 bad units. According to the records maintained in the plant, the reject rate of the device produced is about 5-10%, which is also reflected from the proportion of good and bad units in the data set sampled. Since there are two classes (good and bad units) in the information table (Table 2), only one cut-point is needed for each attribute (Zhai, 2000). Table 3 illustrates the entropy values for the cut-points of the 12 attributes. It can be deduced from the ranking of the entropy values that the device is rather sensitive to lower Q settings. Such an observation is consistent with the physical characteristics of the device (Figure 1). From a mathematical perspective, the entropy value represents how informative an attribute is. The smaller the entropy value, the more informative the said attribute will be. Thus, an attribute with a smaller entropy value will contribute more to classification. This implies that acceptance tests performed using lower Q settings are more important. Thus, the corresponding attributes are more significant compared to others.

The optimal cut-point identified for each of the 12 continuous-valued attributes is able to separate the value space of each attribute into two intervals, which implies that the attributes have been discretized. Optimal cut-points are those cut-points with minimum entropy values (Table 3).

A comparison between the cut-points calculated and the limits of the output parameter imposed by the plant is depicted in Table 4. It is obvious that the all but one (for Attribute A_{12}) cut-points computed are smaller than those imposed by the plant. In other words, the limits defined by these optimal cut-points for acceptance tests are more stringent.

Table 4: Comparison between Optimal Cut-points and Plant Limits

Attributes	A_1	A_2	A_3	A_4	A_5	A_6
Cut-points	82.05	58.25	60.6	82.35	28.7	22.8
Plant's Limit	140	70	70	140	70	35
Attributes	A_7	A_8	A_9	A_{10}	A_{11}	A_{12}
Cut-points	16.2	52.0	29.65	28.20	24.8	131.8
Plant's Limit	35	70	70	35	35	70

Based on these cut-points, the 12 continuous-valued attributes can be, respectively, discretized into binary intervals using the following scheme:

Less than *cut-point value* $\Rightarrow$ 1;
Larger than or equal to *cut-point value* $\Rightarrow$ 2.

As a result, Table 2 can be easily transformed into an information table with discrete attribute values only, which is required by rough set analysis. The result of transformation is shown in Table 5.

Table 5: Information Table after Transformation

Unit	Condition Attributes												Decision
	A_1	A_2	A_3	A_4	A_5	A_6	A_7	A_8	A_9	A_{10}	A_{11}	A_{12}	QLTY
1	1	1	1	1	1	1	1	1	1	1	1	1	0
2	1	1	1	1	1	1	1	1	1	1	1	1	0
3	1	1	2	2	1	1	2	1	1	1	1	1	1
4	2	2	1	1	2	2	1	1	1	1	1	1	1
5	1	1	1	1	1	2	2	1	1	1	1	1	0
6	1	1	1	1	1	1	1	1	1	1	1	1	0
:	:	:	:	:	:	:	:	:	:	:	:	:	:
170	1	1	1	1	1	1	1	1	1	1	1	1	0

Using RClass*, both certain and possible rules can be readily induced. Table 6 shows the best four rules, that is, the first 4 rules with the highest reliability in terms of both probability and coverage for Concept '*good unit*'. The high coverage value attained by these rules shows that they can 'cover' most of the good units listed in Table 6. However, these 4 rules concern only 5 attributes, namely A_1, A_2, A_4, A_5 and A_6, in the lower Q setting range. This is quite consistent with the physical characteristics of the product described in Figure 1. It also reinforces the perception that the device is more sensitive to lower Q settings. Further investigation shows that these 5 attributes have the lowest entropy values and are ranked first to fifth (Table 3). This is consistent with the deduction that attributes with minimum entropy values have a significant impact on classification. Further analysis also shows that if only these 5 attributes are employed to ascertain the quality of the device, good results can still be achieved. Among all the 158 good units listed in Table 5, 152 of them can be correctly recognized if Rule 1(Table 6) comprising Attributes A_1, A_4, and A_5 is used. As for Rule 2 (Table 6) comprising Attributes A_2, A_4, and A_5, 150 out of 158 good units can be correctly recognized. Similar analysis can also be performed to Rules 3 and 4 (Table 6).

Table 6: Rules with the Highest Reliability

Rule		Confidence Level
1	IF $(A_1 = 1)$ & $(A_4 = 1)$ & $(A_5 = 1)$ THEN $QLTY$ = good	152/152=100%
2	IF $(A_2 = 1)$ & $(A_4 = 1)$ & $(A_5 = 1)$ THEN $QLTY$ = good	150/150=100%
3	IF $(A_1 = 1)$ & $(A_4 = 1)$ & $(A_6 = 1)$ THEN $QLTY$ = good	147/147=100%
4	IF $(A_2 = 1)$ & $(A_4 = 1)$ & $(A_6 = 1)$ THEN $QLTY$ = good	144/144=100%

As already mentioned, the objective of this analysis is to identify the most important attributes so as to facilitate the acceptance tests. This would enable quality engineers to distinguish *good* units from *bad* ones. It appears that if a device, which is under acceptance test, satisfies any of the rules listed in Table 6, it is likely to be a good unit. This implies that

IF $\{(A_1 = 1)$ & $(A_4 = 1)$ & $(A_5 = 1)\} \cup \{(A_2 = 1)$ & $(A_4 = 1)$ & $(A_5 = 1)\}$ THEN
$\cup \{(A_1 = 1)$ & $(A_4 = 1)$ & $(A_6 = 1)\} \cup \{(A_2 = 1)$ & $(A_4 = 1)$ & $(A_6 = 1)\}$ $QLTY$ = good

Using such a composite rule that combines Rules 1, 2, 3 and 4, a total of 154 units among the 158 good units can be correctly classified. This gives a confidence level of 97.5%. Further analysis on these five P-Q settings shows that all of them lie in the lower Q setting range, that is, Q_1 and Q_2, and the P settings are P_1, P_2, and P_4 respectively. Thus, the number of acceptance tests to be carried out can be remarkably reduced by omitting other P-Q settings (Table 7). This implies that 7 out of 12 acceptance tests can indeed be eliminated. Thus, a 58% reduction in cost for acceptance tests can be expected.

Table 7: P-Q Settings after Simplification

Physical Characteristic P	Physical Characteristic Q		
	Q_1	Q_2	Q_3
P_1	√	√	×
P_2	√	√	×
P_3	×	×	×
P_4	√	×	×

Note: √ --- Test must be performed.

× --- No test is needed.

At this juncture, it is important to examine the 4 good units, which cannot be detected by the composite rule. For clarity, the acceptance test results attained by the 4 units are tabulated in Table 8.

Table 8: Good Units not Covered by the Composite Rule

Unit	Condition Attributes												Decision
	A_1	A_2	A_3	A_4	A_5	A_6	A_7	A_8	A_9	A_{10}	A_{11}	A_{12}	*QLTY*
1	20.1	21.8	46.8	**82.7**	1.77	1.18	12.0	35.0	5.24	0.19	2.89	12.9	0
2	81.0	**61.7**	41.1	35.9	37.6	23.5	10.9	9.41	8.44	12.7	10.6	4.41	0
3	18.0	4.84	40.7	**91.0**	23.8	20.9	1.05	29.7	4.54	3.31	7.91	5.61	0
4	27.1	30.3	54.8	**85.7**	3.64	4.49	9.66	29.1	19.7	20.0	10.5	1.35	0
Limit	82.05	58.25	60.6	82.35	28.7	22.8	16.2	52.0	29.65	28.2	24.8	131.8	

It is obvious from the table that they fall outside the boundary of a good unit defined by the cut-points (Attribute A_4 for Units 1, 3, and 4; Attribute A_2 for Unit 2). It appears that the cut-points computed using the entropy method are rather stringent, and as a result, a few good units are rejected. Thus, the composite rule appears to rather cautious and would provide added assurance to the quality of the device if implemented. Further examination shows that only a small proportion of the entire family of devices falls under this category. In this case, in the 170 random samples, only 2.35% belongs to this category. Discarding them will not have an impact to the overall production cost.

CONCLUSION

This chapter summarizes the work leading to the establishment of an entropy-based approach to the discretization of continuous-valued attributes for rough set analysis. Using a case study gleaned from the electronics manufacturing industry, the entropy-based approach has demonstrated that the cut-points generated can be employed to discretize the range of continuous-valued attributes into reasonable sub-intervals. The cut-point values except the one defined for Attribute A_{12} are found to be more stringent than the limits pre-determined by the plant. The discretization process transforms the data sampled from the past acceptance test data into a discrete information table as the input to RClass*. RClass* is able to generate a set of rules to characterize the electronics device. It has been demonstrated that a composite rule based on 5 attributes (A_1, A_2, A_4, A_5 and A_6), that is, 5 P-Q settings, can be established. Using such a rule, a total of 154 units among the 158 good units can be correctly classified. This gives a confidence level of 97.5%. This implies that Attribute A_{12}, an attribute with a cut-point value above the acceptance limit pre-determined by the plant can be eliminated from acceptance test. Further analysis on these five P-Q settings shows that all of them lie in the lower range of Q, that is, Q_1 and Q_2, and the settings for P are P_1, P_2, and P_4 respectively. This is consistent with the physical characteristics of the device. Thus, the number of acceptance tests to be carried out can be remarkably reduced by omitting other P-Q settings. This implies that 7 out of 12 acceptance tests can indeed be eliminated. As a result, a 58% reduction in cost for acceptance tests can be expected.

REFERENCES

Fausett, L.V., Fundamentals of Neural Networks: Architectures, Algorithms, and Applications. Englewood Cliffs, NJ: Prentice-Hall, 1994.

Goldberg, D.E., Genetic Algorithms in Search, Optimisation and Machine Learning. Reading, Mass.: Addison-Wesley Pub. Co., 1989.

Khoo, L.P., Tor, S.B. and Zhai, L.Y., "A Rough-set Based Approach for Classification and Rule Induction," International Journal of Advanced Manufacturing Technology, 15, 438-444, 1999.

Khoo, L.P. and Zhai, L.Y., "RClass*: A Prototype Rough-set and Genetic Algorithms Enhanced Multi-concept Classification System for Manufacturing Diagnosis," in Handbook of Computational Intelligence in Design and Manufacturing, Boca Raton: CRC Press LLC, 2000 (in press).

Mitchell, J.S., An Introduction to Machinery Analysis and Monitoring, Tulsa, Oklahoma: PannWell Books Company, 1981.

Pawlak, Z., "Rough Set Approach to Multi-attribute Decision Analysis," European Journal of Operational Research, 72 (3), 443-459, 1994.

Pawlak, Z., Rough Sets - Theoretical Aspects of Reasoning about Data, Kluwer Academic, 1991.

Pawlak, Z., "Rough Set: A New Approach to Vagueness," in Fuzzy Logic for the Management of Uncertainty, pp. 105-108, New York: John Wiley and Sons, 1992.

Pawlak, Z., "Rough Sets," in Rough Sets and Data Mining - Analysis for Imprecise Data, pp. 3-8, Boston, Mass: Kluwer Academic Publishers, 1997.

Pawlak, Z., "Rough Sets," International Journal of Computer and Information Sciences, 11(5), 341-356, 1982.

Pawlak, Z., "Why Rough Sets," in 1996 IEEE International Conference on Fuzzy Systems, pp. 738-743, 1996.

Quinlan, J.R., "Induction of Decision Trees," Machine Learning, 1, 81-106, 1986.

Quinlan, J.R., C4.5: Programs for Machine Learning, Boston: Morgan Kaufmann Publishers, 1992.

Shafer, G., "Belief Functions and Parametric Models," Journal of Royal Statistical Socirety, 44, 322-352, 1982.

Shafer, G., A Mathematical Theory of Evidence, Princeton, NN.JY.: Princeton Univ. Press, 1976.

Wong, S.K.M., Ziarko, W. and Li, Y.R., "Comparison of Rough-set and Statistical Methods in Inductive Learning," International Journal of Man-Machine Studies, 24, 53-72, 1986.

Zadeh, L.A., "Fuzzy Sets," Information and Control, 8, 338-353, 1965.

Zhai, L.Y., Automated Extraction of Diagnostic Knowledge, Master thesis, Nanyang Technological University, Singapore, 2000.

CHAPTER 15

An Evaluation of Sampling Methods for Data Mining with Fuzzy C-Means

K. Josien, G. Wang, T. W. Liao[*], and E. Triantaphyllou
[*]ieliao@lsu.edu
Industrial & Manufacturing Systems Engineering Department
Louisiana State University, Baton Rouge, LA 70803

M. C. Liu
Manufacturing R&D, Boeing Company, Wichita, KS

ABSTRACT

Using fuzzy c-means as the data-mining tool, this study evaluates the effectiveness of sampling methods in producing the knowledge of interest. The effectiveness is shown in terms of the representative-ness of sampling data and both the accuracy and errors of sampled data sets when subjected to the fuzzy clustering algorithm. Two population data in the weld inspection domain were used for the evaluation. Based on the results obtained, a number of observations are made.

D. Braha (ed.), Data Mining for Design and Manufacturing, 355–369.

INTRODUCTION

Data mining is the application of specific algorithms for extracting knowledge from data (Fayyad *et al.*, 1996). Typical kinds of knowledge extracted include association rules, characteristic rules, classification rules, discriminant rules, clustering, etc. Chen *et al.* (1996) surveyed data mining techniques developed in several research communities according to the kinds of knowledge to be mined. This study makes use of a clustering algorithm, specifically the fuzzy c-means (Bezdek, 1987). The possibilistic c-means algorithm (Krishnapuram and Keller, 1993) was tried but eventually not used because of unsatisfactory results.

Clustering or unsupervised classification is the process of grouping physical or abstract objects into classes of similar objects. Consider the partition of a database with N tuples into m clusters. The number of ways in which this can be done, denoted by P(N, m), is as follows (Duran and Odell, 1974):

$$P(N,m) = \frac{1}{m!} \sum_{j=0}^{m} \binom{m}{j} (-1)^j (m-j)^N. \tag{1}$$

As N increases, P(N, m) grows exponentially. Given this huge search space, much effort has been spent to devise better clustering algorithms. Current clustering algorithms can be broadly classified into two categories: partitional and hierarchical. Partitional clustering algorithms attempt to determine m partitions that optimize a clustering criterion. Algorithms in this category include the popular c-means, CLARANS (Ng and Han, 1994), BIRCH (Zhang *et al.*, 1996), and CLIQUE (Agrawal *et al.*, 1998). A hierarchical clustering algorithm performs a nested sequence of partitions by either an agglomerative or divisive approach. The agglomerative approach starts by placing each object in its own cluster and then merges them into larger and larger cluster until all objects are in one cluster (Guha *et al.*, 1998; Loslever *et al.*, 1996). The divisive approach reverses the process.

Use of a distributed framework for parallel data mining offers another alternative to handle large data sets. Rana and Fisk (1999) described a distributed framework employing task and data parallelism using HPJava. A commercial tool that follows this strategy is Darwin of Oracle. The other alternative, called focusing, is to reduce data before applying data mining algorithms. Data reduction can be achieved by reducing the number of tuples and/or attribtues. Using C4.5 (Quinlan, 1993) and IB in MLC++ (Kohavi *et al.*, 1995) as the algorithms for mining classification rules, Reinartz (1999) analyzed the potentials of focusing tuples in data mining. SAS's Enterprise Miner implements most sampling methods.

Our study applies the same methods used by Reinartz (1999) for focusing tuples, but employs clustering instead of classification algorithms on different data sets. In the next section, fuzzy c-means is briefly described. Section 3 presents the sampling methods used, followed by a description of the data set and knowledge sought. Section 5 discusses the results obtained in this study. The paper ends with a conclusion section.

FUZZY CLUSTERING

Fuzzy c-means (FCM) is used to serve as the data mining technique in this study. It is an unsupervised classification method, belonging to the partitional clustering category. It was derived from the hard (or crisp) c-means algorithm.

The hard c-means and its variants (Ball and Hall, 1967) are based on the minimization of the sum of squared Euclidean distances between data ($x_k, k=1, ..., n$) and cluster centers ($v_i, i =1, ..., c$), which indirectly minimizes the variance as follows:

$$Min\ J_1(U,V) = \sum_{i=1}^{c}\sum_{k=1}^{n}(u_{ik})^2 \parallel x_k - v_i \parallel^2. \tag{2}$$

In the above equation, $\mathbf{U} = [u_{ik}]$ denotes the matrix of a hard c-partition and $\mathbf{V}=\{v_i\}$ denotes the vector of all cluster centers. The partition constraints in c-means are: (1) $u_{ik} \in \{0, 1\}\ \forall i, k$, (2) $\sum_{i=1,c} u_{ik}=1, \forall k$, and (3) $0 < \sum_{k=1,n} u_{ik} < n$, $\forall i$. In other words, each x_k either belongs or does not belong to a cluster and it can only belong to one cluster.

Dunn first extended the hard c-means algorithm to allow for fuzzy partition with the objective function as given in Eq. 3 below (Dunn, 1974):

$$Min\ J_2(U,V) = \sum_{i=1}^{c}\sum_{k=1}^{n}(\mu_{ik})^2 \parallel x_k - v_i \parallel^2. \tag{3}$$

Note that $\mathbf{U} = [\mu_{ik}]$ in this and following equations denotes the matrix of a fuzzy c-partition. The fuzzy c-partition constraints are: (1) $\mu_{ik} \in [0, 1]\ \forall i, k$, (2) $\sum_{i=1,c} \mu_{ik}=1, \forall k$, and (3) $0 < \sum_{k=1,n} \mu_{ik} < n, \forall i$. In other words, each x_k could belong to more than one cluster with each belonging-ness taking a fractional value between 0 and 1. Bezdek (1987) generalized $J_2(U, V)$ to an infinite number of objective functions, i.e., $J_m(U, V)$, where $1 \leq m \leq \infty$. The new objective function subject to the same fuzzy c-partition constraints is

$$Min\ J_m(U,V) = \sum_{i=1}^{c}\sum_{k=1}^{n}(\mu_{ik})^m \parallel x_k - v_i \parallel^2. \tag{4}$$

Note that both hard c-means and fuzzy c-means algorithms try to minimize the variance of those data within each cluster.

To solve the above model, an iterative procedure is required. Please refer to the original paper for the solution procedure. The FCM solution procedure was implemented in C language for this study.

SAMPLING METHODS

The sampling methods studied include simple random sampling, systematic sampling, and stratified sampling.

Simple random sampling selects n samples tuple-by-tuple from a population of size N by drawing random numbers between 1 and N. Denote the population of N tuples as the focusing input, F_{in}, and the selected samples as the focusing output, F_{out}. Algorithm RS shows an implementation of simple random sampling.

Algorithm RS(F_{in}, n)
----

begin
$\qquad F_{out} := \emptyset;$
$\qquad$ **while** $|F_{out}| \leq n$ **do**
$\qquad\qquad i := random(1, |F_{in}|);$
$\qquad\qquad F_{out} := F_{out} \cup \{t_i\};$
$\qquad$ **enddo**
$\qquad$ **return** (F_{out});
end;

Note that in this algorithm the sampling is done with replacement. That is, each tuple has the same chance at each draw regardless whether it has already been sampled or not.

To draw n samples, systematic sampling first determine the step size, next draws the first tuple out of the focusing input at a random position, then iteratively adds each tuple with an index which refers to step positions after the selection position in the previous step. Algorithm SS describes an implementation of systematic sampling. Note that in this algorithm $\lfloor \bullet \rfloor$ denotes the largest integer smaller than $\bullet$.

Stratified sampling first uses a stratifying strategy to separate the focusing input into a set of strata $S = \{s_1, ..., s_l, ..., s_L\}$, and then draws samples from each stratum independently by an application of other sampling techniques such as simple random sampling. The stratifying strategy involves the selection of stratified variables, which must be categorical. For the data set

---

Algorithm SS(F_{in}, n)

---

begin

$\quad F_{out} := \varnothing;$

$\quad step := \lfloor |F_{in}|/n \rfloor;$

$\quad i := start;$

$\quad$ **while** $i \leq |F_{in}|$ **do**

$\quad\quad F_{out} := F_{out} \cup \{t_i\};$

$\quad\quad i := i + step;$

$\quad$ **enddo**

$\quad$ **return** $(F_{out});$

end;

---

studied, the stratified variable is binary. There are four variations of stratified sampling: proportional sampling, equal size, Neyman's allocation and optimal allocation. Proportional stratified sampling ensures that the proportion of tuples in each stratum is the same in the sample as it is in the population. Equal size stratified sampling draws the same number of tuples from each stratum. Neyman's allocation allocates sampling units to strata proportional to the standard deviation in each stratum. With optimal allocation, both the proportion of tuples and the relative standard deviation of a specified variable within strata are the same as in the population. Algorithm PSS shows an implementation of proportional stratified sampling that is used in this study. Stratified sampling preserves the strata proportions of the population within the sample. It thus may improve the precision of the fitted models.

For each sampling method, several sampling sizes were obtained at different levels in order to study their effect. The population as well as each

---

Algorithm PSS(F_{in}, n)

begin

$\quad F_{out} := \varnothing;$

$\quad S := \text{stratify}(F_{in});$

$\quad l := 1;$

$\quad$ **while** $l \leq |S|$ **do**

$\quad\quad n_l := \lfloor n|s_l|/|F_{in}| \rfloor + 1;$

$\quad\quad F_{out} := F_{out} \cup RS\{s_l, n_l\};$

$\quad\quad l := l + 1;$

$\quad$ **enddo**

$\quad$ **return**(F_{out});

end;

---

sample data set drawn from it are statistically characterized. The sample characteristics are compared with the population characteristics to show the representative-ness of drawn samples.

Three types of statistical characteristics are distinguished. The first type of characteristics describes the mean and variance of attribute values. The second type considers the distribution of attribute values for simple attributes. The third type takes into account the joint distribution of attribute values for more than one single attribute. The key procedure used to analyze characteristics about focusing outputs in relation to focusing input is hypothesis testing. The null hypothesis, H_0, is that the sample characteristic equal to the population characteristic. The alternative hypothesis, H_1, is that the sample characteristic is not equal to the population characteristic.

To test the mean of attribute j in the focusing output with sample size of n (>30), we compute the test statistic $s_{mj} = n^{1/2}(\mu_j(F_{out}) - \mu_j(F_{in}))/\sigma_j(F_{out})$. H_0 is rejected at confidence level $1-\alpha$ if $s_{mj} > z_{1-\alpha/2}$. For testing the variance of attribute j in the focusing output with sample size of n (>30), we compute the test statistic $s_{Vj} = (n-1)\sigma_j(F_{out})^2/\sigma_j(F_{in})^2$. H_0 is rejected at confidence level $1-\alpha$ if $s_{Vj} < \chi^2_{1-\alpha/2}(n-1)$.

Numeric attributes must be discretized before hypothesis testing for distribution can be performed. Consider attribute j with values in domain dom_j and a set of intervals $I = \{I_1, \ldots, I_l, \ldots, I_L\}$ with $I_l = [b_l, e_l[, b_l < e_l, 1 \le l \le L-1,$ and $I_L = [b_L, e_L]$. I is discretization of dom_j if $dom_j \subseteq I$, $b_1 = \min dom_j$, $e_L = \max dom_j$, and $b_{l+1} = e_l$. This study employs equal-width discretization, as shown in Algorithm EWD.

Algorithm EWD

begin

    $I := \varnothing$;
    $b_1 := \min dom_j$;
    $e_L := \max dom_j$;
    $width := (e_L - b_1)/L$;
    $l := 1$;
    while $l \le L-1$ **do**
        $e_l := b_l + width$;
        $b_{l+1} := e_l$;
        $I := I \cup [b_l, e_l[$;
        $l := l+1$;
    enddo
    $I := I \cup [b_L, e_L]$;
    return(I);

end;

To test the distribution of attribute j in the focusing output with sample size of n after being discretized, we compute the test statistic $s_{Dj} = \sum_{k=1,L}\{[n_{jk}(F_{out})-n \cdot n_{jk}(F_{in})/N]^2/ n \cdot n_{jk}(F_{in})/N\}$. H_0 is rejected at confidence level $1-\alpha$ if $s_{Dj} < \chi^2_{1-\alpha/2}(L-1)$. This test is valid only if $n \cdot n_{jk}(F_{in}) \geq 5$ for all k. A similar test can be performed for joint distribution, but the number of combinations could be high as the number of attributes and the number of discretized intervals increase.

DATA AND KNOWLEDGE

Radiographic testing (RT) is one of several commonly used non-destructive techniques to evaluate welded structures such as off shore oil-drilling plate forms and space shuttle external tanks. With the assistance of a view box, a certified inspector interprets radiographs to determine whether a particular weld is sound or not. Although this is the mode of operation in industries today, human interpretation of weld quality is often subjective, inconsistent, labor intensive, and sometimes biased. Attempts have been made to develop a computer-aided system as an assistant to human inspectors. The key in this effort is to come up with a comprehensive set of interpretation knowledge used by human inspectors. It is our belief that this comprehensive set of interpretation knowledge can be extracted from the huge volumes of radiographic images archived. Because the huge amount of raw data involved, a data reduction operation called feature extraction is usually performed. This operation is critical because good knowledge cannot be obtained without discriminate features. Modeling of interpretation knowledge based on these features is yet another critical task, which is the focus of this work.

This study uses some data extracted from radiographic images of industrial welds that are available to us. Two populations of data organized in the form of tables are used. The first population of data has 2,275 tuples with each tuple having 3 numeric attributes, which were originally extracted for weld identification. Refer to Liao et al. (2000) for more detailed information about feature extraction. The second population of data has 10,500 tuples with each tuple having 25 numeric attributes, which were originally extracted for welding flaw detection (Liao et al., 1999). For both data sets, the categorical value of each record is known, which indicates whether a particular tuple is a weld (for the first data set) or a welding flaw (for the second data set) or not.

The performance measures of interest here are the accuracy of weld identification or welding flaw detection, the false positive rate, the false negative rate, and the accuracy-falsehood ratio that is defined as the ratio between the accuracy and the summation of the false positive rate and the false negative rate.

RESULTS AND DISCUSSIONS

For each data set, we first applied each one of the three sampling methods to generate focusing outputs of different sizes. For each sample size, ten focusing outputs were produced. Each focusing output was then statistically characterized and tested in relation to the population characteristic. Subsequently, we applied fuzzy clustering algorithms to each focusing output. The statistical test results are presented first, followed by the clustering results. In each category, the results are organized by data set.

Statistical Test Results

Weld Identification Data Set

Tables 1-3 summarize the statistical test results of the weld identification data by attribute. For each size of focusing output, the percentage of its passing the test of its representative-ness of the population (or accepting H_0) is shown for some statistical characteristics. Note that each entry corresponding to each statistical characteristic has two numbers with the first number derived from the first five focusing outputs and the second number the second five. The significance of $\alpha = 0.05$ is consistently used throughout all tests.

For each statistical characteristic of each feature, analysis of variance was performed to determine the significance of sampling method, sampling size, and their interaction. The results indicate that:

Table 1. Results of statistical test for feature 1 of the weld identification data set.

Sampling Method	Sample Size	Mean	Variance	Distribution
Random Sampling	50	100, 100	60, 60	60, 20
	100	60, 100	80, 60	0, 20
	200	100, 100	20, 80	0, 20
	300	100, 80	40, 60	0, 0
Systematic	50	100, 80	60, 80	40, 20
Sampling	100	80, 100	20, 80	0, 0
	200	80, 100	60, 80	0, 0
	300	100, 100	40, 60	0, 0
Stratified	50	80, 100	80, 60	0, 20
Sampling	100	100, 100	80, 60	20, 20
	200	100, 100	40, 40	0, 0
	300	80, 100	80, 40	20, 20

Table 2. Results of statistical test for feature 2 of the weld identification data set.

Sampling Method	Sample Size	Mean	Variance	Distribution
Random Sampling	50	60, 100	40, 40	40, 20
	100	80, 80	60, 20	20, 0
	200	60, 100	60, 40	20, 0
	300	100, 100	40, 60	0, 0
Systematic	50	60, 60	40, 40	20, 40
Sampling	100	80, 100	20, 40	0, 0
	200	40, 60	20, 20	0, 0
	300	80, 80	40, 40	0, 0
Stratified	50	60, 100	40, 40	20, 0
Sampling	100	80, 100	0, 20	0, 0
	200	80, 80	20, 0	0, 0
	300	80, 60	0, 40	0, 0

Table 3. Results of statistical test for feature 3 of the weld identification data set.

Sampling Method	Sample Size	Mean	Variance	Distribution
Random Sampling	50	100, 100	80, 40	0, 0
	100	100, 100	60, 60	0, 0
	200	100, 100	80, 60	0, 0
	300	100, 80	100, 60	0, 0
Systematic	50	80, 100	20, 100	20, 20
Sampling	100	100, 100	80, 100	0, 0
	200	100, 100	40, 80	0, 0
	300	100, 80	80, 80	0, 0
Stratified	50	100, 100	40, 80	0, 0
Sampling	100	80, 100	60, 80	0, 0
	200	100, 100	60, 100	0, 20
	300	80, 100	40, 60	0, 0

1) The means are statistically indifferent regardless the sampling method and sample size used.
2) For the variance characteristic, the sampling method factor is significant for feature 2 with p-value = 0.018.
3) For the distribution characteristic, sample size is always significant with p-values of 0.02, 0.003, and 0.044 for features 1, 2, and 3, respectively.
4) The interaction between sampling method and sample size is statistically significant for the distribution characteristic of feature 3 with p-value = 0.005.

Welding Flaw Detection

Because this data set has 25 attributes, it will take up a lot of space to show all of the results. Tables 4-6 summarize the statistical test results of the welding flaw detection data set for three selected attributes. For each size of focusing output, the percentage of its passing the test of its representative-ness of the population (or accepting H_0) is shown for some statistical characteristics. Note that as in Tables 1-3 the first number is derived from the first five focusing outputs and the second number the second five. The significance of $\alpha = 0.05$ is consistently used throughout all tests.

Table 4. Results of statistical test for feature 5 of the welding flaw detection data set.

Sampling Method	Sample Size	Mean	Variance	Distribution
Random Sampling	100	100, 80	100, 60	40, 20
	200	100, 100	100, 80	20, 20
	300	100, 40	100, 60	20, 0
	800	100, 100	80, 60	0, 0
	1000	100, 80	80, 60	0, 0
Systematic Sampling	100	100, 100	100, 100	100, 100
	200	100, 100	100, 100	100, 100
	300	100, 100	100, 100	100, 80
	800	100, 100	100, 100	0, 0
	1000	100, 100	100, 100	0, 0
Stratified Sampling	100	100, 100	80, 100	40, 40
	200	100, 100	80, 100	40, 40
	300	100, 100	100, 100	20, 20
	800	100, 100	80, 100	0, 0
	1000	100, 100	80, 100	0, 0

Table 5. Results of statistical test for feature 15 of the welding flaw detection data set.

Sampling Method	Sample Size	Mean	Variance	Distribution
Random Sampling	100	80, 100	40, 40	20, 20
	200	100, 100	60, 20	0, 0
	300	100, 80	20, 40	0, 0
	800	100, 100	20, 20	0, 0
	1000	100, 100	20, 20	20, 0
Systematic Sampling	100	60, 80	0, 20	60, 20
	200	60, 100	20, 60	20, 0
	300	100, 100	40, 0	0, 0
	800	100, 100	40, 20	20, 0
	1000	100, 100	100, 100	0, 0
Stratified Sampling	100	60, 60	0, 0	40, 60
	200	80, 100	40, 20	0, 20
	300	100, 80	0, 40	0, 0
	800	100, 100	40, 40	0, 0
	1000	100, 100	0, 40	0, 0

Table 6. Results of statistical test for feature 25 of the welding flaw detection data set.

Sampling Method	Sample Size	Mean	Variance	Distribution
Random Sampling	100	100, 100	100, 100	0, 0
	200	60, 100	100, 100	0, 0
	300	100, 100	100, 100	20, 0
	800	100, 100	100, 100	20, 0
	1000	100, 80	100, 100	0, 0
Systematic	100	100, 100	100, 100	0, 0
Sampling	200	100, 100	100, 100	0, 20
	300	100, 100	100, 100	20, 20
	800	100, 100	100, 100	40, 0
	1000	100, 100	100, 100	0, 0
Stratified	100	100, 100	100, 100	0, 20
Sampling	200	100, 100	100, 100	0, 0
	300	100, 100	100, 100	0, 0
	800	80, 100	100, 100	0, 0
	1000	80, 100	100, 100	0, 0

For each statistical characteristic of each one of the above features, an analysis of variance was performed to determine the significance of sampling method, sampling size, and their interaction. The results indicate that:

1) All three statistical characteristics of feature 25 are indifferent regardless the sampling method and sample size used.

2) The sampling method factor is significant for the variance characteristic of feature 5 with p-values = 0.011. In addition, all factors are significant for the distribution characteristic of the same feature with p-values $< 10^{-4}$.

3) The sample size factor is significant for the mean and distribution characteristics of feature 15 with p-values = 0.004 and 0.0002, respectively. In addition, the interaction between sampling method and sample size is significant for the variance characteristic with p-value = 0.015.

Clustering Results

Weld Identification Data Set

Table 7 summarizes the clustering results of the weld identification data set obtained by the FCM algorithm. For each size of focusing output clustered, the mean accurate rate (A), mean false negative rate (FN), mean false positive rate (FP), and mean accuracy-falsehood ratio defined as A/(FN+FP) are shown in each table. Each mean value was computed from ten values corresponding to ten focus output data. A weld not identified is a false negative whereas a non-weld identified as a weld is a false positive.

Table 7. Results of FCM clustering of the weld identification data set.

Sampling Method	Sample Size	Mean Accurate Rate (%)	Mean False Negative Rate (%)	Mean False Positive Rate (%)	Mean Accuracy-Falsehood Ratio
Random Sampling	50	65.8	13.2	56.3	0.95
	100	62.4	14.2	65.3	0.79
	200	59.5	1	80.0	0.74
	300	56.4	5.3	80.7	0.66
Systematic Sampling	50	64.0	20.0	55.1	0.85
	100	58.1	10.4	73.3	0.69
	200	60.0	0.7	79.9	0.74
	300	57.4	0.9	81.0	0.70
Stratified Sampling	50	67.6	5.9	71.5	0.87
	100	61.4	0.2	77.0	0.80
	200	58.8	0.5	81.9	0.71
	300	60.0	0.5	79.8	0.75
Focusing Input	2275	59.2	0.6	80.5	0.73

For each performance measure, an analysis of variance was performed to determine the significance of sampling method, sampling size, and their interaction. The results indicate that sample size is statistically significant for all performance measures and other factors are all insignificant. Overall, the accuracy, false negative rate, and accuracy-falsehood ratio decrease whereas false positive rate increases as sample size increases. It was surprised to find that for all performance measures except false negative rate, sample sizes of 50 and 100 generally fare better than the population. The performance of sampled data sets with size larger than 200 are more comparable with that of the population for this particular data.

Welding Flaw Detection Data Set

Table 8 summarizes the clustering results of the welding flaw detection data obtained by the FCM algorithm. For each size of focusing output clustered, the mean accurate rate (A), mean false negative rate (FN), mean false positive rate (FP), and mean accuracy-falsehood ratio defined as A/(FN+FP) are shown in each table. Each mean value was computed from ten values corresponding to ten focus output data. A welding flaw not detected is called a false negative. On the other hand, a non-flaw called as a flaw is a false positive.

Table 8. Results of FCM clustering of the welding flaw detection data set.

Sampling Method	Sample Size	Mean Accurate Rate (%)	Mean False Negative Rate (%)	Mean False Positive Rate (%)	Accuracy-Falsehood Ratio
Random Sampling	100	58.8	33.0	42.8	0.78
	200	58.3	29.6	43.6	0.80
	300	55.6	28.3	47.1	0.74
	800	57.6	37.3	43.3	0.72
	1000	58.5	28.4	43.6	0.81
Systematic Sampling	100	60.5	33.8	40.3	0.82
	200	58.2	25.1	44.3	0.84
	300	58.8	24.9	44.0	0.85
	800	55.0	21.6	49.0	0.80
	1000	55.9	35.4	45.7	0.70
Stratified Sampling	100	60.0	24.0	40.5	0.93
	200	61.1	31.7	40.2	0.85
	300	61.0	25.2	41.4	0.92
	800	60.6	26.7	41.6	0.89
	1000	56.2	25.7	46.9	0.77
Focusing Input	10,500	64.5	19.0	38.3	1.13

For each performance measure, an analysis of variance was performed to determine the significance of sampling method, sampling size, and their interaction. The results indicate that sampling method is statistically significant for the accuracy and false positive rate. It seems that stratified sampling produces better results than random sampling and systematic sampling in all performance measures. However, no sampling method gives better results than the population for this particular data set.

CONCLUSION

This paper evaluated three sampling methods with respect to the representative-ness and performance of the sampled data. The representative-ness is tested based on three statistical characteristics: mean, variance, and distribution. The performance is measured by using four indices: the accuracy rate, false negative rate, false positive rate, and accuracy-falsehood ratio based on the clustering results of fuzzy c-means. Two population data sets taken from the domain of radiographic testing of welds were used.

It is observed that:

1. Sample means are generally statistically indifferent from the population mean regardless the sampling method and sample size used.

2. The sampling method factor is significant for the variance characteristic for two out of six features tested (feature 2 of weld identification data and feature 5 of welding flaw detection data). It is also significant for the

distribution characteristic for one feature (feature 5 of welding flaw detection data).

3. The sample size factor is significant for the distribution characteristic for five out of six features tested (feature 25 of welding flaw detection data is the only insignificant one).

4. The interaction factor is significant for the variance characteristic of one feature (feature 3 of weld identification data) and for the distribution characteristic of two features (features 5 and 15 of welding flaw detection data).

5. For the weld identification data set, sample size is statistically significant for all performance measures and other factors are all insignificant. It was surprised to find that for all performance measures except false negative rate, sample sizes of 50 and 100 generally fare better than the population.

6. For the welding flaw detection data set, sampling method is statistically significant for the accuracy and false positive rate. In addition, stratified sampling seems to produce better results than random sampling and systematic sampling but worse than the population in all performance measures.

Depending upon the data, one factor might be more important than another. More tests on widely different data are needed to reach any definite conclusion. It is also desirable to determine if there is any correlation between the statistical characteristics of drawn samples and the performance measures of interest.

REFERENCES

Agarwal, R., Gehrke, J., Gunopulos, D., and Raghavan, P., "Automatic Subspace Clustering of High Dimensional Data for Data Mining Applications," *SIGMOD '98*, Seattle, WA, 94-105, 1998.

Ball, G. H. and Hall, D. J., ISODATA, an iterative method of multivariate analysis and pattern recognition, *Behavior Science*, 153, 1967.

Bezdek, J. C., *Pattern Recognition with Fuzzy Objective Function Algorithms* (Plenum Press, New York and London, 1987).

Chen, M.-S., Han, J., and Yu, P. S., "Data Mining: An Overview from a Database Perspective," *IEEE Transactions on Knowledge and Data Engineering*, 8(6), 866-883, 1996.

Dunn, J. C., A fuzzy relative of the ISODATA process and its use in detecting compact well-separated clusters, *J. Cybernet.*, 3, 1974, 32-57.

Duran, B. S. and Odell, P. L., *Cluster Analysis: a Survey*, Volume 100 of *Lecture Notes in Economics and Mathematical Systems*. Springer-Verlag, 1974.

Fayyad, U., Piatetsky-Shapiro, G., and Smyth, P., "From Data Mining to Knowledge Discovery in Databases," *AI Magazine*, 37-54, Fall 1996,

Guha, S., Rastogi, R., and Shim, K., "CURE: An Efficient Clustering Algorithm for Large Databases," *SIGMOD '98*, Seattle, WA, 73-84, 1998.

Kohavi, R., Sommerfield, D., and Dougherty, J., *Data Mining Using MLC++: A Machining Learning Library in C++*, http://robotics.stanford.edu/~ronnyk.

Krishnapuram, R. and Keller, J. M., "A Possibilistic Approach to Clustering," *IEEE Trans. on Fuzzy Systems*, 1(2), 1993, 98-110.

Liao, T. W., Li, D.-M., and Li, Y.-M., "Extraction of Welds from Radiographic Images Using Fuzzy Classifiers," *Information Sciences*, 126, 21-42, 2000.

Liao, T. W., Li, D.-M., and Li, Y.-M., "Detection of Welding Flaws from Radiographic Images with Fuzzy Clustering Methods", *Fuzzy Sets and Systems*, 108(2), 145-158, 1999.

Loslever, P., Lepoutre, F. X., Kebab, A., and Sayarh, H., "Descriptive multidimensional statistical methods for analyzing signals in a multifactorial biomedical database," *Med. & Biol. Eng. & Compt.*, 34, 13-20, 1996.

Ng, R. T. and Han, J., "Efficient and Effective Clustering Methods for Spatial Data Mining," in *Proc. of the VLDB Conference*, Santiago, Chile, 144-155, 1994.

Quinlan, J. R., *C4.5: Programs for Machine Learning*, San Mateo, CA: Morgan Kaufmann, 1993.

Rana, O. F. and Fisk, D., "A Distributed Framework for Parallel Data Mining Using HPJava," *BT Technology Journal*, 17(3), 146-154, 1999.

Reinartz, T., *Focusing Solutions for Data Mining*, Springer, 1999.

Zhang, T., Ramakrishnan, R., and Livny, M., "BIRCH: An Efficient Data Clustering Method for Very Large Databases, " in *Proc. of the ACM SIGMOD Conference on Management of Data*, Montreal, Canada, June 1996.

CHAPTER 16

Colour Space Mining For Industrial Monitoring

K.J. Brazier
kbrazier@liv.ac.uk
Centre for Intelligent Monitoring Systems, Department of Electrical
Engineering, University of Liverpool, Brownlow Hill, Liverpool, Merseyside,
UK, L69 3GJ

A.G. Deakin,
Centre for Intelligent Monitoring Systems, Department of Electrical
Engineering, University of Liverpool, Brownlow Hill, Liverpool, Merseyside,
UK, L69 3GJ

R.D. Cooke
Centre for Intelligent Monitoring Systems, Department of Electrical
Engineering, University of Liverpool, Brownlow Hill, Liverpool, Merseyside,
UK, L69 3GJ

P.C. Russell
Centre for Intelligent Monitoring Systems, Department of Electrical
Engineering, University of Liverpool, Brownlow Hill, Liverpool, Merseyside,
UK, L69 3GJ

G.R. Jones
Centre for Intelligent Monitoring Systems, Department of Electrical
Engineering, University of Liverpool, Brownlow Hill, Liverpool, Merseyside,
UK, L69 3GJ

ABSTRACT

The effectiveness of colour spaces for the encapsulation of multivariate and
distributed source information has led to increasing interest in their
deployment in industrial monitoring technology. Other advantageous features
are the availability of several hardware technologies that allow early
transformation of a variety of monitored quantities into colour information,

D. Braha (ed.), Data Mining for Design and Manufacturing, 371–400.

the compression of data at an early stage, and the familiarity of colour as a means for human assimilation of quantitative information.

To fully realise the potential of colour space representation for enhancing information gathering and enrichment demands its integration with the data mining process. In monitoring applications this must include a capacity for real-time distributed processing and delivery. This contribution describes the relevant characteristics of the representation of information in colour spaces and some of the ways in which it is being combined with data mining methods to develop industrial monitoring solutions. These are illustrated by particular examples from the energy and manufacturing industries of clustering, visualisation and statistical analysis for colour space information.

INTRODUCTION

Industrial monitoring frequently requires the fusion of distributed (eg. spatially, temporally, spectrally) sensor data to derive emergent information characterising the status or progress of the monitored process. This allows the process operator to assimilate more readily key aspects of system behaviour, dispensing with the need to interpret detailed low level sensor data. Chromatic processing methods, which fuse sensor data into composite representations in colour spaces, are becoming increasingly established as a solution for this, finding application initially in the monitoring of electrical arcs and plasmas [Jones (1993), Jones et al. (1994), Russell et al. (1994), Jones (1995), Cosgrave et al. (1997), Yokomizu et al. (1998)].

Given the similarity between the aims of such methods and those of the data mining process, it is natural to consider whether monitoring might benefit from the establishment of a synergistic relationship between them. This contribution first makes some observations regarding the characteristics of chromatic methods as a means for distilling and representing distributed sensor information. It then goes on to describe examples of new applications in which the synergism between chromatic and data mining methods is being developed.

CHROMATIC PROCESSING

An understanding of the following discussion of the chromatic processing that gives rise to colour space representations of data will benefit from some familiarity with the language and conventions of signal processing. Readers unfamiliar with this area may find it useful to consult an introductory text on this subject, for example Connor (1982). Readers wishing to consult a more advanced text might try Gasquet and Witomski (1999).

Chromatic methods fuse distributed data by the application of a number of weighted integrals of the measurand with respect to the variable in which it is distributed. For example, in the optical domain, in which the methods originate, the measurand and distribution variable are typically amplitude of illumination and wavelength, respectively. The weighted integrals thus correspond to the wavelength dependent responses of photodetectors on which a polychromatic optical signal is incident. It can be shown [Jones *et al.* (2000)] that where these responses are Gaussian, their interaction with the signal generates values that correspond to coefficients in its Gabor transform [Gabor (1946), Stergioulas *et al.* (2000)]

$$\varphi(t) = \sum_{m=-\infty}^{\infty} \sum_{n=-\infty}^{\infty} a_{mn} g(t - mD) e^{inWt} \tag{1}$$

where $\varphi(t)$ is the signal, a_{mn} are the coefficients, m and n are integers, and W and D give the intervals between the centre points of the Gabor basis functions in the frequency and time domains, respectively. $g(t)$ is chosen to be

$$g(t) = \left(\frac{\sqrt{2}}{D} \right)^{1/2} e^{-\pi \left(\frac{t}{D} \right)^2} \tag{2}$$

causing the basis function to be a sine wave ($\mathrm{Re}(e^{inWt})$) subject to a Gaussian envelope ($g(t)$) [Porat and Zeevi (1988)]. The coefficients a_{mn} are given by the integral of the signal weighted by the Gabor basis function,

$$a_{mn} = \int_{-\infty}^{\infty} (g(t - mD) e^{inWt})^* \phi(t) \mathrm{d}t \tag{3}$$

The frequency domain form of the Gabor basis functions is similarly a Guassian-enveloped sine wave, so the Gaussian detector responses correspond to the zero-order (in frequency, ie. $n=0$) Gabor basis functions.

The non-orthogonality of the Gaussian basis functions means that, while, owing to the integral nature of the frequency response functions, it is not entirely unambiguously separable, information contained in the relative amplitudes at different frequencies is retained for all sampled frequencies. It has been shown [Stergioulas *et al.* (2000), Stergioulas (1997)] that it is maximal where the slopes of the Gaussians are greatest and that the number of terms in the Gabor expansion may be substantially truncated with only a small loss of information. For practical purposes truncation is usually to three terms and Stergioulas (1997) describes experiments with a test signal that suggest this is sufficient to achieve useful information retention at the same time as maintaining high robustness to noise.

In the optical domain an approximation to these three Gaussian basis functions is provided by the wavelength response of the sensor elements used in colour photodetectors (eg. in CCD cameras). This is known as a tristimulus sensor system. Observations may therefore be represented as data points in a colour space, the most straightforward of which is a Cartesian colour cube having an axis for each of the three sensor elements. The three co-ordinates of a point therefore give a separate measure of each of the familiar red, green and blue components of visible light.

Transformations of this space to highlight potentially interesting characteristics of colours are well established in colour science, many of which operate to confine the overall intensity of the light (proportional to the sum of the three detector outputs) to a single axis. This is often a useful step as signal variations affecting chiefly the overall intensity are frequently attributable to physical causes the influence of which we would like to disregard. For example, where we are monitoring optically in the presence of ambient daylight, we would expect the thickness of cloud cover to affect primarily our intensity component and its effect to be, therefore, readily removable. A more detailed examination of the redistribution of signal to noise ratios among chromatic parameters is given in Yu *et al.* (1997).

The transformation most frequently used in current work is that to the HLS colour system, which re-parameterises a colour (or its analogue, where a non-spectral distribution provides the raw data) as hue, lightness and saturation [Levkowitz and Herman (1993)]. The transformation is

$$
H = \begin{cases}
60(G-B)/(\max(R,G,B)-\min(R,G,B)) & \text{if } \max(R,G,B)=R \\
60(2+(B-R)/(\max(R,G,B)-\min(R,G,B)) & \text{if } \max(R,G,B)=G \\
60(4+(R-G)/(\max(R,G,B)-\min(R,G,B)) & \text{if } \max(R,G,B)=B
\end{cases}
\tag{4}
$$

$$
L = \frac{R+G+B}{3}
\tag{5}
$$

$$
S = \frac{\max(R,G,B)-\min(R,G,B)}{\max(R,G,B)+\min(R,G,B)}
\tag{6}
$$

where R, G, and B are the tristimulus values supplied by the Gabor sensors. The lightness (L) is the intensity parameter mentioned above, while the saturation quantifies the spread of the distribution. Hue quantifies the value of the distribution variable around which the measured data is distributed. Note that this is not necessarily the same thing as the mean of the measured distribution and is frequently referred to as the dominant value of the distribution variable to distinguish it from this. In terms of colour mixing it amounts to a real-valued quantification of that property of colour categorically described by colour name (red, yellow, green etc.). It is important to distinguish this from the real values associated with colour names in the spectral domain (ie., prior to processing with Gabor sensors), in which the description "yellow" and the corresponding wavelength range imply the presence of photons having wavelengths falling in this range. Post Gabor processing, a yellow colour and corresponding dominant wavelength may, for example, be indicated where only red and green photons are present.

This may be interpreted as revealing a "virtual signal" in the yellow wave band, which exists only by virtue of such covariation as exists between the red and green detector responses. The reason for wishing to view the signal in this way is that it is the information conveyed by this virtual signal with which we are familiar from our colour vision, the detectors in our eyes crudely approximating Gabor detectors and sharing the property of non-orthogonality. Humans are thus able to interpret the hue of the HLS colour space effectively without the need to quantify in detail the uncertainties arising from compression losses. Where the distribution is other than a visible light spectrum, interpretation is eased by reference to the optical domain (for example we might talk about a "yellow" noise).

The HLS colour space is most readily represented in cylindrical polar co-ordinates, in which it has the geometry of a cone (Fig. 1). Here, the vertical

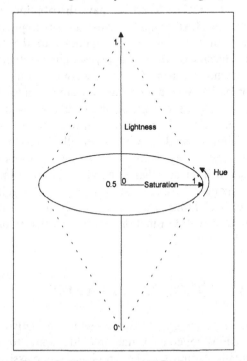

Figure 1: The geometry of HLS colour space

axis gives the lightness value represented by a point in the space, the radial axis its saturation and the angular location its hue. The maximum value on the lightness axis (1) indicates all Gabor sensors saturated (white), while the minimum value (0) indicates no signal (black). A zero saturation value (ie. a point on the L axis) indicates data equally distributed between the Gabor sensors (grey), while a maximum value (1) indicates a "pure" colour. Hue values are arranged such that pure red, green or blue (ie. symmetric distributions about the points of maximum detector sensitivity) give values of 0°, 120° and 240°, respectively. Note that this leaves us with a region between 240° and 360° in which there is no physically realisable dominant value of the distribution variable and which corresponds to magenta in colour mixing terms. Points in this region are generated by distributions having low values of the data in the middle part of the detectors' range and higher values at both ends. Of course, this is only true where our distribution is not around a closed path, which is not always the case. Yokomizu *et al.* (1998) give an example of an application in which pseudo Gabor sensors are applied to a spatial distribution that itself has a polar geometry, so that all angles in the circle correspond to a real physical angle.

Colour science is based around hardware implementations of tristimulus sensors applied to optical spectra and we can also apply it to analogous hardware sensors for other distributed data, for example filters used to process acoustic or electrical signals. Where we have distributions in the form digital electronic data, we can flexibly apply virtual sensors implemented in software to a distribution. In the most direct analogue to the frequency domain hardware sensors discussed, we can apply weighted integrals to the Fourier transforms of distributions, using Guassian weighting functions if we wish to maintain a Gabor basis. We can also apply weighted integrals directly to the distribution itself, giving a set of zero-order Gabor terms in the domain of the distribution variable. There is no obligation to maintain the Gabor basis, of course. While Gabor showed it to be maximally information preserving for the general signal, we may know that we have signals with particular characteristics that will cause other integral weighting functions to be more informative. The methods of colour science may just as well be applied to the values of these weighting functions or even to single data items (corresponding to a weighting function that has a value of one for a particular item and is zero elsewhere).

STATISTICAL COLOUR COMPARISON

This section describes an application in which a statistical approach was applied to observations collected as data in colour space to solve a problem in manufacturing quality control. The problem is a classical application for colour science, however the addition of a statistical comparison method has allowed the cost of the solution to be much reduced.

A great many manufactured products are constructed from a number of components produced in different batches and/or by different suppliers. Even where components are manufactured in the same batch, conditions may vary throughout the process. The result is frequently variation in the finished components. Where a component feature of interest is its colour, some detectable variation is often acceptable, provided this variation is consistent with that of the other components comprising an example of the finished product. In other words, the criterion on which quality is assessed is the ability to discriminate between the colours of the components making up an individual product. If it is not possible to discriminate between the colours of its components, then the quality control test is passed.

Conventionally, one of two methods is used to apply this test of quality:

i.   Human inspection. This has the advantage of being a good model for discrimination by the end user (ie. a human being is modelled by another human being), but the disadvantages of subjectivity, lack of quantitivity and a limited supply of necessary expertise (ie. people with a "good eye").

ii.   Inspection using colorimetric instrumentation. This is quantitative, but less amenable to the use of a range of light sources, which is important to ensure discrimination is impossible under all expected conditions of product deployment. Furthermore, colorimetric instrumentation is a high cost solution.

To improve on this situation, development has been undertaken of a quantitative, flexible and low cost approach to the problem. A prototype has been successfully tested at the premises of a manufacturer of domestic cookers.

Input Data

To fulfil the requirements of low cost and flexibility, input to the system was in the form of image data obtained from a colour CCD camera and video capture card of the types aimed at the domestic PC market. Software was written in C++ to carry out image capture and processing in a single streamlined package. The generic Microsoft video capture programming interface for Windows 9x operating system was used for the image capture to allow the software to be used with capture cards from a wide range of manufacturers.

For the purposes of testing, the camera and pairs of samples to be compared were placed in an enclosure, along with a source of diffuse illumination. The samples were of a material extensively used in the manufacture of domestic cookers, namely back-painted glass. The illumination was from a 3200K dichroic lamp, chosen to be representative of illumination likely to be encountered by the deployed final product. The interior of the enclosure was painted matt black to minimise stray reflections, but it was open at one end. This allowed convenient access, but meant that external lighting could contribute to some small difference between the illumination experienced by the two samples. This was deemed acceptable as preliminary tests had shown that, even where strenuous efforts were made to illuminate only with a single diffuse source, differences in measured colour parameters for the samples were liable to be more strongly influenced by differences in the illumination falling on them than by differences in their pigmentation. The elements were positioned in the enclosure in such a way as to avoid specular reflections from the samples being visible to the camera.

The software allows the user to select an equally sized rectangular region on the image of each sample and the mean of the tristimulus pixel values for each region is calculated. From this, hue, lightness and saturation values are calculated according to eqns.(4)-(6) and the differences between them for the two sampling regions are stored. This is repeated for a number of frames set by the user. The sample positions are then exchanged (in the test enclosure this is conveniently done by opening a hatch in the back of the enclosure, thus avoiding disturbance to the camera and illumination) and an equal number of values are stored for the new arrangement.

Statistical Processing

Time varying offset errors occurring across the image frame (as can occur when, for example, power fluctuations affect the source of illumination) are eliminated by consideration only of the differences between the measured values for the two regions. These differences are then ascribable to the combination of the spatial variation in the illumination, the difference in pigmentation in the sampled regions and random noise. Consideration of the change in these differences when the sample positions are exchanged allows us to account for the spatial variation in lighting. That is, there will, on average, be a change in the measured differences between the sample regions only if there is a difference between the pigmentation of the sample objects. The size of the change is monotonic with the difference in colour of the samples under the chosen illumination (assuming a reasonable approximation of the camera sensor frequency responses to those of the human eye). This is therefore an appropriate measure to quantify the differences between the colours of the samples.

Allowance is made for the random errors by application of a double-sided t-test to the two series of difference values recorded. The user sets a required confidence level and this is used to determine upper and lower limits on the

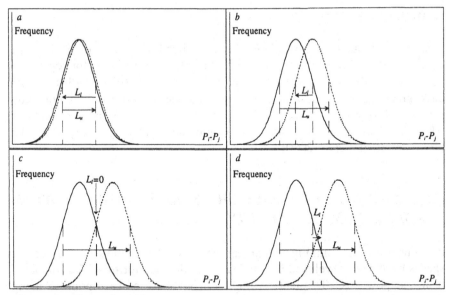

Figure 2: The trade-off between confidence and discriminating power

size and direction of the change in mean difference on exchange of sample positions. Thus, if the upper limit is positive and the lower limit negative, it cannot be claimed with the required confidence that there is a perceptible difference between the colours of the sample objects.

The principle is illustrated in Fig. 2, in which the difference in the colour parameter of interest is plotted on the x-axis and the frequency of occurrence of this difference in the sample sets is plotted on the y-axis. The two distributions of the values are idealised as Gaussian and represent the results before and after the exchange of object positions. Visual inspection of hue values in histogram form has suggested that the Gaussian approximation, assumed by the statistical processing, is reasonable.

In the figure, confidence limits on the separation of the distribution means, L_l and L_u, have been superimposed, and different degrees of similarity between the data set means before and after object exchange are illustrated. In Fig. 2a the means of the distributions are identical and $L_l = -L_u$. This remains true even if the spreads of the distributions about the mean are different. As the means of the distributions become increasingly different, the difference in colour parameter values continues to be indistinguishable with the required confidence as long as $L_l < 0$ and $L_u > 0$ (Fig. 2b). The threshold of discrimination is reached when $L_l = 0$ (Fig. 2c) and when the confidence limits are both positive discrimination with the required confidence is possible (Fig. 2d). For a change in the opposite sense, L_u falls to zero, then both limits become negative.

Achievable Resolution

Tests carried out to compare samples of painted glass panels used in the manufacture of domestic cookers allowed resolution thresholds for separation of the hue value means to be estimated at 3.3° and 0.26° for dark blue and dark green panels, respectively. These correspond to nominal dominant wavelength (see Russell *et al.* 1998) resolutions of about 3.3nm and 0.26nm, which are comparable with the discrimination achievable by an expert human observer (Brazier *et al.* 2000).

CLUSTERING IN COLOUR SPACE FOR DIGITAL INFORMATION RECOVERY

A number of data mining techniques based on the exploitation of colour spaces have been developed by application to the monitoring of the combined heat and power (CHP) plant serving the University of Liverpool. Among these is the use of cluster analysis in the three dimensional space defined by the tristimulus parameters to recover information known *a priori* to be binary from the near-continuous data collected by virtual sensors monitoring multiple regions of a camera image. The substantial data compression achieved in comparison with transmission of a full (or even compressed) image will be of substantial benefit to remote plant monitoring using limited bandwidth and display client equipment (eg. mobile internet telephones). The binary categorical nature of the recovered information makes it readily amenable to further processing by decision-based induction methods.

Input Data

A colour CCD camera was positioned to cover the existing indicator panel used by the operators to monitor the performance of the CHP plant. The purpose of this was to exploit data already being collected by existing instrumentation without the need to make intrusive connections to the existing systems.

A key contributor to information about the plant's behaviour is the bank of indicator lights associated with the turbine generator (Fig. 3). This divided into three groups of lights, each group being of a different colour and indicating a different level of severity of problem with the generator's functioning. White lights give status indications that are part of the normal correct functioning of the generator, yellow ones warnings of potential failures and red ones failures that have actually occurred (Fig. 4). It was considered that by the time a red indicator light was on a problem had progressed too far to take effective remedial action and so only the white and yellow lights will be considered further.

Figure 3: Image of indicator panel as seen in colour by the camera. The interference lines are caused by the electromagnetically noisy environment in the CHP pump room.

	21	0	2	4	6	8	10	12	14	16	18	42	46
20	22	1	3	5	7	9	11	13	15	17	19	43	47
	23	25	27	28	30	31	33	35	36	38	40	44	48
	24	26		29		32	34		37	39	41	45	49

Figure 4: Arrangement of status (white), warning (yellow) and failure (red) lamps:
□ - white lamp, ▨ - yellow lamp, ▤ - red lamp, ■ - no lamp

The camera, video capture card and programming interface were as used in the colour comparison example described above, with the exception that the video signal was boosted by an RF line driver to allow it to be transmitted over a longer distance. The monitoring PC could thus be located well away from the relatively hostile environment of the pump room where the control panel is situated. The computer also controlled a system of RS-232 data acquisition and control modules to allow switching between this camera and cameras monitoring other equipment, and adjustment of camera settings (eg. shutter speed, white balance etc.). This will not be described further here as we are concerned only with the data from one particular camera, the settings of which remained fixed. Images were acquired approximately every thirty seconds.

Clustering

There was some experimentation with the dimensions of the virtual sensors placed over each panel light in the image. A substantial sensing area for each would allow random noise effects to be averaged out, the image quality being particularly poor owing to electromagnetic interference in the camera's operating environment. However, the small size of the panel lights in the image, resulting from constraints on the choice of camera location, meant that the size of sensors was limited. This limitation was compounded by the significant bloom generated in the CCD array. Time series plots of the sensor outputs, both in RGB and HLS forms showed that there was no sensor size that could allow lights in the on and off states to be consistently distinguished. Furthermore, the parameters exhibited substantial variation with the ambient lighting, greater than would allow distinction on the basis of a fixed threshold, even using a combination of parameters.

It was noticed that the problem could readily be cast as a cluster analysis task, each sampled frame having multiple observations (one for each of 50 lights) in multiple variables (the colour parameters) and requiring those observations to be placed in one of two categories (on or off). Given that the number of categories required was fixed, k-means clustering (see eg. Spath, 1980) was deemed an appropriate method. Clustering, of course, leaves the problem of labelling the categories, but it was noticed that this could readily be solved by applying a simple piece of domain knowledge, namely that there would be a greater number of lights that were off than were on.

Clustering was performed in the RGB colour space, this having the advantage over HLS that the angular and periodic nature of the hue parameter would not give rise to scaling difficulties. Two sets of weightings were applied in the software to the R, G and B values, one set for the white lights and one for the yellow. This allowed for compensation for the difference in colours between the lights (otherwise the main distinction between them would have been their colour, rather than whether they were on or off, and the clustering algorithm would have just separated white lights from yellow). Acceptable values for these were found by trial and error.

As regards the dimensions of the virtual sensors, a horizontal row of three pixels for each was found to work well. The clustering algorithm proved more robust to the limited suppression of random noise resulting from averaging over such a small area than to the greater contamination by bloom produced by larger sensors.

Applications of Clustering Results

As well as providing a real time display of the results on the PC hosting the monitoring, the software supplies an output to the web-based user interface described below.

To provide some indication of the success of the method, it was applied to process a number of recorded images of the indicator panel and the results compared with those of visual inspection of the images. These were taken at the rate of one per hour for a 24 hour period. The images selected were taken on a sunny day. This subjects the processing to its maximum stress as during the afternoon strong sunlight illuminates the monitored area, reducing the contrast between on and off lights. Some effort had been made to mitigate this by the erection of a shade over the panel, similar to that frequently seen over LED displays, however any more extensive measures to deal with this were dismissed as potentially too intrusive with respect to the work of plant operators in the area. Fig. 5 shows the results of the test.

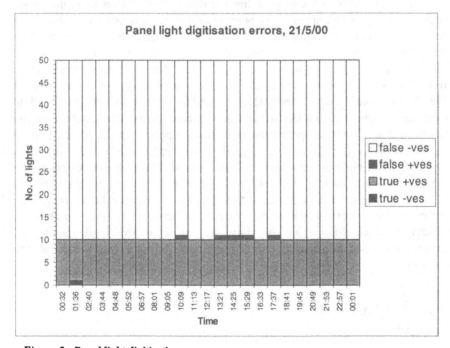

Figure 5: Panel light digitisation errors

The vertical blocks represent numbers of lights and each vertical column represents the information extracted from one image. Throughout the day, ten of the fifty lights were illuminated, although the configuration of illuminated lights was not constant. True positives are the number of correctly reported on lights. There is only one false positive (an on light reported as off) and this persists only for a single frame. Of the off lights, only one is misreported. This is intermittent, but chiefly during the afternoon, when the sunlight was causing the most stress to the processing. Furthermore, this "lamp" was actually a dummy that was included in the processing for historical reasons. Unlike all the others, it had no writing on its cover and this is believed to have produced a tendency for it to be distinct from the other lamps, and hence occasionally misreported. Even continuing to include the results for this lamp in the assessment, we have

Maximum instantaneous true negatives	10%
Maximum instantaneous false positives	2.5%
Maximum instantaneous total errors	2%
Mean level of true negatives	0.43%
Mean level of false positives	0.54%
Mean total errors	0.52%

To demonstrate the utility of the recovered information in higher level processing, some examples of its output were used to construct a tutorial set for a simple ID3-style (Quinlan, 1986) decision tree induction algorithm. The recovered binary state of each individual lamp was presented as a separate attribute and a set of values for these obtained from a single image frame treated as a single observation. Categorical natural language descriptions of the CHP turbine's state of operation provided by domain experts (ie. the plant operators) were used as the dependent variable in the construction of the decision tree. The examples used were from a period during which normal operation of the turbine was temporarily suspended to allow it to perform a scheduled self-cleaning operation. This ensured that a number of operating states and, hence, illumination configurations of the indicator panel occurred.

The decision tree produced, using Quinlan's simple relative frequency estimate for prior probabilities ($p_i = n_i/n$, where n_i is the number of observations of class i in the data set and n is the total number of observations) and information gain proportionate to the reduction in Shannon entropy,

$$S = -\sum_{i=1}^{N} p_i \log_2 p_i$$

to select decision variables, is shown in Fig. 6. Several alternative decision variables could equally well have been selected at many of the nodes, having equal information gain. In these circumstances the algorithm simply defaulted to the rightmost variable in its attribute-value table, the order of the variables corresponding to the clustering-induced states of the panel lights as indicated in Fig. 5.

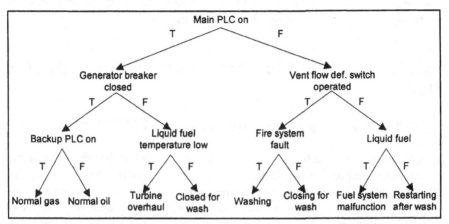

Figure 6: Decision tree formed on panel light tutorial set

The decision variable selected for the root node was the condition of the "enabled" indicator lamp for the main programmable logic controller for the system. While it is not entirely clear why this should be on when the turbine is closed down in preparation for washing or being overhauled, it does seem reasonable that a high contribution to information gain should be associated with the status of this controller. Certainly, we expect it to be enabled when the turbine is running, as the decision tree confirms.

Consideration of just one further variable is then required to isolate both of the running states of the turbine from all others, namely the status of the circuit breaker associated with the generator, which, as we would expect, is closed when the generator is running.

A prolonged shutdown for overhaul clearly means that the turbine is not being maintained in a state of readiness for operation and, hence, there is no good reason to maintain the fuel at a temperature at which it is readily combustible. The decision variable distinguishing closure for washing from closure for overhaul appears to reflect this.

Considering the other main branch of the tree, it is unclear why the ventilation flow deflector of the fire system should play the roles that they do, although it is equally possible to conceive of reasons why they might. Clearly, there is an association between the type of fuel in use (liquid or gas) and a malfunction of the fuel system, as indicated by the decision variable at the final branching node of the tree.

This tree covers only a small part of the space of operating conditions and status indications of the CHP turbine and generator, and was constructed purely to illustrate the applicability of earlier results. Nonetheless, we can clearly see that the values recovered by clustering in colour space are sufficient to begin to allow appropriate decision variables to be selected. It is envisaged that results of clustering in colour space will eventually be tested in real time against trees produced in this way to give summary statements, both diagnostic and prognostic, about the system condition.

PROGNOSTIC TREND MONITORING IN COLOUR SPACE

Acoustic and Temperature Monitoring of CHP Plant

In addition to having information as to the current state of health of the plant, the operators expressed the desire to know prognostically when there should be intervention in a preventative manner, for example when a motor bearing should be changed – too early and some of the bearing's value is wasted, too late and a pump goes out of service, degrading the system's ability to respond to demand. Thermochromic patches were applied to the hot water pumps, indicating pump temperature changes through colour changes. In addition to camera-based monitoring, further non-imaging sensors were applied. The hot water pumps were monitored with optical fibre-based acoustic sensors and with thermocouples, and an air compressor and a gas compressor were monitored with fibre-based acoustic sensors [Russell *et al* (1999)].

The acoustic data is collected by means of optical fibres along which laser signals are propagated. Acoustic compression waves alter the refractive index of the fibre and modulate the interference patterns emerging from the fibre. The patterns are collected from the fibre and converted to voltage signals, which are filtered in the acoustic frequency domain by the application of chromatic filters into notional red, green and blue bands. Each band has a standard deviation of 0.5KHz and centred on 1 KHz, 2 KHz and 3 KHz frequencies, respectively. In this instance, the filters are implemented in hardware and their outputs are converted into the colour space parameters of hue, lightness and saturation in software. One optical fibre is attached to both main pumps ('daisy-chained'). A second fibre is loosely attached to a framework near to the air compressor so that more general background signals can be obtained.

The progressive deterioration of a pump motor bearing was acoustically tracked over a period of several weeks [Russell *et al* (2000)] as illustrated in Fig. 7. Here, for convenient two-dimensional visualisation, lightness is shown radially on a second polar plot.

Over time the bearing noise became 'peakier' as seen in the increase in signal purity (saturation in Fig. 7 (a)) and strength (lightness in Fig. 7 (b)) until the bearing failed. The plots can be used as early warning indicators of deteriorating bearing condition.

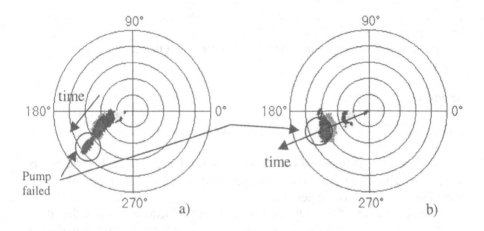

Figure 7: Chromatic plots of the acoustic signals from a failing pump. (a) Hue-Saturation. (b) Hue-Lightness.

Four Stimulus Processing

To allow for the visual fusion of temperature information from four pumps in a hue-saturation plot, an extension of the technique for the transformation of tristimulus signals to process signals from four sources was required.

Derivation of the hue calculation follows the reasoning of Levkowitz and Herman (1993). The hue circle is divided into eight equal sectors of 45° starting at hue angles 0-45° as sector 0, then numbering in an anti-clockwise direction so that hue angles 315-360° are in sector 7. When a particular stimulus is dominant, the hue angle will be located in one of two sectors (which one of the two sectors depends on the values of the other stimuli relative to the dominant stimulus), Table 1.

Table 1: Primary sector selection

Dominant stimulus	Hue located in sector
1	0 or 7
2	1 or 2
3	3 or 4
4	5 or 6

The final hue value, in degrees, is calculated as:

$$H = 45 \times (\text{sector}(s_1, s_2, s_3, s_4) + \text{adjustment}(s_1, s_2, s_3, s_4)) \qquad (7)$$

where s_i are the values returned by each of the four sensors. The sector function identifies the appropriate sector and the adjustment function gives the location within the sector.

For example, if the four stimulus values are such that $s_1 > s_2$, $s_2 >= s_3$ and $s_3 > s_4$:

$\text{sector} = 0$

$\text{adjustment} = (s_2 - s_4)/(\max(s_1, s_2, s_3, s_4) - \min(s_1, s_2, s_3, s_4))$

For instance, if the stimulus values s_1-s_4 are 0.6, 0.4, 0.25, 0.1 the hue is $45.0° \times 0.6$ giving $27.0°$. The full specification of the four stimulus hue function is given in the Appendix.

Lightness (eqn. (5)) for the four stimulus transformation is just the mean of all four values, and saturation is calculated as before (eqn.(6)), considering all four values in the determination of the maximum and minimum.

HUMAN-COMPUTER INTERFACES

Operator Requirements

One of the broad aims for data mining is the integration of decision-making capabilities with intelligence gathering. In the context of an industrial process, this aspiration might ultimately take the form of autonomous, self-regulating machines. Apart from considerations as to how far this level of automation is realisable, there are many reasons why it might not be generally acceptable and why people are involved in the process control 'loop'. Nevertheless, the level of automation that is achievable is helping to free the process operator from the role of passively waiting at the plant site to react to events. For example, the CHP system uses a customised hard-wired system that enables the plant to be remotely started and stopped, alarms to be directed to the operator's home or mobile telephone, and a log of events to be accessed via a computer. Hence, it is possible for the plant to run without being continuously manned. The integration of visual reporting tools and data mining techniques has the potential to further enhance decision support for the remote operator.

Data mining can be used to discover patterns of behaviour in data [Fayyad (1997)] and one can envisage this being used to extend the time horizon of the operator from monitoring the immediate state of the plant process to looking ahead to the time when an operator intervention will be required. At the same time, intelligent information compression can focus the multiple sensor data up through, for example, decision tree hierarchies to the level of a yes/no intervention decision, reducing complexity for the operator.

Thus the requirements of any operator-centred, semi-autonomous system will be indications as to when (1) routine maintenance is due; (2) the current condition warrants intervention earlier than planned maintenance is due; (3) immediate intervention is required. In connection with interventions in general and requirement (1) in particular, it is to be noted that maintenance itself can have unforeseen side-effects (e.g. [Atlas and Jones (1996)]).

Development of Solutions

In summary, operator requirements demanded consideration of timescale and timeliness, levels of accuracy and detail provided/available, accessibility and distribution, and the forms of information presentation and visualisation. These requirements inevitably involve some tradeoffs. In terms of information presentation and visualisation, it was considered that a variety of media would be appropriate – images and colour codes, digital representations, textual data, graphs and plots. Iterative development was driven by consultation with the plant operators, the richest and most accessible source of domain expertise in this area. System integrity was enhanced, wherever possible, by the use of integrating sensors overlapping in their distribution variables. Data was then converted to low level chromatic information at the earliest possible stage, providing partial redundancy and, hence, alleviating the problem of sensor or monitoring system malfunctions. This helps to maintain operator confidence, which was further supported by the redundancy achieved through supply of picture images of the plant from the cameras.

Digital representation of the control panel lights, achieved through clustering in colour space, described above, gives a clear identification of which lights are on/off. This information is not always evident from the image alone. For example at night the turbine hall is not lit and, although the overall pattern of lights can be seen against a black background, the mapping of this visual light pattern to the actual lights on the panel is not straightforward as both the frame and the other unlit lights are not visible. Another advantage of the digital representation is that it enables a digital replay of light patterns to be presented to the operator. Discussions with CHP system operators indicated that a *visual* replay of the light patterns is more assimilable for diagnostic purposes than a textual list of alerts and alarms in the traditional log. Furthermore, from the image it is possible to visually monitor the behaviour of key analogue indicators when initiating a plant re-start remotely. The operators' requirements for visual and visually-derived information coincided well with the overall approach of cost-effective, non-intrusive, camera-based monitoring. It should be emphasised that the image of the control panel was supplied at the operators' request, but this is not necessary to allow mining of information from the image and its transmission to the operator. Reconstruction of a visual representation at the operator's location can optionally provide for familiar cues in the presentation.

A number of prototype user interfaces were developed for the CHP application. They are based on world wide web technology using web pages written in Html and JavaScript, with ActiveX components and Java applets. This technology has the advantages of being standard and accessible 'world-wide' through internet technology. Some prototypes include ActiveX controls, linked to ActiveX servers, for switching between three cameras, one on the control panel, one on the heating pumps and a black/white/infrared camera in the turbine room. Another control enables a light in the turbine hall to be switched on and off. Access to these controls is password restricted to authorised personnel. They may be activated via a dial-in connection or via the Internet. Fig. 8 depicts the range of control (client) - server pairs deployed in the second generation of prototype interfaces.

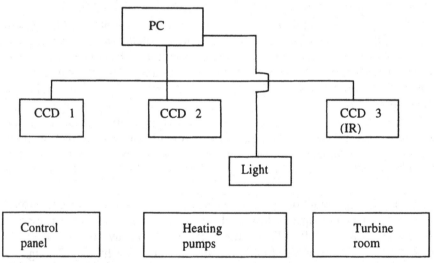

Figure 8: From the PC, ActiveX controls enable three cameras in the CHP to be alternated and a light in the Pump hall to be operated.

Another ActiveX control was implemented to provide a visual alarm for pump motor temperature, indicating green when the temperature is in range and red when out of range. The control is illustrated in Fig. 9. The temperature range limits are set in the html page. The approach using controls and indicators may be likened to providing a virtual dashboard.

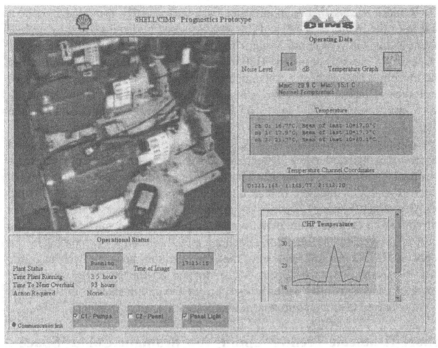

Figure 9: The camera has been switched to the pump hall, the light is on, and pump temperature is within the normal range (as specified in the web page source html script)

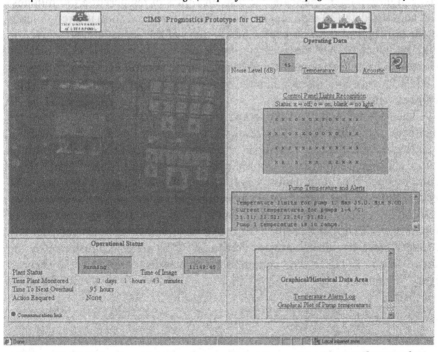

Figure 10: Web-based user interface for monitoring the combined heat and power plant.

Remote access to the monitoring data was initially provided by direct dial-in to the monitoring PC and this was accessed by landline and by mobile phone using GSM (Global System for Mobile communication), which was successfully tested between Holland and the UK. Subsequently, the monitoring data has been provided by a web server. The most recent user prototype is publicly available, using MS Internet Explorer 4 and above, at www.cims.org.uk/chp2. Fig. 10 illustrates a typical operational status of the CHP plant.

The main camera image in Fig. 10 shows the control panel with dials and lamps. It is updated every 30 seconds and is in Jpeg format to provide for a practical level of compression. The latest temperature data for the pumps, obtained every 5 minutes, is also displayed, together with the status for the primary pump (pump 1). At the suggestion of the operators, a procedure was introduced to raise an alert if the pump temperature goes out of range. The alert is relayed to the operator telephonically. An entry is recorded in the legacy CHP system log requesting that the operator check the web page for further information. The event is also logged on the web site so that the operator can inspect the permanent record of the circumstances in which the event occurred (Temperature Alarm Log).

Short Timescale Trend Mapping in Chromatic Colour Space

On a slightly longer timescale, the last ten values of temperature (50 minutes) and acoustic data (30 minutes) are available in graphical format through icons at the top right of the web page. An example of the acoustic data is given in Fig. 11 which shows the acoustic data from a sensor arranged to pick up vibrations from both pumps 1 and 2, and for the general background noise in the CHP plant room (predominantly from an air compressor) in terms of hue, lightness and saturation.

The vertical scales on the graph represent 0-360° hue, and 0.0-1.0 for lightness and saturation. The results show a fairly constant Hue value of ~200°, corresponding to a dominant acoustic frequency of ~1.75 KHz, and a fairly constant saturation of approximately 0.9, indicating the signal to be quite monochromatic in nature, covering a relatively narrow waveband. The graphs indicate that one of the main pumps is running (lightness>0) and also that they are the main contributory factor in the general background level of sound in the CHP plant room. The background signal becomes considerably stronger when the air compressor comes on. This effect is illustrated by the dotted line in Fig. 11. The compressor operates for a period of 50 seconds every 50 minutes.

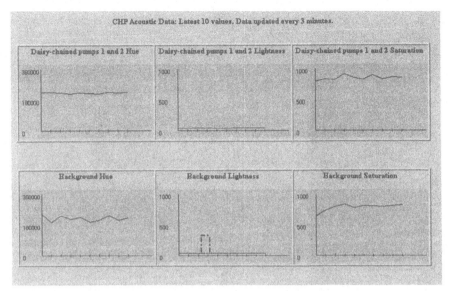

CHP Acoustic Data: Latest 10 values. Data updated every 3 minutes.

Figure 11: Acoustic data for the main pumps and background noise from an air compressor. The dotted line in background lightness illustrates the compressor switching on and then off.

Long Timescale Trend Mapping in Chromatic Colour Space

An alternative approach to displaying the status of the system is to show the Hue-Saturation values on a chromatic map. This enables trends over longer timescales to be conveniently displayed. Fig. 7 showed an example of this for tristimulus signals. An example using the four stimulus processing described is shown in Fig. 12 for the temperatures of the four CHP pumps. In this case the filtering stage is not applicable and each pump's temperature is taken as a direct input to a four-colour space. They are then compressed by the four stimulus transformation to a single point in the HS plane (lightness is disregarded to provide for common noise rejection, as described for tristimulus chromatic processing). In the absence of a signal from the three other pumps the active pump is apparent from a marker displayed in one of four axis directions. Thus the location of a system status point at a given angular position is indicative of the relative influence of each pump output i.e. showing which pump has a dominating temperature effect. The extent to which the pump(s) dominate is given by the distance of the point from the origin. It should be noted that this chromatic map is based on a set of Cartesian coordinates, $ScosH$ vs. $SsinH$, rather than simply being an HS polar plot. This allows contributions from the pumps to be more readily re-scaled relative to one another, distorting the usual HS circle into elliptical or oval shapes.

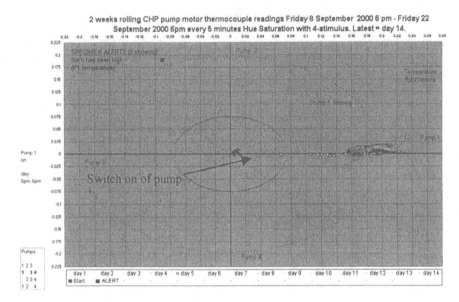

Figure 12: Cartesian Hue-Saturation plot for the CHP pumps showing pump 1 switch-on

Fig. 12 presents temperature patterns for the fortnight after a Summer break, during which main pump number 1 was switched on. The Hue-Saturation values start near the origin (all pumps off) then over a period of a few hours the combined temperature can be seen to move in the direction of the dominant pump, here pump 1. Data from the pumps were collected over a year, so that their behaviour in Hue-Saturation terms, as in Fig. 12, could be seen in a long-term context. The inner and outer ellipses bound the region in which indicators for the active pumps normally fall. Here, only pump 1 is running. Other trajectories involving pump switchovers were observed as illustrated in Fig. 13.

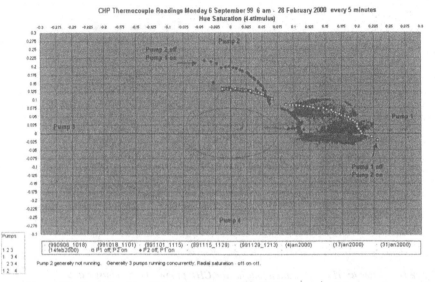

Figure 13: Cartesian Hue-Saturation plot for the CHP pumps: switch-overs from pump 1 to 2 and back .

Fig. 13 indicates the situation where pump 1 has been switched off and pump 2 switched on. The trace moves from pump 1 quadrant to pump 2 quadrant, starting at '3 o'clock' and following the white markers to '12 o'clock'. After several hours, pump 2 was switched off and pump 1 was switched back on again. The trace moves back from quadrant 2 to quadrant 1, following the black markers back to '3 o'clock'. In each case, the saturation increases when the pump is switched off because of the residual heat in the motor and the loss of cooling by the motor's fan blades.

A further advantage of expressing sensor data in compressed colour terms is that in addition to fusion of distributed sensor data (eg pump temperatures), it enables data from different sources to be fused and plotted on Hue-Saturation and Hue-Lightness plots whilst maintaining access to the raw data if required [Russell *et al.* (1999)]. As a preliminary step, acoustic data and temperature data have been initially plotted on the same Hue-Saturation plot to establish behaviour patterns as depicted in Fig. 14.

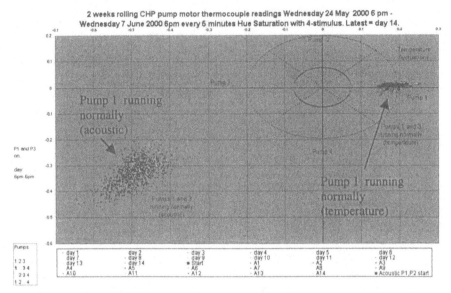

2 weeks rolling CHP pump motor thermocouple readings Wednesday 24 May 2000 6 pm -
Wednesday 7 June 2000 6pm every 5 minutes Hue Saturation with 4-stimulus. Latest = day 14.

Figure 14: Cartesian Hue-Saturation plot for the CHP pumps: both acoustic and temperature data are prior to fusion.

The longest timescale of data availability is represented by the log of temperature alerts that is maintained on the web site (Temperature Alarm Log). As noted, it records, by way of example, all events with main pump 1 and it is a permanent record of exceptional events.

SUMMARY and CONCLUSIONS

The fusion of distributed measurements by the application of weighted integrals into parameter values in low dimensional colour spaces and the transformation of these spaces according to the established methods of colour science forms the basis of the methodology of chromatic processing. Analysis in terms of signal processing theory, in particular where the weighted integrals form the basis functions of a truncated Gabor expansion, provides an interpretation of the information preservation characteristics of the lossy compression of the dimension of a problem. Conventional colour space transformations have been shown to be amenable to extension to deal with cases having a natural low dimensionality that differs from the three dimensions of classical chromatic processing.

   A number of examples have been described that show how the integration of this methodology with the techniques of data mining, particularly statistical and classification tools, can contribute to industrial process and quality control. Additionally, the use of colour spaces in the low dimensional presentation of information contributes new visualisation tools to industrial monitoring and to data mining.

As regards product quality, the amalgamation of chromatic and statistical techniques has placed a previously subjective monitoring regime on a quantitative footing oriented towards the in-service criteria that influence customer satisfaction with the product. Furthermore, it has allowed this to be done at low cost and using mass-market hardware.

Chromatic and data mining techniques have also been brought together to provide for the monitoring of several aspects of plant performance and their operator-centred visual fusion. Video cameras are deployed to allow for exploitation of new and existing monitoring systems with minimal disturbance to the existing systems. Diverse monitoring information can be mined from each camera to serve multiple purposes. The resulting information extends and reinforces existing information and can be combined with other new sources of information that extend monitoring to equipment that has not previously been monitored out of consideration for cost, such as the pump motors and compressors.

Standard communications technologies allow for access to a visual reporting interface through a landline, GSM mobile telephone or the internet. This can be to multiple remote operators in different locations if required. Remote interaction is also possible such that aspects of the monitoring (eg. cameras, lights) can be controlled. The existing facility for remote plant operation has been enhanced by the availability of images and monitoring data that assist the operator in diagnostics when alerts occur and in decision-making – principally whether the plant can safely be remotely re-started or whether a visit on-site is necessary.

The operator's reach is also extended in the time domain. An information picture is constructed. Supplementing the status information from the control panel lights are graphs that cover periods of several minutes of equipment operation, as with acoustic signatures from the pumps. An early warning of abnormal trends may be obtained from these graphs. As an alternative, tolerances are also placed on equipment performance, such as with the pump temperature range, so that alarms are raised whenever actual values go out of range. Over longer timescales, the colour space compression of sensor data in terms of hue, saturation and lightness, provides assistance with tracking of equipment integrity, as with the pump bearings, and plant response patterns, as with the pump temperatures, supporting prognosis and planning. The final visual fusion of monitoring information in the reporting interface provides a holistic picture of plant health, with facilities to drill down to more detailed focus as required.

ACKNOWLEDGEMENTS

The authors gratefully acknowledge the contributions of Stoves Ltd., the University of Liverpool Energy Company Ltd., Shell UK plc and the support of ERDF funds from the European Union.

APPENDIX - FOUR STIMULUS HUE CALCULATION

The full specification of the sector function (eqn. (7)) depends on the relative sizes of all four stimuli, Table A1. The adjustment functions are given in Table A2.

Table A 1: Full specification of sector selection function

Condition	Sector
$\max(s_1, s_2, s_3, s_4)=s_1 \wedge (\max(s_2, s_3, s_4)=s_2 \vee (\max(s_2, s_3, s_4)=s_3 \wedge \max(s_2, s_4)=s_2))$	0
$\max(s_1, s_2, s_3, s_4)=s_2 \wedge (\max(s_1, s_3, s_4)=s_1 \vee (\max(s_1, s_3, s_4)=s_4 \wedge \max(s_1, s_3)=s_1))$	1
$\max(s_1, s_2, s_3, s_4)=s_2 \wedge (\max(s_1, s_3, s_4)=s_3 \vee (\max(s_1, s_3, s_4)=s_4 \wedge \max(s_1, s_3)=s_3))$	2
$\max(s_1, s_2, s_3, s_4)=s_3 \wedge (\max(s_1, s_2, s_4)=s_2 \vee (\max(s_1, s_2, s_4)=s_1 \wedge \max(s_2, s_4)=s_2))$	3
$\max(s_1, s_2, s_3, s_4)=s_3 \wedge (\max(s_1, s_2, s_4)=s_4 \vee (\max(s_1, s_2, s_4)=s_1 \wedge \max(s_2, s_4)=s_4))$	4
$\max(s_1, s_2, s_3, s_4)=s_4 \wedge (\max(s_1, s_2, s_3)=s_3 \vee (\max(s_1, s_2, s_3)=s_2 \wedge \max(s_1, s_3)=s_3))$	5
$\max(s_1, s_2, s_3, s_4)=s_4 \wedge (\max(s_1, s_2, s_3)=s_1 \vee (\max(s_1, s_2, s_3)=s_2 \wedge \max(s_1, s_3)=s_1))$	6
$\max(s_1, s_2, s_3, s_4)=s_1 \wedge (\max(s_2, s_3, s_4)=s_4 \vee (\max(s_2, s_3, s_4)=s_3 \wedge \max(s_2, s_4)=s_4))$	7

Table A 2: Full specification of adjustment function

Sector	Adjustment function
0	$(s_2 - s_4)/(\max(s_1, s_2, s_3, s_4) - \min(s_1, s_2, s_3, s_4))$
1	$(1-(s_1 - s_3))/(\max(s_1, s_2, s_3, s_4) - \min(s_1, s_2, s_3, s_4))$
2	$(s_3 - s_1)/(\max(s_1, s_2, s_3, s_4) - \min(s_1, s_2, s_3, s_4))$
3	$(1-(s_2 - s_4))/(\max(s_1, s_2, s_3, s_4) - \min(s_1, s_2, s_3, s_4))$
4	$(s_4 - s_2)/(\max(s_1, s_2, s_3, s_4) - \min(s_1, s_2, s_3, s_4))$
5	$(1-(s_3 - s_1))/(\max(s_1, s_2, s_3, s_4) - \min(s_1, s_2, s_3, s_4))$
6	$(s_1 - s_3)/(\max(s_1, s_2, s_3, s_4) - \min(s_1, s_2, s_3, s_4))$
7	$(1-(s_4 - s_2))/(\max(s_1, s_2, s_3, s_4) - \min(s_1, s_2, s_3, s_4))$

REFERENCES

Atlas, L. and Jones, D., "National Science Foundation Workshop on Signal Processing for Manufacturing and Machine Monitoring", Draft 1.2 (11/15/96) of Final Report to the National Science Foundation, Alexandria, VA, March 14-15, 1996 http://isdl.ee.washington.edu/wkshp/nsf/manmon.html

Brazier, K.J., Russell, P.C., Jones, G.R. and Shankland, I., "A statistical approach to economical camera-based colorimetric comparison", Sensor Review, 20(1), 43-49, 2000

Connor, F.R., Signals, Edward Arnold (Publishers) Ltd. (2nd ed.), 1982

Cosgrave, J.A., Russell, P.C., Hall, B.S.B. and Jones, G.R., "Optoacoustic monitoring of electric arcs in high voltage circuit breakers", Int. Conf. on Switching Arcs, Xian Jiaotong University, People's Republic of China, pp. 598-605, 1997

Fayyad, U.M., "Editorial", in Data Mining and Knowledge Discovery, pp. 5–10, Kluwer, 1997

Gabor, D., "Theory of Communication", J. IEE, 93(3), 429-457, 1946

Gasquet, C. and Witomski, P., Fourier analysis and applications, Springer-Verlag Inc., 1999

Jones, G.R, "Plasma monitoring using chromatically processed optical signals", Proc. XX1 Int. Conf. on Phenomena in Ionised Gases Proc. III (Invited Lectures) (Bochum), pp. 24-33, 1993

Jones, G.R., Russell, P.C. and Khandaker, I., "Chromatic interferometry for an intelligent plasma processing system", Meas. Sci. Technol., 5, 639-647, 1994

Jones, G.R., "Electric arc monitoring using intelligent optical fibre systems", Proc. XI Int. Conf. on Gas Discharges and their Applications (Tokyo), vol. 2, pp. 504-512, 1995

Jones, G.R., Russell, P.C., Vourdas, A., Cosgrave, J., Stergioulas, L.K. and Haber, R., "The Gabor transform basis of chromatic monitoring", Meas. Sci. Technol., 11, 1-8, 2000

Levkowitz, H. and Herman, G.T., "GLHS: A generalized lightness, hue, and saturation color model" CVGIP: Graphical Models and Image Processing, 55(4), 271-285, 1993

Porat, M. and Zeevi, Y.Y., "The generalized Gabor scheme of image representation in biological and machine vision", IEEE Trans. Pattern Analysis and Machine Intelligence, 10(4), 452-468, 1998

Quinlan, J.R., "Induction of decision trees", Machine Learning, 1, 81-106, 1986

Russell, P.C., Spencer, J.W. and Jones, G.R., "Optical fibre sensing for intelligent monitoring using chromatic methodologies", Sensor Review, 18(1), 44-48, 1998

Russell, P.C., Craven, A., Deakin, A., Cooke, R., Furlong, S., Spencer, J.W. and Jones, G.R., "Intelligent chromatic monitoring of industrial plant", Sensors and their Applications X , Eds. N M White and J T Augousti, Proceedings of the Tenth Conference on Sensors and their Applications, Cardiff, Institute of Physics Publishing, Bristol, pp. 245-250, 5-8 September 1999

Russell, P.C., Brazier, K.J., Deakin, A., Cooke, R. and Jones, G.R., "Holistic Monitoring of Power Plant", Intelligent Systems and Applications (ISA 2000), Wollongong, Australia, 11-15 Dec. 2000

Spath, H., Cluster analysis algorithms, Ellis Horwood Ltd., Chichester, UK, 1980

Stergioulas, L.K., "Time-frequency methods in optical signal processing", PhD Thesis, University of Liverpool, UK, 1997

Stergioulas, L.K., Vourdas, A. and Jones, G.R., "Gabor representation of a signal using a truncated von Neumann lattice and its practical implementation", Opt. Engng., 39(7), 1965-1971, 2000

Yokomizu, Y., Spencer, J.W. and Jones, G.R., "Position location of a filamentary arc using a tristimulus chromatic technique", J. Phys. D: Appl. Phys., 31(23), 3373-3382, 1998

Yu, R.J., Lisboa, P.J.G., Russell, P.C. and Jones, G.R., "Resolution capabilities of chromatic sensing in the monitoring of semiconductor plasma processing systems", Nondestr. Test. Eval., 13, 347-360, 1997

CHAPTER 17

Non-Traditional Applications of Data Mining

Andrew Kusiak
andrew-kusiak@uiowa.edu
Intelligent Systems Laboratory, Department of Industrial Engineering,
4312 Seamans Center, The University of Iowa, Iowa City, IA 52242 – 1527

ABSTRACT

Machine learning offers algorithms for extraction of knowledge in an understandable form based on historical data. It is viewed as a key tool in development of autonomous systems. This chapter shows that learning algorithms can be used for novel problem solving in engineering design and manufacturing. Decomposition is a key problem in the latter two application areas. A data mining approach is used for matrix decomposition for the case with unknown and known decisions associated with each object in the matrix. Of particular interest to control of manufacturing processes is the case when decisions, e.g., product quality, are ill-defined. Data mining is a viable tool for solving problems with ill-defined outcomes.

D. Braha (ed.), Data Mining for Design and Manufacturing, 401–416.

INTRODUCTION

Data mining techniques and tools explore patterns and associations in databases. As a new approach, it awaits applications in diverse areas. This chapter discusses a new class of potential applications of data mining that have not been discussed in the literature, such as:

❑ Matrix decomposition with learning algorithms; Machine learning algorithms produce interesting matrix structures.
❑ Learning from data sets with ill-defined decisions; Decisions assigned to objects in a training data set may be ill-defined, e.g., the object may be classified as category B rather than C for the set {A, B, C} of decision values.
❑ Using learning from data sets with ill-defined outcomes to enhance decision-making accuracy; Approximate rules identify objects of a training set with conflicting outcomes which are assigned new decision values thus improving classification accuracy of decion rules.

Learning (classification) systems fall into five general categories:

A. Classical statistical methods (e.g., linear, quadratic, and logistic discriminant analyses) (see Michie *et al.* 1994).
B. Modern statistical techniques (e.g., projection pursuit classification, density estimation, k-nearest neighbor, casual networks, Bayes theorem [Domingos and Pazzani 1996]).
C. Neural networks (e.g., backpropagation, Kohonen, linear vector quantifiers, and radial function networks) [see Michie *et al.* 1994]).
D. Decision tree algorithms (e.g., ID3 [Quinlan 1986], CN2 [Clark 1989], C4.5 [Quinlan 1993], T2 [Auer *et al.* 1995], Lazy decision trees [Friedman *et al.* 1996], OODG [Kohavi 1995], OC1 [Aha 1992], AC, BayTree, CAL5, CART, ID5R, IDL, TDIDT, and PROSM [all discussed in Michie *et al.* 1994]).
E. Decision rule algorithms (e.g., AQ15 [Michalski *et al.* 1986], LERS [Grzymala-Busse 1997, and numerous other algorithms based on the rough set theory [Pawlak 1991]).

In this chapter the algorithms of the last category are used for rule extraction from data sets.

MATRIX DECOMPOSITION

Learning algorithms can be used to explore clusters of objects and features in data sets. This task has been traditionally performed with cluster analysis. Clustering algorithms tend to use all features while creating clusters, e.g., based on distances between objects (Kusiak 2000). Machine learning algorithms usually use a few features at a time.

Clustering algorithms support unsupervised learning, which implies that the objects in a data set are not assigned decision values. On the other hand, machine learning algorithms follow the principle of supervised learning and therefore decision values have to be specified.

In this chapter two ways of dealing with decision values in matrices are considered:

❑ Dummy object approach; All objects in a matrix (data set) are assigned identical decision values and a dummy object (row) is added to the matrix with a different value of the decision. The feature values of the dummy object should differentiate it from all other objects in the matrix.
❑ Filling undesirable entry approach; To increase discernity among objects of a sparse data set, random values are assigned to the blank entries.

The following types of decomposition are considered:

❑ Mutually separable matrix
❑ Non-decomposable matrix
❑ Matrix with conflicting outcomes

Mutually Separable Matrix

Consider the data in Table 1, which could represent a machine–part incidence matrix in the group technology problem. Matrix decomposition has applications beyond group technology and therefore the rows of the matrix in Table 1 are called objects and the columns are referred to as features. The entry '1' in the object-feature matrix indicates that the feature is present and 'blank' denotes absence of the corresponding feature.

The matrix in Table 1 is sparse and blank entries have to be addressed before a learning algorithm is used. Data mining offers numerous algorithms for filling missing data. In this case it was assumed that a blank implies that a corresponding feature is not present rather than it is missing.

Table 1. Object-feature incidence matrix

	F1	F2	F3	F4	F5	F6
1		1		1		
2			1			
3	1				1	
4			1			1
5		1		1		
6	1				1	
7					1	

As it is done in the group technology matrix, the blanks in Table 1 are replaced with zeros. In addition, dummy object 8 with decision value Two is introduced as shown in Table 2.

Table 2. Matrix with dummy object 8

	F1	F2	F3	F4	F5	F6	D
1	0	1	0	1	0	0	One
2	0	0	1	0	0	0	One
3	1	0	0	0	1	0	One
4	0	0	1	0	0	1	One
5	0	1	0	1	0	0	One
6	1	0	0	0	1	0	One
7	0	0	0	0	1	0	One
8	1	1	1	1	1	1	Two

A machine learning algorithm applied to the data set in Table 2 has produced the decision rules in Figure 1.

```
Rule 1. (F6 = 0) THEN (D = One); [6, 85.71%,
100.00%][6, 0][1, 2, 3, 5, 6, 7]
Rule 2. (F5 = 0) THEN (D = One); [4, 57.14%,
100.00%][1, 2, 4, 5]
Rule 3. (F1 = 1) AND (F6 = 1) THEN (D = Two); [1,
100.00%, 100.00%][8]
```

Figure 1. Decision rules extracted from the data set in Table 2

These decision rules in Figure 1 are presented in the following format:
(Condition) THEN (Outcome); [Rule support, Relative rule strength, Discrimination level] [Objects represented by the rule] (For the definitions of these terms see Stefanowski 1998).

For example, Rule 3 `(F1 = 1) AND (F6 = 1) THEN (D = Two);`
`[1, 100.00%, 100.00%][8]`
reads

IF (The value of feature F1 equals 1) AND (The value of F6 equals 1) THEN (The decision D is Two); [This rule represents one object; In this case the one object makes up 100.00% of all objects in the training data set with the decision D = Two; The two objects match the conditions and decision of this rule with the accuracy of 100%] [The object represented by this rule is 8].

Each of the three rules in Figure 1 group the objects as shown in Table 1. The matrix corresponding to Rule 1 is shown in Table 3. The cluster of objects {1, 2, 3, 5, 6, 7} shown in Table 3 has been created based on the value of feature F6 and is not interesting. Rule 2 of Figure 1 has resulted in two clusters shown in Table 4.

Table 3. Clusters corresponding to Rule 1 of Figure 1

	F1	F2	F3	F4	F5	F6
1		1		1		
2			1			
3	1				1	
5		1		1		
6	1				1	
7					1	
4			1			1

Table 4. Clusters corresponding to Rule 2 of Figure 1

	F1	F2	F3	F4	F5	F6
1		1		1		
2			1			
4			1			1
5		1		1		
3	1				1	
6	1				1	
7					1	

The matrices in Tables 3 and 4 indicate that the clusters may be formed based on '0' (previously 'blank') or '1' entries as opposed to some clustering algorithms that create clusters based on the entries '1' only. A way to 'force' the rules to select features with entries '1' is to assign random numbers to all 'blank' entries.

The '0' entries of the incidence matrix in Table 3 have been filled with random integer numbers in the interval [3, 9] as shown in Table 5.

Table 5. The data of Table 3 with entries '0' replaced by random numbers

	F1	F2	F3	F4	F5	F6	D
1	4	1	3	1	8	5	One
2	3	0	1	2	2	4	One
3	1	0	4	5	1	2	One
4	6	0	1	8	3	1	One
5	7	1	5	1	4	3	One
6	1	0	6	9	1	8	One
7	8	0	7	4	1	5	One
8	1	1	1	1	1	1	Two

The rule set extracted from the data in Table 5 is shown in Figure 2.

```
Rule 4.   (F2  =   0)   THEN    (D   =   One);    [5,   71.43%,
100.00%] [2, 3, 4, 6, 7]
Rule 5.   (F1  =   4)   THEN    (D   =   One);    [1,   14.29%,
100.00%] [1]
Rule 6.   (F1  =   7)   THEN    (D   =   One);    [1,   14.29%,
100.00%] [5]
Rule 7.  (F1 = 1) AND  (F6 = 1)  THEN  (D = Two);   [1,
100.00%, 100.00%] [8]
```

Figure 2. Rules extracted from the data in Table 5

The clusters corresponding to Rules 4 – 6 of Figure 2 are shown in Tables 6 and 7.

Table 6. Clusters corresponding to Rule 4 of Figure 2

	F1	F2	F3	F4	F5	F6
1		1		1		
5		1		1		
2			1			
3	1				1	
4			1			1
6	1				1	
7					1	

Table 7. Clusters corresponding to Rules 5 and 6 of Figure 2

	F1	F5	F2	F4	F3	F6
3	1	1				
7		1				
6	1	1				
1	4		1	1		
5	7		1	1		
4					1	1
2					1	

Non-Decomposable Matrix

Most of the time a matrix does not decompose into mutually exclusive submatrices. Consider the matrix in Table 8 and its transformed form in Table 9 with '0s' replacing 'blanks', dummy row 9, and decision values {One, Two}.

Table 8. Matrix with eight objects

	F1	F2	F3	F4	F5	F6	F7	F8	F9	F10
1		1	1	1				1		
2			1				1			
3	1				1					1
4					1					1
5		1		1						
6		1				1	1		1	
7					1					
8	1					1	1		1	

Table 9. Expanded Table 8

	F1	F2	F3	F4	F5	F6	F7	F8	F9	F10	D
1	0	1	1	1	0	0	0	1	0	0	One
2	0	0	1	0	0	0	1	0	0	0	One
3	1	0	0	0	1	0	0	0	0	1	One
4	0	0	0	0	1	0	0	0	0	1	One
5	0	1	0	1	0	0	0	0	0	0	One
6	0	0	1	0	0	1	1	0	1	0	One
7	0	0	0	0	1	0	0	0	0	0	One
8	1	0	0	0	0	1	1	0	1	0	One
9	1	1	1	1	1	1	1	1	1	1	Two

```
Rule 8.  (F8  =  0)   THEN   (D  =  One);   [7,   87.50%,
100.00%][2, 3, 4, 5, 6, 7, 8]
Rule 9.  (F5  =  0)   THEN   (D  =  One);   [5,   62.50%,
100.00%][1, 2, 5, 6, 8]
Rule 10. (F1 = 1) AND (F8 = 1) THEN (D = Two); [1,
100.00%, 100.00%][0, 1][9}]
```

Figure 3. Rules from the data in Table 9

The clusters corresponding to Rules 8 and 9 of Figure 3 are shown in Tables 10 and 11.

Table 10. Clusters associated with Rule 8 of Figure 3

	F5	F10	F1	F4	F2	**F8**	F3	F6	F7	F9
3	1	1	1							
7	1									
4	1	1								
1				1	1	1	1			
5					1	1				
6							1	1	1	1
2							1		1	
8			1					1	1	1

Table 11. Clusters associated with Rule 9 of Figure 3

	F5	F10	F1	F4	F2	F8	F3	F6	F7	F9
3	1	1	1							
7	1									
4	1	1								
1				1	1	1	1			
5				1	1					
6							1	1	1	1
2							1		1	
8			1					1	1	1

Clustering matrices based on the rules extracted with machine learning algorithms proves to be useful. The fact that machine learning algorithms work on different principles than traditional clustering algorithms may provide interesting insights into structure of matrices, especially when considered in a broad context of autonomous systems.

The next section will show that machine learning algorithms can be applied to structure matrices containing objects with conflicting (ill-defined) outcomes. In some applications, e.g., in group technology, the conflicting outcome approach is equivalent to grouping with alternative process plans (Kusiak 2000).

Matrix with Ill-Defined Outcomes

In the previous section, the exact rules were considered for matrix decomposition. The concept of approximate rules will be applied to discover objects with alternative outcomes. The approximate rules 5 involve two objects 4 and 9 as well as 5 and 10 with identical features and different outcomes.

Consider the data set in Table 12 and the corresponding rule set in Figure 4. As the set in Table 12 contains two objects with conflicting decisions, the rule set in Figure 4 contains approximate rules. The exact and approximate rules extracted from the data in Table 12 are shown in Figure 4.

Table 12. Matrix containing two objects 4 and 5 with conflicting decisions

	F1	F2	F3	F4	F5	F6	F7	F8	F9	F10	D
1	0	1	1	1	0	0	0	1	0	0	Two
2	0	0	1	0	0	0	1	0	0	0	Three
3	1	0	0	0	1	0	0	0	0	1	One
4	0	0	0	0	1	0	0	0	0	1	One
5	0	1	0	1	0	0	0	0	0	0	Two
6	0	0	1	0	0	1	1	0	1	0	Three
7	0	0	0	0	1	0	0	0	0	0	One
8	1	0	0	0	0	1	1	0	1	0	Three
9	0	0	0	0	1	0	0	0	0	1	Two
10	0	1	0	1	0	0	0	0	0	0	Three

Exact rules

```
Rule 1. (F1 = 1) AND (F9 = 0) THEN (D = One); [1,
33.33%, 100.00%][3]
Rule 2. (F4 = 0) AND (F7 = 0) AND (F10 = 0) THEN (D =
One); [1, 33.33%, 100.00%][7]
Rule 3. (F8 = 1) THEN (D = Two); [1, 33.33%,
100.00%][1]
Rule 4. (F7 = 1) THEN (D = Three); [3, 75.00%,
100.00%[2, 6, 8]
```

Approximate rules

```
Rule 5. (F1 = 0) AND (F10 = 1) THEN (D = One) OR (D =
Two); [2, 100.00%, 100.00%][4, 9]
Rule 6. (F3 = 0) AND (F4 = 1) THEN (D = Two) OR (D =
Three); [2, 100.00%, 100.00%][5, 10]
```

Figure 4. Exact and approximate rules for the data in Table 12

The exact rules of Figure 4 formed the clusters illustrated in Table 13 while the approximate rules discovered that objects (4 and 9) and (5 and 10) have conflicting outcomes.

Table 13. Clusters corresponding to the exact rules of Figure 4

	F5	F10	F1	F4	F2	F8	F3	F6	F7	F9
3	1	1	1							
7	1									
1				1	1	1	1			
6							1	1	1	1
2							1		1	
8			1					1	1	1
4	1	1								
5				1	1					

CLASSIFICATION ACCURACY ENHANCEMENT

The property of approximate rules discussed in the previous section will be applied to an industrial data set with ill-defined outcomes. It will be shown that approximate rules will identify objects in the training set that have been assigned incorrect outcomes. Rules are extracted from the same data set for two different feature sets.

Feature Set 1

Case 1: One approximate rule involving two conflicting objects

The quality of predictions with the rules extracted from a data set is usually evaluated by a cross-validation scheme (Stone 1974). The k-fold (here $k = 10$) cross-validation scheme is often recommended. In this scheme, a training data set is partitioned into $k = 10$ folds (subsets) and one fold of objects is removed from the training data set and the rules are extracted from the remaining $k - 1$ = 9 folds.

Table 14. Absolute classification accuracy

	N	Z	P	None
N	3	2	11	2
Z	3	11	12	1
P	2	3	30	6

The diagonal numbers in Table 14 represent the number of outcomes $D = Y$, Z, P that have been correctly predicted. In this case 3 of the 17 (= 3 + 2 + 11 + 2) decisions $D = N$ have been correctly predicted along with 11 decisions $D = Z$, and 30 decisions $D = P$. The numbers off the diagonal indicate the incorrectly predicted decisions. For the decision $D = N$ (row N in Table 14) 2 objects were incorrectly classified as Z, 11 objects as P, and two objects were placed in the 'None' category.

Table 15 reports the percentage of objects (known in data mining as classification accuracy) that have been classified correctly, incorrectly, or fall into the 'None' category for the three decision values $D = N$, Z, P.

Table 15. Classification accuracy

	Correct	Incorrect	None
N	12.50%	57.50%	10.00%
Z	50.83%	44.17%	5.00%
P	79.17%	10.69%	10.12%
Av	51.39%	38.33%	10.28%

Note that the numbers in each row of Table 15 do not necessarily add up to 100%.

Case 2: The outcome of one of the two conflicting object of Case 1 changed from $D = P$ to $D = Z$

Table 16. Absolute classification accuracy

	N	Z	P	None
N	9	1	7	1
Z	3	11	10	4
P	2	3	33	2

Table 17. Classification accuracy

	Correct	Incorrect	None
N	36.37%	40.83%	2.50%
Z	50.00%	32.50%	17.50%
P	86.48%	9.60%	3.92%
Av	61.67%	30.00%	8.33%

Case 3: The outcomes of the two conflicting objects of Case 1 changed from $D = P$ to $D = Z$ and from $D = N$ to $D = Z$.

Table 18. Absolute classification accuracy

	N	Z	P	None
N	3	3	8	3
Z	4	7	15	3
P	2	4	27	7

Table 19. Classification accuracy

	Correct	Incorrect	None
N	12.50%	47.50%	20.00%
Z	25.83%	56.67%	17.50%
P	73.38%	15.33%	11.29%
Av	43.06%	42.36%	14.58%

Case 4: The two conflicting objects of Case 1 removed from the training set

Table 20. Absolute classification accuracy

	N	Z	P	None
N	4	5	4	4
Z	1	7	10	9
P	1	5	24	10

Table 21. Classification accuracy

	Correct	Incorrect	None
N	21.67%	43.33%	15.00%
Z	25.00%	35.50%	39.50%
P	63.69%	12.62%	23.69%
Av	41.39%	31.25%	27.36%

Before the results of Cases 1 through 4 will be summarized classification accuracy for Feature Set 2 is computed. This data set contains features that differ from the ones in Feature Set 1.

Feature Set 2

Case 5: Training data set producing seven approximate rules

The support of each approximate rule varied from one to four objects.

Table 22. Absolute classification accuracy

	N	Z	P	None
N	4	1	10	3
Z	3	8	6	0
P	3	2	32	4

Table 23. Classification accuracy

	Correct	Incorrect	None
N	15.83%	49.17%	15.00%
Z	42.50%	57.50%	0.00%
P	84.45%	9.60%	5.95%
Av	51.39%	40.56%	8.05%

The outcomes of the 15 objects included in the approximate rules were arbitrarily modified $D = P$ to $D = Z$ and from $D = N$ to $D = Z$.

Case 6: Training data set with 15 modified outcomes

Table 24. Absolute classification accuracy

	N	Z	P	None
N	1	5	6	2
Z	3	16	10	7
P	2	9	14	11

Table 25. Classification accuracy

	Correct	Incorrect	None
N	5.00%	65.00%	10.00%
Z	46.26%	27.90%	25.83%
P	45.83%	28.00%	26.17%
Av	36.39%	40.56%	23.05%

Rather than changing values of 15 objects, their removal from the training set could be considered. This action would certainly improve classification accuracy. However, the generality of the rules might be sacrificed.

Table 26 summarizes the correctly predicted outcomes of the six cases for two deferent feature sets.

Table 26. Summary of correctly predicted outcomes for cases 1 – 4 and 5 – 6

	Feature Set 1				Feature Set 2	
	Case 1	Case 2	Case 3	Case 4	Case 5	Case 6
N	3	9	3	4	4	1
Z	11	11	7	7	8	16
P	30	33	27	24	32	14

The most desirable values of absolute accuracy in Table 26 are highlighted. Case 2 of Feature Set 1 has produced the most desirable values of classification accuracy with 53 of 86 correctly predicted outcomes D. Case 6 with modified decision values for 15 objects resulted in improvement in the D = Z category.

It should be stressed that the modifications of decision values for all six cases were rather random and no attempt was made to identify assignments maximizing classification accuracy.

The average classification accuracy in Table 27 reinforces the above observations.

Table 27. Summary of average classification accuracy for cases 1 – 4 and 5 – 6

	Feature Set 1				Feature Set 2	
	Case 1	Case 2	Case 3	Case 4	Case 5	Case 6
Correct	51.39%	61.67%	43.06%	41.39%	51.39%	36.39%
Incorrect	38.33%	30.00%	42.36%	31.25%	40.56%	40.56%
None	10.28%	8.33%	14.58%	27.36%	8.05%	23.05%

The results presented in Tables 28 and 29 present interesting insights into classification accuracy. The feature set of each training data set used in Cases 1 – 6 was expanded to the same base feature set. The results of the $k = 10$ fold cross-validation are reported in Tables 28 and 29.

Table 28. Summary of absolute classification accuracy of the data set with expanded features

	Case B1	Case B2	Case B3	Case B4	Case B5	Case B6
N	2	3	5	6	2	2
Z	9	7	11	10	9	15
P	29	29	19	25	29	14

Table 29. Summary of average classification accuracy of the data set with expanded features

	Feature Set 1				Feature Set 2	
	Case B1	Case B2	Case B3	Case B4	Case B5	Case B6
Correct	46.25%	45.28%	40.97%	48.61%	46.25%	36.25%
Incorrect	45.97%	35.28%	31.53%	31.25%	45.97%	31.39%
None	7.78%	19.44%	27.50%	20.14%	7.78%	32.36%

The highlighted cells in the two tables indicate the most preferred accuracy values. The computational results summarized in Tables 26 – 29 prove that changing values of ill-defined outcomes impacts classification accuracy. The classification accuracy can be controlled by changing values of objects with conflicting outcomes, eliminating some of these objects from the training data set, and using decision making rules extracted for different data sets, i.e., the rules corresponding the categories highlighted in Tables 26 – 29.

CONCLUSIONS

The interest in data mining is growing and its full potential is awaiting realization. While some applications of data mining have been discussed in the literature, many are still to come. In this chapter it was shown that machine learning algorithms could be applied to problems traditionally solved with operations research, mathematical programming, or cluster analysis tools. Machine learning may provide unknown insights into problem structure and discover other interesting properties. Besides discovering patterns, learning algorithms can be used to modify data sets in order to enhance decision making accuracy.

REFERENCES

Auer, P., R. Holte and W. Maass, "Theory and application of agnostic PAC-learning with small decision trees," in A. Prieditis and S. Russell, Eds, ECML-95: Proceedings of 8th European Conference on Machine Learning, Springer Verlag, New York, 1995.

Clark, P. and R. Boswell, "The CN2 induction algorithm," Machine Learning, 3(4), pp. 261-283, 1989.

Domingos, P. and M. Pazzani, "Beyond independence: conditions for the optimality of the simple Bayesian classifier," Machine Learning: Proceedings of the Thirteenth International Conference, Morgan Kaufmann, Los Altos, CA, pp. 105-112, 1996.

Friedman, J., Y. Yun and R. Kohavi, "Lazy decision trees," Proceedings of the Thirteenth National Conference on Artificial Intelligence, AAAI Press and MIT Press, 1996.

Grzymala-Busse, J., "A new version of the rule induction system LERS," Fundamenta Informaticae, Vol. 31, pp. 27-39, 1997.

Kohavi, R., "Wrappers for Performance Enhancement and Oblivious Decision Graphs," Ph.D. Thesis, Computer Science Department, Stanford University, Stanford, CA, 1995.

Kusiak, A., "Computational Intelligence in Design and Manufacturing," John Wiley, New York, 2000.

Kusiak, A., "Engineering Design: Products, Processes, and Systems," Academic Press, San Diego, 1999.

Kusiak, A., "Decomposition in data mining; An industrial case study," IEEE Transactions on Electronics Packaging Manufacturing, 23(4), pp. 345-353, 2000a.

Kusiak, A., J.A. Kern, K.H. Kernstine, and T.L. Tseng, "Autonomous decision-making: A data mining approach," IEEE Transactions on Information Technology in Biomedicine, 4(4), pp. 274-284, 2000.

Michalski, R.S., I. Mozetic, J. Hong, and N. Lavrac, "The multi-purpose incremental learning system AQ15 and its testing application to three medical domains," Proceedings of the 5th National Conference on Artificial Intelligence, AAAI Press, Palo Alto, CA, pp. 1041-1045, 1986.

Michie, D., D.J. Spiegelhalter, and C.C. Taylor, "Machine Learning, Neural, and Statistical Classification," Ellis Horwood, New York, 1994.

Pawlak, Z., "Rough Sets: Theoretical Aspects of Reasoning About Data," Kluwer, Boston, MA, 1991.

Quinlan, J.R., "C4.5: Programs for Machine Learning," Morgan Kaufmann, Los Altos, CA, 1993.

Quinlan, J.R., "Induction of decision trees," Machine Learning, 1(1), pp. 81-106, 1986.

Stefanowski, J., "On rough set approaches to induction of decision rules," in L. Polkowski and A. Skowron (Eds), Rough Sets in Knowledge Discovery 1: Methodology and Applications, Springer-Verlag, New York, pp. 501-529, 1998.

Stone, M., "Cross-validatory choice and assessment of statistical predictions," Journal of the Royal Statistical Society, Vol. 36, 111-147, 1974.

CHAPTER 18

Fuzzy-Neural-Genetic Layered Multi-Agent Reactive Control of Robotic Soccer

Andon V. Topalov
topalov@mbox.digsys.bg
Department of Control Systems, Technical University of Sofia - branch
Plovdiv, 61, Sankt Petersburg Blvd., 4000, Plovdiv, Bulgaria

Spyros G. Tzafestas
tzafesta@softlab.ece.ntua.gr
Intelligent Robotics and Automation Laboratory, Computer Science Division,
Department of Electrical and Computer Engineering, National Technical
University of Athens, Zografou, GR-157 73, Athens, Greece

ABSTRACT

Robotic soccer belongs to the class of multi-agent systems and involves many challenging sub-problems. Teams of robotic players have to cooperate in order to put the ball in the opposing goal and at the same time defend their own goal. This paper is concerned with the problem of learning two basic reactive behaviors of robotic agents playing soccer, namely: (i) learning to intercept the moving ball while avoiding collisions with other players and play field walls, and (ii) learning to shoot the ball toward the goal or pass it in a desired direction. The approach adopted has a "layered structure", i.e. the *ball interception/obstacle avoidance* (BIOA) behavior is first learned, and the skills obtained are then employed to learn the *shooting ball* (SB) behavior at a higher layer.

The proposed control scheme involves a *fuzzy-neural trajectory generator* (FNTG), which supplies data to a *trajectory-tracking controller* (TTC) consisting of a *conventional PD feedback controller* (CFC) followed by a *fuzzy-neural controller* (FNC). This allows the implementation of the robot behaviors (tasks) at a trajectory generator level using off-line learning and the robot kinematics model only. The complete dynamics of the mobile base is taken into account by the TTC, and it's learning is performed on a real mobile robot. Considering the advantages of genetic algorithms (GAs), a GA approach is employed to perform the learning process of the FNTG layers. The overall system (including the play field) was simulated in the MATLAB®

D. Braha (ed.), Data Mining for Design and Manufacturing, 417–442.

environment and the results obtained are very encouraging showing the effectiveness of the proposed layered fuzzy-neural-genetic learning control scheme.

INTRODUCTION

One of the ultimate dreams in robotics is to create life-like robotic systems. However, number of challenges exists before such robots could be fielded in the real world. Robots have to be reasonably intelligent, maintain certain level of agility, and be able to engage in some collaborative behaviors. Robotic soccer is an ideal challenge to foster robotic technologies for small personal and mobile robotic systems. The research in this area and the soccer games carried out with robotic players can help to create the necessary knowledge and know-how that will facilitate the use of robots not only for entertainment but also for such non-entertainment purposes like rescue robots or dangerous works carried instead of humans. It is believed that this will become effective in the future, when the robots will be able to coexist and collaborate with humans and to perform things that the human is unable to do and to offer a mobility which will bring additional value to the human society.

In this paper, control architectures, learning techniques and a design methodology for the development of some reactive behaviors of robotic agents playing soccer are proposed. It is our believe that the proposed approach could be also easily applied to more generalized multi-robotics environments.

Robotic soccer, as first proposed by [Mackworth, 1993], is a challenging emerging research domain that is particularly appropriate for studying a large spectrum of issues of relevance to the development of complete autonomous agents [Asada et al., 1998], [Arkin, 1998], and [Tzafestas, 1999]. The behaviors and decision making processes range from the most simple reactive behaviors, such as moving directly towards the ball, to arbitrarily complex reasoning procedures that take into account the actions and perceived strategies of teammates and opponents. Since a few years robotic soccer tournaments are held worldwide between institutes working in the field of mechatronics, multi-agent systems, intelligent robotics or AI in general. There exist different leagues for various sizes of robots and a simulator league organized in two federations, the RoboCup, (see [Asada et al., 1999], [Kitano, 1998], http://www.robocup.org), and the FIRA, (see [Kim, 1996], [Stonier et al., 1999], http://www.fira.net). Regulations for international tournaments are provided by these federations. A set of benchmark tests for robot soccer ball control skills has been also adopted by FIRA [Jonson et al., 1999].

The competing robotic teams are considered as multi-agent systems coupled with collaborative and adversarial learning in an environment that requires real-time dynamic planning. The learning opportunities in the robotic soccer domain could be broken down into several types [Kitano *et al.*, 1997]:

- *Off-line skill learning by individual agents*: learning to intercept the ball or learning to kick the ball with the appropriate power when passing.
- *Off-line collaborative learning by teams of agents*: learning to pass and receive the ball.
- *On-line skill and collaborative learning*: learning to play positions. Although off-line learning methods can be useful in the above cases, there may also be advantages to learn incrementally as well.
- *On-line adversarial learning*: learning to react to predicted opponent actions.

In this paper control architectures, learning approaches and design methodology are proposed for the development of following two basic reactive ball handling behaviors for a robotic agent playing soccer:

1) Learning to intercept the moving ball while avoiding collisions with other players and playground walls;

2) Learning to shoot the ball toward the goal (or to pass the ball in a specified direction).

These two low-level behaviors are crucial for successful action in a robotic soccer multi-agent domain. They also equip soccer robots with the skills necessary to learn higher-level collaborative and adversarial behaviors.

Previously, Asada [Asada *et al.*, 1994a, 1994b] investigated several low-level behaviors such as shooting and avoiding using the Reinforcement Learning (RL) technique. The authors have developed a real-world robotic system with many advanced capabilities and learned a robot to shoot a stationary ball into a goal. A separate learned decision mechanism was also proposed by them for combining different learned behaviors. Several other researchers have used robotic soccer simulators for their experiments. Stone and Veloso [Stone *et al.*, 1996] used Memory-based Learning to allow a player to learn when to shoot and when to pass the ball. The same problem was investigated by Matsubara [Matsubara *et al.* 1996]. In another work Stone and Veloso proposed a supervised learning technique for a robotic player to learn shooting a moving ball [Stone *et al.*, 1998]. The behavior to be learned consisted of the decision of when the shooter should begin to accelerate so that it intercepts the ball's path and redirects it into the goal. Once deciding to start, the shooter then steered along an imaginary line, continuously adjusting it's heading until it was moving in the right direction along this line. The influence of the other robotic players of which the paths could take crosses with the shooter's path and prevent the shooter to intercept the ball was not investigated. In addition, the supervised learning approach adopted by these authors leaded them to the need of prediction of all possible situations in the environment and gathering representative training data in order to achieve robust and situation-independent skills which is not an easy task. Further development of the above research was reported lately by Veloso *et al.* in

[Veloso *et al.*, 1999a], [Veloso *et al.*, 1999b]. In addition to the behavior developed in [Stone *et al.*, 1998] and called by the authors *ball interception* a new behavior named *ball collection* was developed. When using it, the robot considers a line from the target position (e.g. the adversary's goal, or a teammate position) to the ball's current or predicted position, depending on whether or not the ball is moving. The robot then plans a path to a point on the line and behind the ball such that it does not hit the ball on the way and such that it ends up facing the target position. Finally, the robot accelerates to the target, thus hitting the ball toward it. The robot chooses from between its two ball handling routines based on whether the ball will eventually cross its path at a point such that the robot could intercept it towards the goal. It gives precedence to the ball interception routine, only using ball collection when necessary. An obstacle avoidance feature was also incorporated in the both behaviors. In the event that an obstacle blocks the direct path to the goal location, the robot aims to one side of the obstacle until it is in a position such that it can move directly to its original goal. However, it was mentioned by the authors that even with obstacle avoidance in place, their robots can occasionally get stuck against other robots or against the wall. They made also a conclusion that if the opponent robots do not use obstacle avoidance, collisions with their robots are inevitable.

In this paper, a control scheme consisting of a trajectory generator supplying data to a trajectory-tracking controller is proposed for the implementation of the basic reactive behaviors. This allows the robot behaviors to be implemented completely on the trajectory generator level and thus to be learned off-line from simulations that use only the robot kinematics model. It is the tracking controller level that takes into account the complete dynamics of the mobile base. Learning at this level is made on a real mobile robot. Genetic algorithms (GAs) are adopted as a learning technique where the robot is expected to self-improve its behavior to fit the different situations that it experiences, through the robot-environment interaction. By defining a fitness function as the selective pressure, the control system that drives the robot to perform the desired behavior well will emerge over generations through competition between members of populations.

In the proposed approach increasingly complex layers of learned behaviors are developed from the bottom up using the local learning method [Izumi *et al.*, 1999]. To be able to control the robots to intercept a moving ball is one of the real challenges in robotic soccer. This capability is essential for a high-level ball-passing behavior. So the *ball interception and obstacle avoidance* (BIOA) behavior is learned first. The robot learns to avoid reliably the obstacles while generating a path to the location of the ball. It is not required for the adversaries to be equipped with obstacle avoidance capability too. Then the acquired skills are incorporated into the next higher-level multi-agent learning scenario that gives the robot the ability to direct the ball in a specified direction - the *shooting ball* (SB) behavior. We developed only one

SB behavior instead of two behaviors as proposed in [Veloso *et al.*, 1999a,b], thus avoiding the need of an additional behavior arbitration mechanism at this level.

For the implementation of the above idea a *fuzzy-neural trajectory generator* (FNTG), with a layered structure consisting of two layers, is developed. During the learning of the BIOA behavior the first layer of the generator is trained without using the SB behavior layer. The learning of the later is then performed together with the use of the outputs of the layer already trained. By layering learned behaviors one on the top of the other, as in this case, all levels of learning that involve interaction with other agents contribute to, and are a part of, multi-agent learning.

Although in this paper the focus is on a single agent that learns to do the shooting of the moving ball, it is considered as a multi-agent learning scenario since the ball is typically moving as the result of a pass from a teammate: it is possible only because other agents are present [Stone *et al.*, 1998].

THE TRAJECTORY TRACKING CONTROL PROBLEM OF ROBOTIC SOCCER AGENTS

While the main goal of this investigation is to study the principles of learning basic reactive behaviors in robotic soccer in general, a close conformity with the FIRA MiroSot Game Rules (see http://www.fira.net) is followed. The size of the robotic players in this category is limited to 7.5cm × 7.5cm × 7.5cm, the playground size is 130cm × 150cm and the goal is 40cm wide. The allowance for a global vision system to be used with a video camera overlooking the complete field gives the opportunity to get a global view of the world state (see Fig. 1).

Figure 1: An overall View of the Robotic Soccer System in the MiroSot Category

It is accepted here that the soccer players are wheeled mobile robots with a differential drive mechanism. Their motion and orientation is achieved by independent actuators (geared DC motors) providing the necessary torque to the driving wheels. This means that robots are non-holonomic.

Fig. 2 shows the robot's geometric model. Its position in an inertial Cartesian frame is completely specified by the posture $\mathbf{q} = \begin{bmatrix} x & y & \theta \end{bmatrix}^T$ where x, y and θ are the coordinates of the reference point C (center of gravity), and the orientation of the mobile basis $\{C, X_C, Y_C\}$ with respect to the inertial basis, respectively.

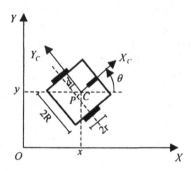

Figure 2: The Geometric Model of a Non-Holonomic Mobile Platform

The vehicle's motion is controlled by its linear (υ) and angular (ω) velocities. The robot's kinematics is defined by a Jacobian matrix $\mathbf{J}(\theta)$ that transforms velocities $\mathbf{v} = \begin{bmatrix} \upsilon & \omega \end{bmatrix}^T$ expressed in mobile base coordinates into velocities $\dot{\mathbf{q}}$ expressed in Cartesian coordinates, where the system (1) is called the steering system of the vehicle [Fierro *et al.*, 1995]:

$$\begin{bmatrix} \dot{x} \\ \dot{y} \\ \dot{\theta} \end{bmatrix} = \dot{\mathbf{q}} = \mathbf{J}\mathbf{v} = \begin{bmatrix} \cos\theta & -d\sin\theta \\ \sin\theta & d\cos\theta \\ 0 & 1 \end{bmatrix} \begin{bmatrix} \upsilon \\ \omega \end{bmatrix} \qquad (1)$$

The trajectory-tracking problem for non-holonomic vehicles can be stated as follows: *Given a prescribed reference cart* $\mathbf{q}_r = \begin{bmatrix} x_r & y_r & \theta_r \end{bmatrix}^T$, $\mathbf{v}_r = \begin{bmatrix} \upsilon_r & \omega_r \end{bmatrix}^T$ *with* $\upsilon_r > 0$ *for all* t, *find a smooth velocity control* $\mathbf{v}_c$ *such that* $\lim_{t \to \infty}(\mathbf{q}_r - \mathbf{q}) = 0$, *and compute the torque input* $\tau(t)$ *by some means such that* $\mathbf{v} \to \mathbf{v}_c$ *as* $t \to \infty$.

The implemented structure for the robot tracking control system is shown in Fig. 3.

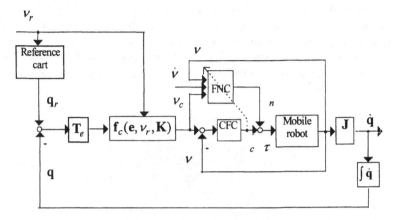

Figure 3: Block Diagram of the Tracking Control of Mobile Robot

The tracking error posture $e(t) = [e_1(t) \quad e_2(t) \quad e_3(t)]^T$ is expressed on the basis of the frame linked to the mobile platform:

$$\begin{bmatrix} e_1 \\ e_2 \\ e_3 \end{bmatrix} = \begin{bmatrix} \cos\theta & \sin\theta & 0 \\ -\sin\theta & \cos\theta & 0 \\ 0 & 0 & 1 \end{bmatrix} \begin{bmatrix} x_r - x \\ y_r - y \\ \theta_r - \theta \end{bmatrix} \quad (2)$$

$$\mathbf{e} = \mathbf{T}_e(\mathbf{q}_r - \mathbf{q})$$

It is assumed that the solution to the steering system tracking problem, as found in [Kanayama *et al.*, 1990], is available, and that the auxiliary velocity control input $\mathbf{v}_c(t)$ which achieves tracking for (1) is given by

$$\mathbf{v}_c = \begin{bmatrix} \upsilon_r \cos e_3 + k_1 e_1 \\ \omega_r + k_2 \upsilon_r e_2 + k_3 \upsilon_r \sin e_3 \end{bmatrix} \quad (3)$$

$$\mathbf{v}_c = \mathbf{f}_c(\mathbf{e}, \mathbf{v}_r, \mathbf{K})$$

where k_1, k_2 and k_3 are positive constants.

Given the velocity vector $\mathbf{v}_c$, the auxiliary velocity tracking error can be defined as $\mathbf{e}_c = \mathbf{v}_c - \mathbf{v}$. The velocity control scheme does not assume preliminary knowledge of the cart dynamics. The feedback-error-learning approach [Gomi *et al.*, 1993] is applied, where *conventional PD feedback controller* (CFC) is used both as an ordinary feedback controller to guarantee global asymptotic stability during the learning period, and as a reference model for the responses of the controlled object. To obtain the desired response during tracking-error convergence movement by compensating for the nonlinear object dynamics, a *fuzzy-neural controller* (FNC) is trained to become a nonlinear regulator. The output of the CFC is fed to the FNC as the error signal used for learning. A detailed description of the trajectory-tracking level is given in [Topalov *et al.*, 1998], [Topalov *et al.*, 1999].

An experimental micro-robot that has been developed in TU Plovdiv in conformity with the MiroSot size restrictions is considered as a soccer agent in this work (see Fig. 4).

Figure 4: The Experimental Micro-Robot

Completely made with Lego® parts this robot is equipped also with an on-board control unit based on Intel® 80C196 16-bit, 20 MHz microcontroller.

The trajectory-tracking control scheme has been implemented in a C++ program working under MS Windows 95/98® (see Fig. 5). The program communicates with the robot's on-board control unit via the RS 232 interface. It has also a built in capability to import data from the Matlab® environment. This allowed us, at this stage, to develop and learn the robot's trajectory generator level within Matlab®, while still keeping the possibility for a real-time tracking of the generated trajectory by the laboratory micro-robot.

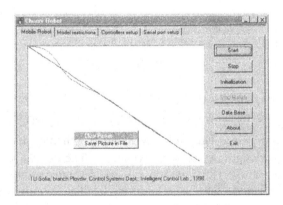

Figure 5: The Main Window of the Trajectory-Tracking Control Program

THE FUZZY-NEURAL TRAJECTORY GENERATOR

The vision system in MiroSot robotic soccer category supplies the agents with the current coordinates of the ball and the postures of the players. It can provide also additional data, including the team to which they belong to and their numbers inside of the team. The input data used by the implemented behaviors must be easily calculated from this information.

The developed BIOA and SB behaviors have been completely implemented on the trajectory generator level.

The BIOA Layer

The BIOA behavior has been built as a first layer of the robot's FNTG. It uses input data that include only the relative positions of the ball and the closest obstacle. Thus the number of the situations that the robotic agent could encounter in its environment has been significantly reduced. The entire layer was built as a fuzzy-neural network shown on Fig. 6

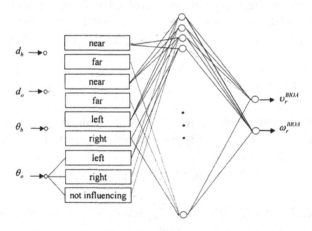

Figure 6: The BIOA Behavior Fuzzy-Neural Network

Its inputs are:
- The distance between the ball and the mobile robot - d_b;
- The distance between the robot and the closest obstacle (it could be the nearest robotic agent or a playground wall depending which of them is closer to the controlled agent) - d_o;
- The relative angle between the azimuth of the robot and the direction of the ball - θ_b;
- The relative angle between the azimuth of the robot and the direction of the closest obstacle - θ_o.

The fourth input signal θ_o is associated with three fuzzy labels: "left", "right" and "not influencing" in accordance with the three different situations considered:

1) θ_o has small positive value which means that the obstacle is on the left side, so the robot should move to the right in order to avoid collision;

2) θ_o has small negative value meaning that the obstacle is on the right side and the robot should turn to the left;

3) The absolute value of the angle θ_o is big enough and there is no danger of collision with the obstacle.

There are two outputs providing the linear reference velocity υ_r^{BIOA} and the angular reference velocity ω_r^{BIOA}. The reference velocities $\mathbf{v}_r^{BIOA} = \begin{bmatrix} \upsilon_r^{BIOA} & \omega_r^{BIOA} \end{bmatrix}^T$, expressed in mobile base coordinates, are further transformed into velocities $\dot{\mathbf{q}}_r^{BIOA} = \begin{bmatrix} \dot{x}_r^{BIOA} & \dot{y}_r^{BIOA} & \dot{\theta}_r^{BIOA} \end{bmatrix}^T$, expressed in Cartesian coordinates by using equation (1). The later are used for the calculation of the robot's posture at the next time step.

A rule base of Takagi - Sugeno type initially consisting of 24 fuzzy "if-then" rules where the fuzzy sets are involved only in the premise part has been implemented. The $ijkl$-th rule can be expressed as follows:

R_{ijkl} : IF d_b is A_i and d_o is B_j and θ_b is C_k and θ_o is D_l THEN $f_{\upsilon_{ijkl}} = V_{ijkl}$

$$\text{and } f_{\omega_{ijkl}} = \Omega_{ijkl}$$

where $i, j, k = 1, 2$; $l = 1, 2, 3$; A_i, B_j, C_k, D_l are fuzzy sets and V_{ijkl} and Ω_{ijkl} are constants.

The firing strength of the $ijkl$-th rule is obtained as a T-norm of the membership values of the premise part (by using a multiplication operator):

$$W_{ijkl}^{BIOA} = \mu_{A_i}(d_b)\,\mu_{B_j}(d_o)\,\mu_{C_k}(\theta_b)\,\mu_{D_l}(\theta_o) \qquad (4)$$

The overall outputs of the BIOA fuzzy-neural network are computed as a weighted average of each rule's output

$$\upsilon_{rl}^{BIOA} = \frac{\sum_{i=1}^{2}\sum_{j=1}^{2}\sum_{k=1}^{2}\sum_{l=1}^{3} W_{ijkl}^{BIOA} f_{\upsilon_{ijkl}}}{\sum_{i=1}^{2}\sum_{j=1}^{2}\sum_{k=1}^{2}\sum_{l=1}^{3} W_{ijkl}^{BIOA}};$$

$$\omega_r^{BIOA} = \frac{\sum_{i=1}^{2}\sum_{j=1}^{2}\sum_{k=1}^{2}\sum_{l=1}^{3} W_{ijkl}^{BIOA} f_{\omega_{ijkl}}}{\sum_{i=1}^{2}\sum_{j=1}^{2}\sum_{l=1}^{2}\sum_{k=1}^{3} W_{ijkl}^{BIOA}} \qquad (5)$$

The membership functions $\mu_{A_i}(d_b)$, $\mu_{B_j}(d_o)$ and $\mu_{C_k}(\theta_b)$ are sigmoid functions expressed as follows:

$$\mu_{A_i}(d_h) = \frac{1}{1 + e^{-a_{A_i}(d_h - c_{A_i})}}; \quad \mu_{Bj}(d_o) = \frac{1}{1 + e^{-a_{B_j}(d_o - c_{B_j})}};$$

$$\mu_{C_k}(\theta_h) = \frac{1}{1 + e^{-a_{C_k}(\theta_h - c_{C_k})}} \qquad (6)$$

where a and c are the parameters to be tuned.

Gaussian functions are used for the membership functions $\mu_{D_l}(\theta_o)$, where:

1) The membership functions for "left" or "right" ($l = 1$ or 2) are chosen to be as follows:

$$\left|
\begin{array}{ll}
\mu_{D_l}(\theta_o) = e^{\frac{-(\theta_o^2)}{2\sigma_{D_l}^2}} & \text{if } l = 1 \text{ and } \theta_o > 0 \text{ or if } l = 2 \text{ and } \theta_o \leq 0 \\
\mu_{D_l}(\theta_o) = 0 & \text{otherwise}
\end{array}
\right. \qquad (7)$$

2) The membership function for "not influencing" ($l = 3$) is defined as follows:

$$\mu_{D_3}(\theta_o) = 1 - e^{\frac{-(\theta_o^2)}{2\sigma_{D_3}^2}} \qquad (8)$$

where σ_{D_l} is the parameter to be tuned.

The fuzzy-neural network so proposed had initially 63 tunable parameters. In order to decrease their number, several logical relations have been taken further into consideration:

1) It is obvious that with increasing the distance to the closest obstacle d_o its influence on the generated trajectory will decrease. So the topology of the layer can be simplified taking into account only part of the combinations between d_o and θ_o as shown on Fig. 7. In this way the number of the fuzzy rules has been decreased to 16.

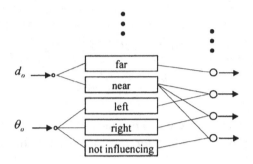

Figure 7: The Implemented Combinations Between d_o and θ_o

2)   Symmetry between the robot's movements to the left and to the right has been incorporated into the trajectory generation algorithm. This could be illustrated by the following example: suppose for simplicity there are no obstacles in the robot's working area and consider two cases when the positions of the ball are defined on equal distances from the robot and symmetrically to the robot's orientation. The idea is that the algorithm should in these two cases generate two symmetrical trajectories. When applied, it turned out that there are only 16 different outputs (regarded as absolute values) of the fuzzy "if-then" rules that have to be learned.

So the entire number of the tunable parameters has been finally decreased to 31.

The SB Layer

The learned individual skill at the first layer has been used as a basis for the next level multi-agent learning - learning to shoot the moving ball toward the goal. It was required that the robotic player not only could intercept the ball while avoiding collisions, but it should attempt also to redirect the ball into the goal. If a robot is to accurately direct the ball towards a target position, it must be able to approach the ball from a specified direction.

The SB behavior has been implemented as a second layer of the trajectory generator. The entire layer was built again as a Takagi – Sugeno type fuzzy-neural network presented in Fig. 8.

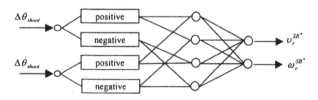

Figure 8: The Shooting Ball Behavior Fuzzy-Neural Network

Two easily computable coordinate-independent predicates have been used as inputs to the second layer:

- The heading offset ($\Delta\theta_{shoot}$): the relative angle between the agent's current heading and its desired heading in the time of shooting toward the goal (see Fig. 9);
- The rate of change of the heading offset - $\Delta\dot{\theta}_{shoot}$.

The network has two outputs, providing corrective signals for the vehicle's linear ($\upsilon_r^{SB^*}$) and angular ($\omega_r^{SB^*}$) reference velocities.

Since in general the ball's momentum is initially across the front of the goal, the shooter must compensate by kicking the ball towards another appropriate direction so that finally it goes in the goal. Several considerations were taken into account when determining the appropriate robot's heading at the time of shooting:

1) The presence of the other players on the field which paths could take crosses with the shooter's path makes unrealistic the expectation that the shooter could determine in advance a point on the ball's trajectory at which to strike the ball. Instead, the desired robot heading is changing continuously with the change of the coordinates of the contact point.

2) The direction in which the ball moves after an impact with the shooter depends on many parameters: the speed, the radius and the mass of the ball, the speed and the mass of the robot, the construction of robot's catching-shooting mechanism, the elasticity of the bodies and the duration of the contact, the influence of the forces, originating from the robot's motors and applied during the contact, the robot and the ball slippage, etc. Several simplifying assumptions have been made in order to perform simulations:

- The robot is not equipped with a ball catching mechanism and its front side is flat.
- Both the robot and the ball are rigid bodies.
- The ball is a dimensionless particle and its diameter is neglected.
- The mass of the robot is much larger than the mass of the ball.
- The duration time of the robot-ball contact is negligible.
- The velocity of the robot at the contact point is close to zero.

After applying the laws of momentum conservation and the kinetic energy conservation together with these simplifications it has been accepted that the relative angle between the direction of the movement of the ball before the impact and the heading of the robot at the time of shooting and the angle between this robot's heading and the direction of movement of the ball after the impact are approximately equal. (see Fig. 9). So the case of pure reflection is considered. It should be mentioned that no principal changes or relearning of the robot's SB behavior should be made if some of the above assumptions do not apply. Only the desired direction that the robot should aim should be specified appropriately (e.g. by using a vector summation of the velocities of the robot, of the kicking mechanism and of the ball at the contact point).

During the robot's motion the heading offset was practically measured as the angle between the current robot's azimuth and the bisection line of the ball-agent angle. The later is defined as angle between the ball's path and the line connecting the shooter's position with the center of the goal.

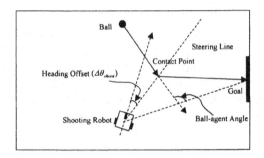

Figure 9: Model of the Shooting Behavior for a Robotic Player

The *mn*-th rule for the above fuzzy-neural network can be expressed as follows:

R_{mn}: IF $\Delta\theta_{shoot}$ is E_m and $\Delta\dot{\theta}_{shoot}$ is F_n THEN $f_{\upsilon_{mn}} = V_{mn}$ and $f_{\omega_{mn}} = \Omega_{mn}$

where $m, n = 1, 2$; E_m and F_n are fuzzy sets and V_{mn} and Ω_{mn} are constants.

The firing strength of the *mn*-th rule is obtained as in the first layer i. e. as a T–norm of the membership values of the premise part (by using a multiplication operator):

$$W_{mn}^{SB} = \mu_{F_m}(\Delta\theta_{shoot})\mu_{F_n}(\Delta\dot{\theta}_{shoot}) \tag{9}$$

The corrective outputs for the SB behavior level are computed as the weighted average of each rule's output:

$$\upsilon_r^{SB^*} = \frac{\sum_{m=1}^{2}\sum_{n=1}^{2} W_{mn}^{SB} f_{\upsilon_{mn}}}{\sum_{m=1}^{2}\sum_{n=1}^{2} W_{mn}^{SB}}, \qquad \omega_r^{SB^*} = \frac{\sum_{m=1}^{2}\sum_{n=1}^{2} W_{mn}^{SB} f_{\omega_{mn}}}{\sum_{m=1}^{2}\sum_{n=1}^{2} W_{mn}^{SB}} \tag{10}$$

The membership functions $\mu_{F_m}(\Delta\theta_{shoot})$ and $\mu_{F_n}(\Delta\dot{\theta}_{shoot})$ are sigmoid functions, namely:

$$\mu_{F_m}(\Delta\theta_{shoot}) = \frac{1}{1 + e^{-a_{F_m}(\Delta\theta_{shoot} - c_{E_m})}},$$

$$\mu_{F_n}(\Delta\dot{\theta}_{shoot}) = \frac{1}{1 + e^{-a_{F_n}(\Delta\dot{\theta}_{shoot} - c_{F_n})}} \tag{11}$$

where a and c are the parameters to be tuned.

The proposed fuzzy-neural network for learning the SB behavior has 18 tunable parameters.

The overall structure of FNTG is shown in Fig. 10. When the robot is executing the shooting behavior its trajectory is calculated by using the corrective outputs of the shooting layer together with the outputs of the interception and avoidance layer. If only the BIOA behavior is fulfilled, then only the outputs of this layer are used.

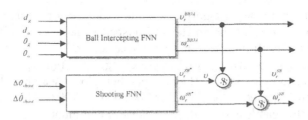

Figure 10: The Fuzzy-Net Trajectory Generator

The symbol S in Fig. 10 denotes a suppression unit that implies the cooperation and competition for the output results of the two behavioral layers. If two reasoning results are expressed by a and b then the suppression logic can be written by

$$c = (1-s)a + sb \qquad (12)$$

where c denotes the result of competition or cooperation and s is defined as follows:

$$s = \frac{|m+n|}{2}, \qquad m = \frac{2}{1+e^{-\gamma a}} - 1, \qquad n = \frac{2}{1+e^{-\gamma b}} - 1 \qquad (13)$$

where γ was taken to be equal to 2.5.

In this problem, the case in which the output results of the two behavioral fuzzy-neural networks have the same sign is defined as "cooperation", whereas the case in which these results have opposite sign is defined as "competition". The reference velocities $v_r^{SB} = \left[v_r^{SB} \quad \omega_r^{SB}\right]^T$, expressed in mobile base coordinates, are further transformed into velocities $\dot{q}_r^{SB} = \left[\dot{x}_r^{SB} \quad \dot{y}_r^{SB} \quad \dot{\theta}_r^{SB}\right]^T$, expressed in Cartesian coordinates by using equation (1). The later are used for determining the robot's posture at the next time step.

Once developed, the shooting template can be further built in both collaborative and adversarial directions. In particular, from a collaborative standpoint, the learned shooting skill can be used to develop a skill to pass the ball to a teammate. The passer can redirect a moving ball in exactly the same way as the shooter, only aiming at a point in front of the teammate instead of a point within the goal. On the other hand, adversarial issues can be studied by introducing a defender, which tries to block the shooter's attempts.

GENETIC LEARNING OF THE FUZZY-NEURAL TRAJECTORY GENERATOR

The classical optimization algorithms are not suitable for finding the optimal (global) solution in multi-parameter search spaces that are typically subject to discontinuities, multi-modality and nonlinear constraints. Genetic algorithms are adaptive methods widely used in searching the optimal point in a highly nonlinear and complex space. They are parallel, global search techniques that emulate natural genetic operations [Michalewicz, 1992]. GAs need not to assume that the search space is differentiable or continuous.

Considering the advantages of GAs, a genetic-based approach has been developed to provide learning of the layers of the FNTG. A local learning method was applied and the learned behaviors were developed from the bottom up. The ball interception and obstacle avoidance behavior was learned first. Then the acquired skills were incorporated into the shooting ball behavior. During the learning of the BIOA behavior the first layer of the trajectory generator was trained without using the SB behavior layer. The learning of the later was then performed together with the use of the outputs of the already trained layer.

Each parameter of the fuzzy-neural networks from the both layers was encoded by 15 bits. The real values of the parameters were limited to (-5, 5). For each fuzzy-neural network the population size was taken to be 60.

The binary strings used for the representation of the candidate solutions for the BIOA layer had a length of 465 bits and the corresponding strings for the SB layer had a length of 270 bits.

Each string from the BIOA layer was encoded in the following form:

$$a_{A_i}^b c_{A_i}^b a_{B_j}^b c_{B_j}^b a_{C_k}^b c_{C_k}^b \sigma_{D_l}^b f_{\upsilon_{1111}}^b \ldots f_{\upsilon_{2223}}^b f_{\omega_{1111}}^b \ldots f_{\omega_{2223}}^b b_{21}^b b_{22}^b \qquad (14)$$

where the superscript "b" stands for binary value and b_{21} and b_{22} were the biases of the output neurons .

Similarly the strings of candidate solutions for the SB layer were represented as follows:

$$a_{F_m}^b c_{F_m}^b a_{F_n}^b c_{F_n}^b f_{\upsilon_{11}}^b \ldots f_{\upsilon_{22}}^b f_{\omega_{11}}^b \ldots f_{\omega_{22}}^b b_{21}^b b_{22}^b \qquad (15)$$

The performance evaluation was based on the following requirements:

1) For each candidate solution the robot was allowed to move during 220 time steps (the time step was set to be 0.03s) and some data were collected during the 100-th, 170-th, 200-th and 220-th step. The purpose of the robot was defined to try to intercept the ball for not more than 200 time steps. During the training of the SB fuzzy-neural network an additional requirement was set up: at the time when the robot intercepts the ball it has to have a heading offset close to zero. For each of the above candidate solutions four trial motions from different starting postures were allowed and the average data were taken as representative.

2) The robot should not collide and should not be in a dangerous proximity with the obstacles during its motion.

3) The robot must stop after a contact with the ball occurs.

The fitness function for the evaluation of the BIOA behavior was defined as follows:

$$F = \frac{10}{\gamma + (6d)^2 + d_1 + d_2 + 2\upsilon_{fil} + 0.5\omega_{fil} + \delta + 60\phi} \qquad (16)$$

where F is the fitness function, γ is a small positive constant (γ was chosen to be 0.001), d, d_1 and d_2 is the average distance (based on four trials) from the robot to the ball at the time step 200, 170 and 220 respectively, $\upsilon_{fil} = \dfrac{\upsilon + \upsilon_1 + \upsilon_2}{3}$ and $\omega_{fil} = \dfrac{|\omega| + |\omega_1| + |\omega_2|}{3}$ where υ and ω, υ_1 and ω_1, υ_2 and ω_2 are the robot's linear and angular velocity at time step 200, 170 and 220 respectively. $\delta = (20(0.145 - H))^2$, where 0.145m is the minimal allowed distance between the robot and the obstacle, $H = \min(h, 0.145)$ and h is the shortest distance between the robot and the obstacle achieved during the robot motion. ϕ is a boolean variable where $\phi = 0$ if linear velocity has achieved during the motion of the robot a 150mm/s and $\phi = 1$ otherwise. The constant value 10 is introduced to enlarge the fitness difference between any two offspring. The variable ϕ is included to force the robot to move during the evaluation period. The variables υ_{fil} and ω_{fil} were included in the denominator to force the robot to stop when the ball is intercepted.

The fitness function used for performance evaluation of the SB behavior was very similar to that used for the evaluation of the BIOA behavior. It is distinguished from the later only by the average absolute values of the robot's heading offset, at the time step 100, 200 and 220 based on four trial motions, added to the denominator.

Initial populations of candidate solutions for both fuzzy-neural networks were generated randomly. Evaluation was executed for the fitness value of each solution to rank the trial solutions. The candidate with higher value had the higher opportunity to reproduce a new population by crossover and mutation. During the evolution, the elite preserving strategy was adopted. Candidates were chosen from both of the offspring cluster and the former solution cluster if their fitness values were higher than the other chromosomes. Thus a new group of candidate solutions was formed at each iteration whose size remained the same as that before the reproduction.

The purpose of crossover lies in the combination of useful string segments from different individuals to form new and better performing offspring. A three-point crossover technique was used. During each crossover operation, two parent strings were chosen from the population randomly and three crossover points were generated randomly. The parent strings were cut from the crossover points to produce two "head" segments and two "tail" segments. The "tail" segments were swapped over to produce two new full-length strings.

Mutation randomly alters each gene in the string with a small probability. Therefore mutation helps ensure that no point in the search space has a zero probability of being explored. It was set to 0.001.

EXPERIMENTAL RESULTS

The entire structure of FNTG and the playground were simulated in the MATLAB® environment. Several moving obstacles representing the other robotic players were also introduced. The following parameters of the geometric model were chosen to be adequately near to those of the laboratory micro-robot: r=0.025 m, R= 0.036 m, d=0.02 m. The maximal linear velocity was limited to 0.7 m/s and the maximal angular velocity – to 4.5 rad/s. Genetic learning with reproduction process iterating until the 95-th population, as explained in the previous section, was implemented consecutively for each of the developed behaviors.

The BIOA behavior was learned first. Fig. 11 shows the change of the average fitness and the best fitness function in the population during the generations.

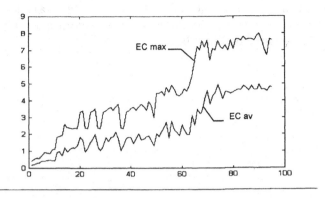

*Figure 11: The Change of the Average Fitness of the Population (EC av) and the Best Fitness
Function in the Population (EC max) During the Generations*

Fig. 12 shows the experimental setup during the genetic learning of BIOA behavior. The generated trajectories correspond to the four trial motions from different starting postures using the parameters having the best fitness function in 95-th population. The positions of the ball, marked with the * symbol, and the other robots were static and randomly changing after 15 generations. A point model of the controlled robot was used and simultaneously the dimensions of the other robots were isotropicaly increased in order to allow correct obstacle avoidance learning.

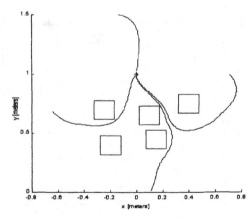

Figure 12: The Experimental Setup During the Genetic Learning of the BIOA Behavior

Fig. 13 shows two examples of the trajectories generated on a simulated playground when the BIOA behavior has been already learned. The robot is intercepting the moving ball while avoiding collisions with three moving robots. Fig. 13b shows that the robot can avoid collisions with the playground walls too.

Fig. 14 shows the changes of the linear velocity (14a) and the changes of the angular velocity (14b) during the time steps corresponding to the motion of the robot pictured on Fig. 13a.

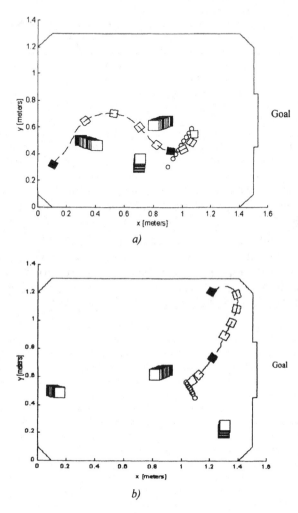

a)

b)

Figure 13: Generated Trajectories that Show the Robot's BIOA Behavior in the Presence of Three Moving Obstacles (the Trajectory of the Ball is Marked with o)

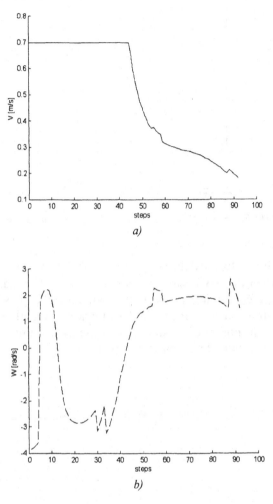

Figure 14: The Changes of the Linear Velocity (14a) and the Angular Velocity (14b) Corresponding to the Movement of the Robot Pictured on Fig. 13a

The learning of the SB behavior was then performed together with the use of the outputs of the already trained BIOA layer. Fig.15 shows the experimental setup. The generated trajectories correspond again to the four trial motions from different starting postures. The parameters having the best fitness function in the 95-th population were used. It can be seen that the robot is moving toward the ball and trying to maintain at the same time an appropriate heading for shooting. The trajectory that starts from the upper area of the playground shows that after surpassing the ball, the robot turns back toward it and tries again to adjust its heading.

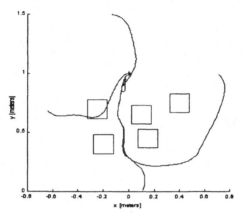

Figure 15: The Experimental Setup During the Genetic Learning of the SB Behavior

Fig. 16 shows two examples of trajectories generated on a simulated playground when the SB behavior has been already learned. It can be seen that at the moment of impact with the ball the robot has the appropriate heading that allows it to redirect the ball toward the goal. The initial world state presented on Fig. 16b is exactly the same as that previously showed in Fig. 13b, but the robot's behavior is different.

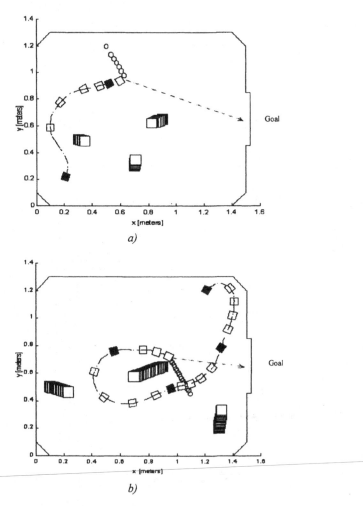

Figure 16: Two Examples of Generated Trajectories when the SB Behavior Has Been Already Learned

Fig. 17 shows the changes of the linear velocity (17a) and the changes of the angular velocity (17b) during the time steps corresponding to the motion of the robot pictured on Fig. 16b.

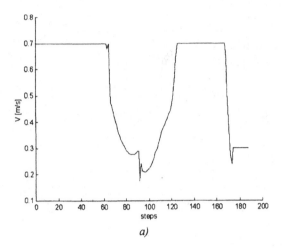

a)

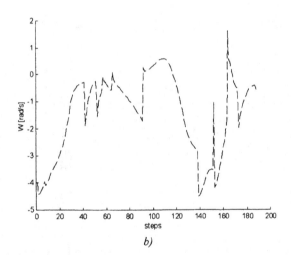

b)

*Figure 17: The Changes of the Linear Velocity (17a) and the Angular Velocity (17b)
Corresponding to the Movement of the Robot Pictured on Fig. 16b*

By layering learned behaviors one of the top of the other, as in this case, all levels of learning that involve interaction with other agents contribute to, and are part of, multi-agent learning. The applicability of the generated trajectories for the two learned behaviors was also tested by using the laboratory micro-robot presented in Fig. 4.

CONCLUSIONS

Control architectures and learning techniques were developed for two basic reactive behaviors of robotic agents playing soccer: (i) interception the moving ball while avoiding collisions with other players and playground walls, and (ii) shooting the ball toward the goal. The two behaviors have been entirely implemented into a trajectory generator level that has a layered structure. This level supplies data to a trajectory-tracking controller. The acquired skills can be further incorporated into higher-level collaborative and adversarial behaviors. The experiments carried out showed the effectiveness of the proposed approach.

ACKNOWLEDGEMENTS

The authors would like to acknowledge the fellowship provided to the first author during his sabbatical leave at the Intelligent Robotics and Automation Laboratory, Department of Electrical and Computer Engineering, NTUA, Greece by the Greek Ministry of Education and the Institute of Communication and Computer Systems, NTUA, Greece.

REFERENCES

Arkin, R. C., Behavior-Based Robotics. The MIT Press, Cambridge, Mass., 1998.

Asada, M., Noda, S., Tawaratsumida, S., and Hosoda, K., "Purposive Behavior Acquisition on a Real Robot by Vision-based Reinforcement Learning," in Proceedings of MLC COLT Workshop on Robot Learning, pp. 1-9, 1994a.

Asada, M., Uchibe, E., Noda, S., Tawaratsumida, S., and Hosoda, K., "Coordination of Multiple Behaviors Acquired by Vision-Based Reinforcement Learning," in Proceedings of IEEE/RSJ/GI International Conference on Intelligent Robots and Systems 1994 (IROS'94), pp. 917-924, 1994b.

Asada, M., Kuniyoshi, Y., Drogoul, A., Asama, H., Mataric, M., Duhaut, D., Stone, P., and Kitano, H., "The RoboCup physical agent challenge: Phase-I," Applied Artificial Intelligence, 12, pp. 127-134, 1998.

Asada, M., and Kitano H., (Eds.), RoboCup-98: Robot Soccer World Cup II, Springer, Berlin, 1999.

Fierro, R., and Lewis, F. L., "Control of a Nonholonomic Mobile Robot: Backstepping Kinematics into Dynamics," in Proceedings of the 34th Conf. On Decision &Control, pp. 381-385, 1995.

Gomi, H., and Kawato, M., "Neural Network Control of a Closed-Loop System Using Feedback-Error-Learning," Neural Networks, Vol. 6, pp. 933-946, 1993.

Izumi, K., and Watanabe, K., "Fuzzy Behavior-based Control with Local Learning," in Tzafestas, S. G., (Ed.), Computational Intelligence in Systems and Control Design and Applications, Kluwer, Boston/Dordrecht, 1999.

Johnson, J., de la Rosa Evista, P., and Kim, J.-H., Benchmark Tests of Robot Soccer Ball Control Skills, (*http://www.fira.net*), 1999.

Kanayama, Y., Kimura, Y., Miyazaki, F., and Noguchi, T., "A Stable Tracking Control Method for an Autonomous Mobile Robot," in Proceedings of IEEE Int. Conf. On Robotics and Automation, Vol. 1, pp. 384-389, 1990.

Kim, J.-H. (Ed.), Proceedings of the Micro-Robot World Cup Soccer Tournament, Taejon, Korea, 1996.

Kitano, H., Veloso, M., Matsubara, H., Tambe, M., Coradeschi, S., Noda, I., Stone, P., Osawa, E., and Asada, M., "The RoboCup Syntethic Agent Challenge 97," in Proceedings of the Fifteen International Joint Conference on Artificial Intelligence, San Francisco, CA, Morgan Kaufman, 1997.

Kitano, H. (Ed.), RoboCup-97: Robot Soccer World Cup I. Springer, Berlin, 1998.

Mackworth, A. K., On seeing robots, in A. Basu and X. Li, (Eds.), Computer Vision: Systems, Theory, and Applications, Word Scientific Press, Singapore, pp. 1-13, 1993.

Matsubara, H., Noda, I., and Hiraki, K., "Learning of Cooperative Actions in Multi-agent Systems: a Case Study of Pass Play in Soccer," in Adaptation, Coevolution and Learning in Multiagent Systems: Papers from the 1996 AAAI Spring Symposium, Menlo Park, CA. AAAI Press, pp. 63-67, 1996.

Michalewicz, Z., Genetic Algorithms + Data Structures = Evolution Programs, Springer, Berlin, 1992.

Stone, P., Veloso, M., and Achim, S., "Collaboration and Learning in Robotic Soccer," in Proceedings of the Micro-Robot World Cup Soccer Tournament, Taejon, Korea, 1996.

Stone, P., and Veloso, M., "Towards Collaborative and Adversarial Learning: A Case Study in Robotic Soccer," International Journal of Human Computer Studies, 48, 1998.

Stonier, R, and Kim, J.-H., (Eds.), FIRA Robot World Cup France'98 Proceedings, FIRA, 1999.

Topalov, A. V., Kim, J.-H., and Proychev, T. Ph., "Fuzzy-Net Control of Non-Holonomic Mobile Robot Using Evolutionary Feedback-Error-Learning," Robotics and Autonomous Systems, 23, pp. 187-200, 1998.

Topalov, A. V., and Tsankova, D. D., "Goal-Directed, Collision-Free Mobile Robot Navigation and Control," in Proceedings of First IFAC Workshop on Multi-Agent Systems in Production, pp. 31-36, 1999.

Tzafestas, S. G., (Ed.), Advances in Intelligent Autonomous Systems, Kluwer, Boston/Dordrecht, 1999.

Veloso, M., Stone, P., and Han, K., "The CMUnited-97 Robotic Soccer Team: Perception and Multi-Agent Control," Robotics and Autonomous Systems, 29, pp. 133-143, 1999a.

Veloso, M., Bowling, M., Achim, S., Han, K., and Stone, P., "The CMUnited-98 Champion Small-Robot Team," in Asada, M., and Kitano, H. (Eds.), RoboCup98: Robot Soccer World Cup II, Springer Verlag, 1999b.

CHAPTER 19

Method-Specific Knowledge Compilation

J. William Murdock
murdock@cc.gatech.edu
Intelligent Systems Group, College of Computing, Georgia Institute of Technology, Atlanta, GA 30332-0280

Ashok K. Goel
goel@cc.gatech.edu
Intelligent Systems Group, College of Computing, Georgia Institute of Technology, Atlanta, GA 30332-0280

Michael J. Donahoo
Jeff_Donahoo@baylor.edu
School of Engineering and Computer Science, Baylor University, P.O. Box 97356, Waco, TX, 76798-7356

Shamkant Navathe
sham@cc.gatech.edu
Database Systems Group, College of Computing, Georgia Institute of Technology, Atlanta, GA 30332-0280

ABSTRACT

Generality and scale are important but difficult issues in knowledge engineering. At the root of the difficulty lie two challenging issues: how to accumulate huge volumes of knowledge and how to support heterogeneous knowledge and processing. One approach to the first issue is to reuse legacy knowledge systems, integrate knowledge systems with legacy databases, and enable sharing of the databases by multiple knowledge systems. We present an architecture called HIPED for realizing this approach. HIPED converts the second issue above into a new form: how to convert data accessed from a legacy database into a form appropriate to the processing method used in a legacy knowledge system. One approach to this reformed issue is to use method-specific compilation of data into knowledge. We describe an experiment in which a legacy knowledge system called INTERACTIVE KRITIK is integrated with an ORACLE database. The experiment indicates the computational feasibility of method-specific data-to-knowledge compilation.

D. Braha (ed.), Data Mining for Design and Manufacturing, 443–463.
© 2001 *Kluwer Academic Publishers. Printed in the Netherlands.*

MOTIVATIONS, BACKGROUND, AND GOALS

Generality and scale have been important issues in knowledge systems research ever since the development of the first expert systems in the mid sixties. Yet, some thirty years later, the two issues remain largely unresolved. Consider, for example, current knowledge systems for engineering design. The scale of these systems is quite small both in the amount and variety of knowledge they contain, and the size and complexity of problems they solve. In addition, these systems are severely limited in that their knowledge is relevant only to a restricted class of device domains (e.g., electronic circuits, elevators, heat exchangers) and their processing is appropriate only to a restricted class of design tasks (e.g., configuration, parameter optimization).

At the root of the difficulty lie two critical questions. Both generality and scale demand huge volumes of knowledge. Consider, for example, knowledge systems for a specific phase and a particular kind of engineering design, namely, the conceptual phase of functional design of mechanical devices. A robust knowledge system for even this very limited task may require knowledge of millions of design parts. Thus the first hard question is this: How might we accumulate huge volumes of knowledge? Generality also implies heterogeneity in both knowledge and processing. Consider again knowledge systems for the conceptual phase of functional design of mechanical devices. A robust knowledge system may use a number of processing methods such as problem/object decomposition, prototype/plan instantiation, case-based reuse, model-based diagnosis and model-based simulation. Each of these methods uses different kinds of knowledge. Thus the second hard question is this: How might we support heterogeneous knowledge and processing?

Recent work on these questions may be categorized into two families of research strategies: (i) ontological engineering, and (ii) reuse, integration and sharing of information sources. The well-known CYC project (Lenat and Guha, 1990) that seeks to provide a global ontology for constructing knowledge systems exemplifies the strategy of ontological engineering. This bottom-up strategy focuses on the first question of accumulation of knowledge. The second research strategy has three elements: reuse of information sources such as knowledge systems and databases, integration of information sources, and sharing of information in one source by other systems. This top-down strategy emphasizes the second question of heterogeneity of knowledge and processing and appears especially attractive with the advent of the world-wide-web which provides access to huge numbers of heterogeneous information sources such as knowledge systems, electronic databases and digital libraries. Our work falls under the second category.

McKay, et al. (1990) have pointed out that a key question pertaining to this topic is how to convert data in a database into knowledge useful to a knowledge system. The answer to this question depends in part on the processing method used by the knowledge system. The current generation of knowledge systems are heterogeneous both in their domain knowledge and control of processing. They not only use multiple methods, each of which uses a specific kind of knowledge and control of processing, but they also enable dynamic method selection. Our work focuses on the interface between legacy databases and legacy knowledge systems of the current generation.

The issue then becomes: given a legacy database, and given a legacy knowledge system in which a specific processing method poses a particular knowledge goal (or query), how might the data in the database be converted into a form appropriate to the processing method? The form of this question indicates a possible answer: method-specific knowledge compilation, which would transform the data into a form appropriate to the processing strategy. The goal of this paper is to outline a conceptual framework for the method-specific knowledge compilation technique. Portions of this framework are instantiated in an operational computer system called HIPED (for Heterogeneous Intelligent Processing for Engineering Design). HIPED integrates a knowledge system for engineering design called INTERACTIVE KRITIK (Goel et al., 1996b&c) with an external database represented in Oracle (Koch and Loney, 1995). The knowledge system and the database communicate through IDI (Paramax, 1993).

THE HIPED ARCHITECTURE

To avoid the enormous cost of constructing knowledge systems for design, HIPED proposes the reuse of legacy knowledge and database systems, so that we can quickly and inexpensively construct large-scale systems with capabilities and knowledge drawn from existing systems. To facilitate easy integration which, in effect, increases overall scalability, we restrict ourselves to making few, if any, changes to the participating legacy systems. The long-term goal is to allow a system to easily access the capabilities of a pool of legacy systems.

The architecture of Figure 1 illustrates the general scheme. In this figure, arrowed lines indicate unidirectional flow of information; all other lines indicate bidirectional flow. Annotations on lines describe the nature of the information which flows through that line. Rectangular boxes indicate functional units and cylinders represent collections of data. The architecture presented in this figure is a projected reference architecture and not a description of a specific existing system. In this section, we describe the entire reference architecture. In the following sections we will further elaborate on a particular piece of work which instantiates a portion of this architecture.

Database Integration

An enormous amount of data is housed in various database systems. Unfortunately, the meaning of this data is not encoded within the databases themselves. At best, the database schema has meaningful names for individual data elements, but often it is difficult to infer all, if any, of the meaning of the data from the schema. This lack of metadata about the schema and a myriad of interfaces to various database systems create significant difficulties in accessing data from various legacy database systems. Both of these problems can be alleviated by creating a single, global representation of all of the legacy data, which can be accessed through a single interface.

Common practice for integration of legacy systems involves manual integration of each legacy schema into a global schema. That is, database designers of the various legacy systems create a global schema capable of representing the collection of data in the legacy databases and provide a mapping between the legacy system schemas and this global schema (Batini et al., 1986). Clearly, this approach does not work for integration of a large number of database systems. We propose (see the right side of Figure 1) to allow the database designers to develop a metadata description, called an augmented export schema, of their database system. A collection of augmented export schemas can then be automatically processed by a schema builder to create a partially integrated global schema which can be as simple as the actual database schema, allowing any database to easily participate, or as complicated as the schema builder can understand; see (Navathe and Donahoo, 1995) for details on possible components of an augmented export schema. A user can then submit queries on the partially integrated global schema to a query processor which fragments the query into sub-queries on the local databases. Queries on the local databases can be expressed in a single query language which is coerced to the local database's query language by a database wrapper.

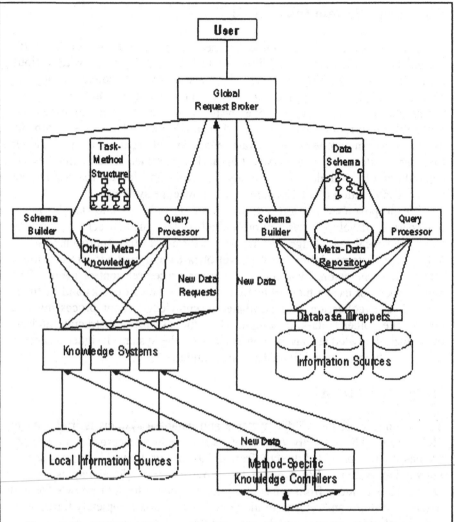

Figure 1: The HIPED architecture. The right side of the figure depicts the integration of databases through partially integrated schema. The left side of the figure shows an analogous integration of knowledge systems. The global request broker at the top provides a uniform interface to both sets of integrated systems so that a request for information may be retrieve from memory and / or deduced by a knowledge system without a specific demand from the user for one approach or the other.

Knowledge System Integration

As with databases, a considerable number of knowledge systems exist for design each with their own abilities to perform certain tasks with various methods. Users wishing to access the capabilities of a collection of such systems encounter problems of different interfaces and knowledge representations. Most knowledge systems do not provide an externally accessible description of the tasks and methods they address. As with the database system, one way to integrate legacy knowledge systems is to gather together the designers and construct an ad hoc interface which combines the capabilities of the underlying systems. Once again, this approach does not work for integration of a large number of knowledge systems.

We propose (see the left side of Figure 1) to allow knowledge system designers to develop a description, called a "task-method schema," of the tasks each local knowledge system can perform . In this approach, a set of knowledge systems, defined at the level of tasks and methods, are organized into a coherent whole by a query processor or central control agent. The query processor uses a hierarchically organized schema of tasks and methods as well as a collection of miscellaneous knowledge about processing and control (i.e., other meta-knowledge). Both the task-method structure and the other meta-knowledge may be constructed by the system designer at design time or built up by an automated schema builder.

Integrated Access

The dichotomy of knowledge systems and database systems is irrelevant to global users. Users simply want answers and are not concerned with whether the answer was provided directly from a database or derived by a process in a knowledge system. We propose the provision of a global request broker which takes a query from a user, submits the query to both knowledge and database systems and returns an integrated result. It is completely transparent to a global user how or from where an answer was derived.

Furthermore, the individual knowledge systems may, themselves, act as users of the integrated access mechanism. The knowledge systems each have their own local repositories of data but may also find that they need information from a database or another knowledge system. When they need external knowledge, they simply contact the global request broker which can either recursively call the general collection of knowledge systems to generate a response or contact the system of databases. When either a database or a knowledge system generates a response to a request from a knowledge system, the resulting answer is then sent through a method-specific knowledge compiler which does whatever specific translation is needed for the particular system.

Method-Specific Knowledge Compilation

In this paper, we are concerned with the compilation of knowledge from external sources into a form suitable for use by a knowledge system method. Recall that we do not want to alter the knowledge system, so the form of the knowledge may be very specific to the particular method which executed the query; consequently, we call this a "method-specific knowledge compilation." We will examine the mechanisms behind this component of the reference architecture in more detail in later sections.

Information Flow

Consider a design knowledge system which spawns a task for finding a design part such as a battery with a certain voltage. In addition to continuing its own internal processing, the knowledge system also submits a query to the global request broker. The broker sends the query to the query processors for both integrated knowledge and database systems. The database query processor fragments the query into subqueries for the individual databases. The data derived is merged, converted to the global representation, and returned to the global request broker. Meanwhile, the knowledge query processor, using its task-method schema, selects knowledge systems with appropriate capabilities and submits tasks to each. Solutions are converted to a common representation and sent back to the global request broker. It then passes the output from both the knowledge and database system query processors through a method-specific knowledge compiler which coerces the data into a form which is usable by the requesting knowledge system. The resulting battery may be an existing battery which satisfies the voltage specification from a knowledge or database system information source or it may be a battery constructed from a set of lower voltage batteries by a knowledge system.

KNOWLEDGE COMPILATION

The architecture described in the previous section raises an enormous variety of issues. The one that we want to focus on more closely here is that of knowledge compilation, i.e. the principled transformation of knowledge from one form to another. Large volumes of information can be found in existing databases of components, design schematics, etc. An intelligent design system can make use of this existing data by compiling it into a form which is suitable to its use. There are several closely interrelated perspectives on knowledge compilation as presented in "Knowledge Compilation: A Symposium" (Goel ed., 1991). A few of these perspectives relevant to this context are:

1. Knowledge implemented in one representational paradigm may be transformed into another. One example of this is the conversion of tuples in a relational database into objects in an object-oriented programming language (Castellanos et al., 1994; Ramanathan, 1994; Fahrner and Vossen, 1995). Another is the collection of information from diverse network resources into a semantically-annotated document in a browsable markup language, as in (Stroulia et al., 2000).

2. Knowledge in a generally useful organization may be transformed into a different organization which is more efficient for a specific application. For example, a model of a device may be transformed into a set of rules for diagnosing faults of that device, as per (Keller, 1991).

3. Declarative knowledge may be transformed into procedural knowledge (Tong, 1991; Anderson and Fincham, 1994), i.e., a specification of a task may be compiled into a procedure for accomplishing that task. This is really an extreme form of the former approach; the result is a knowledge element which typically supports only one application, its execution, but presumably does so in the most efficient way that the compiler can generate.

4. Knowledge of patterns or categories can be inductively inferred from elements. This can be also be seen as an extension of point 2, above; knowledge of instances may be voluminous and difficult to apply to new situations and thus this knowledge is compiled into descriptions or rules which directly enable recognition, classification, etc. Virtually all work done in the field of machine learning on the subject of classification can be viewed as knowledge compilation from this perspective, e.g., (Quinlan, 1986; Mitchell, 1997).

In this work, we have limited our attention to the first of these topics. We believe that all of these approaches to knowledge compilation are interesting and important. We intend to address all of these concerns in our future research. However, for the purposes of supporting large scale, heterogeneous processing, it is clear that the first issue, that of transforming the structural details of the representation, is inherently fundamental; without a framework for such basic transformations, any of the more complex, sophisticated approaches to knowledge compilation are useless because they simply cannot access any knowledge to compile.

INTERACTIVE KRITIK

INTERACTIVE KRITIK is a legacy knowledge system which we have integrated into the HIPED architecture. INTERACTIVE KRITIK is a computer-based design environment. A major component of this system is KRITIK3, an autonomous knowledge-based design system. When completed, INTERACTIVE KRITIK is intended to serve as an interactive constructive design environment. At present, when asked by a human user, INTERACTIVE KRITIK can invoke KRITIK3 to address specific kinds of design problems. In addition, INTERACTIVE KRITIK can provide explanations and justifications of KRITIK3's design reasoning and results, and enable a human user to explore the system's design knowledge.

KRITIK3 evolves from KRITIK, an early multi-strategy case-based design system. Since KRITIK is described in detail elsewhere, e.g., (Goel and Chandrasekaran, 1989; Goel et al. 1996a), in this paper we only sketch the outlines of KRITIK3. One of the major contributions of KRITIK is its device modeling formalism: the Structure-Behavior-Function (SBF) language. The remarkable characteristics of SBF models are: (i) they are functional, i.e. they describe both basic components and complex devices in terms of the function they achieve; (ii) there are causal, i.e. they describe sequences of interactions which constitute the internal behavior of the device; and (iii) they are compositional, i.e. they describe how the function of the device emerges from the functions of the components.

KRITIK3 is a multi-strategy process model of design in two senses. First, while the high-level design process in KRITIK3 is case-based, the reasoning about individual subtasks in the case-based process is model-based; KRITIK3 uses device models described in the SBF language for adapting a past design and for evaluating a candidate design. Second, design adaptation in KRITIK3 involves multiple modification methods. While all modification methods make use of SBF device models, different methods are applicable to different kinds of adaptation tasks.

The primary task addressed by KRITIK3 is the extremely common functions-to-structure design task in the domain of simple physical devices. The functions-to-structure design task takes as input the functional specification of the desired design. For example, the functions-to-structure design of a flashlight may take as an input the specification of its function of creating light when a force is applied on a switch. This task has the goal of giving as output the specification of a structure that satisfies the given functional specification, i.e., a structure that results in the given functions.

KRITIK3's primary method for accomplishing this task is case-based reasoning. Its case-based method sets up four subtasks of the design task: problem elaboration, case retrieval, design adaptation, and case storage.

The task of problem elaboration takes as input the specification of the desired function of the new design. It has the goal of generating a probe to be used by case retrieval for deciding on a new case to use. KRITIK3 uses domain-specific heuristics to generate probes based on the surface features of the problem specification. The task of case retrieval takes as input the probes generated by the problem elaboration component. It has the goal of accessing a design case, including the associated SBF model whose functional specification is similar to the specification of desired design. KRITIK3's case memory is organized in a discrimination tree, with features in the functional specifications of the design cases acting as the discriminants. Its retrieval method searches through this discrimination tree to find the case that most closely matches the probe.

The task of design adaptation takes as input (i) the specification of the constraints on the desired design, and (ii) the specifications of the constraints on and the structure of the candidate design. It has the goal of giving as output a modified design structure that satisfies the specified constraints. KRITIK3 uses a model-based method of design adaptation which divides the design task into three subtasks: computation of functional differences, diagnosis, and repair. The idea here is that the candidate design can be viewed as a failed attempt to accomplish the desired specifications. The old design is first checked to see how its functionality differs from the desired functionality. The model of the design is then analyzed in detail to determine one or more possible causes for the observed difference. Lastly, KRITIK3 makes modifications to the device with the intent of inducing the desired functionality.

The method of repair used by KRITIK3 is generate and test. This method sets up two subtasks of the repair task: model revision and model verification. The task of model revision takes as input (i) the specification of the constraints on the desired design, and (ii) the model of the candidate design. It has the goal of giving as output a modified model that is expected to satisfy the constraints on the desired design. KRITIK3 knows of several model revision methods such as component replication or component replacement. KRITIK3 dynamically chooses a method for model revision at run time based on the results of the diagnosis task. Depending on the modification goals set up by the diagnosis task, the system may also use more than one model-revision method.

The task of model verification takes as input (i) the specification of the constraints on the desired design, and (ii) the specification of the structure of the modified design. It has the goal of giving as output an evaluation of whether the modified structure satisfies the specified constraints. KRITIK3 qualitatively simulates the revised SBF model to verify whether it delivers the functions desired of it.

The task of case storage takes as input (i) a specification of the case memory, and (ii) a specification of a new case. It has the goal of giving as output a specification of the new case memory with the new case appropriately indexed and organized in it. Recall that KRITIK3's case memory is organized in a discrimination tree. The system uses a model-based method for the task of storing a new case in the tree. This method sets up the subtasks of indexing learning and case placement. The SBF model of the new design case enables the learning of the appropriate index to the new case. This directly enables the task of case placement, which stores the model in KRITIK3's local information sources.

AN EXPERIMENT WITH HIPED

We have been conducting a series of experiments in the form of actual system implementations. Some of these experiments have focused on issues most closely related to the data end of the data to knowledge compilation process; these issues include data organization, access, transfer, etc. The experiment we focus on here, however, is more closely connected to the knowledge end of the process. This experiment examines the use of knowledge compiled at run-time in the context of the operation of INTERACTIVE KRITIK.

Figure 2 presents an architectural view of the experiment, in which a legacy knowledge system for design requests and receives information from a general-purpose database system. Since this experiment deals with only one knowledge system and only one database, we are able to abstract away a great many issues and focus on a specific question: method-specific knowledge compilation.

General Method

The overall algorithm developed in this experiment breaks down into four steps which correspond to the four architectural components shown in Figure 2:

Step 1: The knowledge system issues a request when needed information is not available in its local information source.

Step 2: The query processor translates the request into a query in the language of the information source.

Step 3: The information source processes the query and returns data to the query processor, which sends the data to the method-specific knowledge compiler.

Step 4: The method-specific knowledge compiler converts the data into a knowledge representation format which can by understood by the knowledge system.

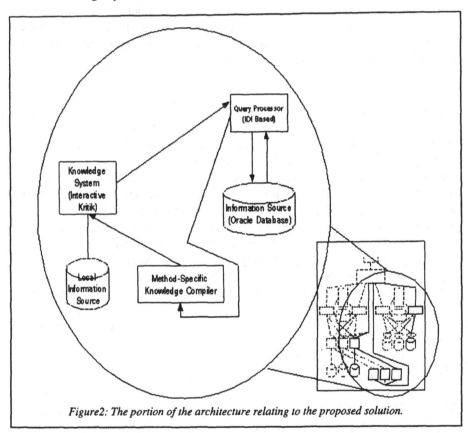

Figure2: The portion of the architecture relating to the proposed solution.

All four of these steps pose complex problems. Executing step one requires that a knowledge system recognize that some element is missing from its knowledge and that this element would help it to solve the current problem. Performing step two requires a mechanism for constructing queries and providing communication to and from the external system. Step three is the fundamental task of databases: given a query produce a data item. Lastly, step four poses a challenging problem because the differences between the form of the data in the information source and the form required by the knowledge system may be arbitrarily complex. We focus on the fourth step:

method-specific knowledge compilation. The algorithm for the method-specific knowledge compiler implemented in our experimental system is as follows:

Substep 4.1: Database data types are coerced into to knowledge system data types.
Substep 4.2: Knowledge attributes are constructed from fields in the data item.
Substep 4.3: Knowledge attributes are synthesized into a knowledge element.

The particular legacy systems for which we have implemented these algorithms are INTERACTIVE KRITIK and a relational database developed under Oracle. Thus the experimental system serves as an interface between INTERACTIVE KRITIK and our Oracle database.

An Illustrative Example

Our experiment takes place during a session in which INTERACTIVE KRITIK is designing an electric light circuit. It has retrieved from its case-memory a model of a circuit which produces light. However, in comparing the functional specification of the retrieved case with the desired functional specification, INTERACTIVE KRITIK determines that the retrieved case does not produce enough light. Consequently, it applies its diagnosis methods to determine components which might be responsible for the amount of light produced. One of the results generated by the diagnosis mechanism is that a higher capacity bulb will lead to the production of more light. Consequently, INTERACTIVE KRITIK may be able to apply the component replacement method of model revision. However, in order to apply this method, it must have knowledge of a light bulb of sufficient capacity. No such bulb is available in its local knowledge base. In earlier versions of this system, it would conclude that replacement of the bulb is impossible and thus either attempt to replace a different component or attempt a different model revision method altogether. However, in this version, INTERACTIVE KRITIK has access to an external source of knowledge via the HIPED architecture.

INTERACTIVE KRITIK sends a request for the desired light bulb to the query processor. The request is made as a LISP function call to a function named lookup-database-by-attribute which takes three arguments: a prototype, an attribute, and a value for that attribute. An example of such a call from the system is a request for a more powerful light bulb for which the prototype is the symbol 'L-BULB which refers to the general class of light bulbs, the attribute is the symbol 'CAPACITY, and the value is the string "capacity-more'" which is internally mapped within INTERACTIVE KRITIK to a quantitative value.   Note that INTERACTIVE KRITIK makes use of both quantitative and qualitative values in its reasoning methods. The details of the interactions between these two kinds of information within the system are moderately complex and beyond the scope of this paper. Obviously, the experiment would be more realistic if the external database used quantitative values.   This would add another step to the method-specific knowledge compilation process (mapping quantitative to qualitative values) but would not significantly affect the process as a whole. The query processor uses IDI to generate an SQL query as follows:

```
SELECT  DISTINCT RV1.inst_name
FROM    PROTO_INST RV1, INSTANCE RV2
WHERE   RV1.proto_name = 'l-bulb'
AND     RV1.inst_name = RV2.name
AND     RV2.att_val = 'capacity-more'
```

IDI sends this query to Oracle running on a remote server.   Oracle searches through the database tables illustrated below.   The first of these tables, INSTANCE, holds the components themselves.   The second table, PROTOINST, is a cross-reference table which provides a mapping from components to prototypes.

Table INSTANCE			Table PROTO_INST	
NAME	ATTRIBUTE	ATT_VAL	INST_NAME	PROTO_NAME
littlebulb	lumens	capacity-less	littlebulb	l-bulb
bigmotor	watts	power-more	bigmotor	motor
bigbulb	lumens	capacity-more	bigbulb	l-bulb

If Oracle finds a result, as it does in this example, it returns it via the method-specific knowledge compiler. In this case, the query generates the string "bigbulb" as the result. The prototype name and the value are also part of the result, but they are not explicitly returned by the database since they are the values used to select the database entry in the first place. The method-specific knowledge compiler converts the raw data from the database to a form comprehensible to INTERACTIVE KRITIK by using the algorithm described earlier. In Substep 4.1, the string "bigbulb" is converted from a fixed length, blank padded string, as returned by Oracle, to a variable length string, as expected by INTERACTIVE KRITIK. In Substep 4.2, the attributes of the new bulb are generated. The values "bigbulb" and 'L-BULB are used as the knowledge attributes name and prototype-comp; the values 'CAPACITY,

'LUMENS, and "capacity-more" are combined into an object of a class named parameter and a list containing this one object is created and used as the parameters attribute of the component being constructed. Finally, in Substep 4.3 these three attribute values are synthesized into a single object of the component class. The end result of this process is an object equivalent to the one defined by the following statement:

```
(make-instance 'component
   :init-name      "bigbulb"
   :prototype-comp 'L-BULB
   :parameters (list (make-instance 'parameter
                      :init-name    'CAPACITY
                      :parm-units   'LUMENS
                      :parm-value   "capacity-more")))
```

These commands generate an object of the component class with three slots. The first slot contains the component name, the second contains the prototype of the component, and the third is a list of parameters. The list of parameters contains a single item, which is, itself, an object. This object is a member of the parameter class and has a parameter name, the units which this parameter is in, and a value for the parameter. This object is then returned to INTERACTIVE KRITIK.

Once INTERACTIVE KRITIK has received the description of the bulb, it is consequently able to apply the component replacement method. It replaces the bulb in the case retrieved earlier with the new bulb returned by the query processor. The effects of this substitution are propagated through the model and INTERACTIVE KRITIK verifies that the adapted model does accomplish the requirements which were initially specified. Finally, INTERACTIVE KRITIK presents the revised model to the user and stores it into the case-memory for further reuse in a later problem-solving session.

RELATED RESEARCH

Knowledge systems for design developed thus far appear incapable of supporting practical design. This critique surely is valid for all laboratory knowledge systems such as AIR-CYL (Brown and Chandrasekaran, 1989), COMADE (Lenz et al., 1996), and our own KRITIK series of systems (Stroulia et al., 1992; Bhatta et al., 1994, Goel et al., 1996). But it is also applicable to systems that have directly led to real applications such as R1 (McDermott,

1982), PRIDE (Mittal et al. 1986), VT (Marcus et al., 1988), CLAVIER (Hennessy and Hinkle, 1992), and ASK-JEF (Barber et al., 1992). The problem is the limited scale and scope of knowledge systems for design.

The ontological engineering approach to extending the scale and scope of knowledge systems has been the focus of some research. CYC (Lenat and Guha, 1990) pursues this approach at a global level, seeking to provide a single, coherent ontology for constructing knowledge systems. Ontolingua (Gruber, 1993) provides another, domain-specific example of ontological engineering. The bottom-up approach focuses on the second question of accumulation of knowledge: domain-specific and domain-independent ontologies may one day enable interactive knowledge acquisition from external sources and autonomous acquisition of knowledge through learning from experience. But ontological engineering requires the building of new systems based on the common ontology.

In contrast, our work falls under the information integration paradigm. This approach emphasizes the reuse of legacy information sources such as databases, integration of the information sources, and sharing of information in one source by other systems. This top-down strategy focuses on one part of the scalability issue, namely, heterogeneity of knowledge and processing. Various projects on information integration have focused on different aspects of information integration. For example, KQML (Finin et al., 1994) provides a protocol language for communication among database systems, and KIF (Genesreth and Fikes, 1991) provides a meta-language for enabling translation between knowledge systems. In contrast, (Brodie, 1988) has emphasized the need for integrating knowledge systems and databases. McKay et al. (1990) in particular have pointed out that a key question is how to convert data in a database into knowledge useful to a knowledge system. The answer to this question clearly depends in part on the problem-solving method used by the knowledge system. The work presented here on method-specific knowledge compilation provides one perspective on this issue.

DISCUSSION

The complexity involved in constructing a knowledge system makes reuse an attractive option for true scalability. However, the reuse of legacy systems is non-trivial because we must accommodate the heterogeneity of systems. The scalability of the HIPED architecture comes from the easy integration of legacy systems and transparent access to the resulting pool of legacy knowledge. Sharing data simply requires that a legacy system designer augment the existing local schema with metadata that allows a global coordinator to relate data from one system to another, providing a general solution to large-scale integration.

While the experiment described here shows how data in a general-purpose design database can be compiled into a specific kind of knowledge required by a particular problem-solving method, it raises an additional issue. If the number of the problem-solving methods is large, and each method requires a knowledge compiler specific to it, then the HIPED architecture would require the construction of a large number of method-specific data to knowledge compilers. In the case of knowledge systems for design, which typically contain many problem-solving methods, this itself would make for significant knowledge engineering.

The issue then becomes whether we can identify primitive building blocks from which we can rapidly construct individual method-specific knowledge compilers. In the example discussed, it appears that the three steps of the specific method for converting data into knowledge described can all reasonably be considered to be relatively generic units of functionality. Consider Substep 4.1 in the example, coercion of database types into knowledge system types: it is not unreasonable to expect that a wide variety of methods might have the same data coercion requirements and thus be able to use the same data coercion routines in their method-specific knowledge compilers. Further, many knowledge systems for design use representations which are characterized as knowledge elements composed of a set of attribute-value pairs. The general framework for Substeps 4.2 and 4.3 of the algorithm (building attribute-value pairs and then combining them to form a knowledge element) probably can be applied to a wide variety of knowledge-based methods. Furthermore, to the extent that some methods have similar forms and mechanisms for constructing these elements, they might be able to share specific routines. Our experiment suggests that it may be possible to abstract generic components of method-specific compilations. Doing so may partially mitigate the problem of constructing large numbers of method-specific knowledge compilers as individual knowledge compilers might be built from a small and parsimonious set of components. But our experiments with HIPED have not yet demonstrated this; more research is required to fully explore this hypothesis.

The specific experiment described here models only a small portion of the general HIPED architecture. In a related experiment, we have worked with another portion of the architecture (Navathe et al., 1996). Here, five types of queries that INTERACTIVE KRITIK may create are expressed in an SQL-like syntax. The queries are evaluated by mapping them using facts about the databases and rules that establish correspondences among data in the databases in terms of relationships such as equivalence, overlap, and set containment. The rules enable query evaluation in multiple ways in which the tokens in a given query may match relation names, attribute names, or values in the underlying databases tables. The query processing is implemented using the CORAL deductive database system (Ramakrishnan et al., 1992). While the experiment described in this paper demonstrates method-specific compilation of data into knowledge usable by INTERACTIVE KRITIK, the other experiment shows how queries from INTERACTIVE KRITIK can be flexibly evaluated in multiple ways. We expect an integration of the two to provide a seamless and flexible technique for integration of knowledge systems with databases through method-specific compilation of data into useful knowledge.

The complexity involved in constructing knowledge systems for practical design makes integration of legacy knowledge systems and legacy databases an attractive option. But, insofar as we know, past research on information integration has almost exclusively focused on heterogeneity of information, not on heterogeneity of processing. However, third-generation knowledge systems for design are heterogeneous in that they use multiple problem-solving methods, each of which uses a specific kind of knowledge and control of processing. Thus the goal of constructing third-generation knowledge systems for practical design requires support for heterogeneity of processing in addition to that of information. This itself is a hard and complex goal that requires long-term research. Building on earlier work on integrating knowledge systems and databases systems (Brodie, 1988; McKay et al., 1990), HIPED identifies some issues in supporting heterogeneous information processing and takes a first step towards achieving the goal.

The issues of generality and scale of intelligent design systems require robust answers to two questions: how to accumulate huge volumes of knowledge, and how to support heterogeneous knowledge and processing? Our approach suggests reuse of legacy design systems, integration of knowledge systems with legacy information sources, and sharing of the information sources by multiple design systems. In addition, given a reusable and sharable information source such as a legacy database, and given a legacy design system that uses a specific processing strategy, the approach suggests the strategy of method-specific knowledge compilers that convert the content of the information into the appropriate form for the processing strategy. HIPED represents a first step towards realizing this strategy.

ACKNOWLEDGEMENTS

This paper has benefited from many discussions with Edward Omiecinski. This work was funded by a DARPA grant monitored by WPAFB, contract #F33615-93-1-1338, and has benefited from feedback from Chuck Sutterwaite of WPAFB.

REFERENCES

Anderson, J. R. and J. M. Fincham: 1994, "Acquisition of procedural skills from examples". Journal of Experimental Psychology: Learning, Memory, and Cognition 20, 1322--1340.

Barber, J., M. Jacobson, L. Penberthy, R. Simpson, S. Bhatta, A. K. Goel, M. Pearce, M. Shankar, and E. Stroulia: 1992, "Integrating Artificial Intelligence and Multimedia Technologies for Interface Design Advising". NCR Journal of Research and Development 6(1), 75--85.

Batini, C., M. Lenzernini, and S. B. Navathe: 1986, "A Comparative Analysis of Methodologies for Database Schema Integration". ACM Computing Surveys 18(4), 325--364.

Bhatta, S., A. K. Goel, and S. Prabhakar: 1994, "Analogical Design: A Model-Based Approach". In: Proceedings of the Third International Conference on Artificial Intelligence in Design. Lausanne, Switzerland.

Brodie, M.: 1988, "Future Intelligent Systems: AI and Database Technologies Working Together". In: Mylopoulos and Brodie (eds.): Reading in Artificial Intelligence and Databases. Morgan Kauffman, pp. 623--641.

Brown, D. and B. Chandrasekaran: 1989, Design Problem Solving: Knowledge Structures and Control Strategies. London, UK: Pitman.

Castellanos, M., F. Saltor, and M. Garcia-Solaco: 1994, "Semantically Enriching Relational Databases into an Object Oriented Semantic Model". In: D. Karagiannis (ed.): Proceedings of the 5th International Conference on Database and Expert Systems Applications - DEXA-94, Vol. 856 of Lecture Notes in Computer Science. Athens, Greece, pp. 125--134.

Codd, E.: 1970, "A Relational Model for Large Shared Data Banks". CACM 13(6).

Fahrner, G. and G. Vossen: 1995, "Transforming Relational Database Schemas into Object Oriented Schemas according to ODMG-93". In: Proceedings of the Fourth International Conference of Deductive and Object Oriented Databases. pp. 429--446.

Finin, T., D. McKay, R. Fritzson, and R. McEntire: 1994, "KQML: An Information and Knowledge Exchange Protocol". In: K. Fuchi and T. Yokoi (eds.): In: Knowledge Building and Knowledge Sharing. Ohmsha and IOS Press.

Genesreth, M. R. and R. Fikes: 1991, "Knowledge Interchange Format Version 2 Reference Manual". Stanford University Logic Group.

Goel, A. K., S. Bhatta, and E. Stroulia: 1996a, "Kritik: An Early Case-Based Design System". In: M. L. Maher and P. Pu (eds.): Issues and Applications of Case-Based Reasoning to Design. Lawrence Erlbaum Associates.

Goel, A. K. and B. Chandrasekaran: 1989, "Functional Representation of Designs and Redesign Problem Solving". In: Proc. Eleventh International Joint Conference on Artificial Intelligence. pp. 1388--1394.

Goel, A. K., A. Gomez, N. Grue, J. W. Murdock, M. Recker, and T. Govindaraj: 1996b, "Explanatory Interface in Interactive Design Environments". In: J. S. Gero and F. Sudweeks (eds.): Proceedings of the Fourth International Conference on Artificial Intelligence in Design. Stanford, California.

Goel, A. K., A. Gomez, N. Grue, J. W. Murdock, M. Recker, and T. Govindaraj: 1996c, "Towards Design Learning Environments - I: Exploring How Devices Work". In: C. Frasson, G. Gauthier, and A. Lesgold (eds.): Proceedings of the Third International Conference on Intelligent Tutoring Systems. Montreal, Canada.

Goel (ed.), A. K.: 1991, "Knowledge Compilation: A Symposium". IEEE Expert 6(2).

Gruber, T. R.: 1993, "A Translation Approach to Portable Ontology Specification". Knowledge Acquisition 5(2), 199--220.

Hennessy, D. and D. Hinkle: 1992, "Applying Case-Based Reasoning to Autoclave Loading". IEEE Expert pp. 21--26.

Keller, R. M.: 1991, "Applying Knowledge Compilation Techniques to Model-Based Reasoning". IEEE Expert 6(2).

Koch, G. and K. Loney: 1995, Oracle: The Complete Reference. Osborne/McGraw Hill/Oracle, 3rd edition.

Lenat, D. and R. Guha: 1990, Building Large Knowledge Based Systems: Representation and Inference in the CYC Project. Addison-Wesley.

Lenz, T., J. McDowell, A. Kamel, J. Sticklen, and M. C. Hawley: 1996, "The Evolution of a Decision Support Architecture for Polymer Composites Design". IEEE Expert 11(5), 77--83.

Marcus, S., J. Stout, and J. McDermott: 1988, "VT: An Expert Elevator Designer that Uses Knowledge-Based Backtracking". AI Magazine 9(1), 95--112.

McDermott, J.: 1982, "R1: A Rule-Based Configurer of Computer Systems". Artificial Intelligence 19, 39--88.

McKay, D., T. Finin, and A. O'Hare: 1990, "The Intelligent Database Interface". In: Proceedings of the Eight National Conference on Artificial Intelligence. Menlo Park, CA, pp. 677--684.

Mitchell, T. M.: 1997, Machine Learning. New York: McGraw-Hill.

Mittal, S., C. Dym, and M. Morjaria: 1986, "PRIDE: An Expert System for the Design of Paper Handling Systems". Computer 19(7), 102--114.

Navathe, S. B. and M. J. Donahoo: 1995, "Towards Intelligent Integration of Heterogeneous Information Sources". In: Proceedings of the 6th International Workshop on Database Re-engineering and Interoperability.

Navathe, S. B., S. Mahajan, and E. Omiecinski: 1996, "Rule Based Database Integration in HIPED: Heterogeneous Intelligent Processing in Engineering Design". In: Proceedings of the International Symposium on Cooperative Database Systems for Advanced Applications.

Paramax: 1993, "Software User's Manual for the Cache-Based Intelligent Database Interface of the Intelligent Database Interface". Paramax Systems Organization, 70 East Swedesford Road, Paoll, PA, 19301. Rev. 2.3.

Quinlan, J. R.: 1986, "Induction of decision trees". Machine Learning 1, 81--106.

Ramakrishnan, R., D. Srivastava, and S. Sudarshan: 1992, "CORAL: Control, Relations, and Logic". In: Proceedings of the International Conference of the Internation Conference on Very Large Databases.

Ramanathan, C.: 1994, "Providing object--oriented access to a relational database". In: D. Cordes and S. Vrbsky (eds.): Proceedings of the 32nd Annual Southeast Conference held in Tuscaloosa, Alabama. New York.

Stroulia, E. and A. K. Goel: 1995, "Functional Representation and Reasoning in Reflective Systems". Journal of Applied Intelligence 9(1). Special Issue on Functional Reasoning.

Stroulia, E., M. Shankar, A. K. Goel, and L. Penberthy: 1992, "A Model-Based Approach to Blame Assignment in Design". In: J. S. Gero (ed.): Proceedings of the Second International Conference on Artificial Intelligence in Design.

Stroulia, E., J. Thomson, and Q. Situe: 2000, "Constructing XML-speaking wrappers for WEB Applications: Towards an Interoperating WEB". In: Proceedings of the 7th Working Conference on Reverse Engineering - WCRE-2000. Brisbane, Queensland, Australia.

Tong, C.: 1991, "The Nature and Significance of Knowledge Compilation". IEEE Expert 6(2).

CHAPTER 20

A Study of Technical Challenges in Relocation of a Manufacturing Site

Guangming Zhang
zhang@isr.umd.edu
Department of Mechanical Engineering and Institute for Systems Research
University of Maryland at College Park, MD 20742, USA.

Sameer Athalye
samathalye@hotmail.com
Department of Mechanical Engineering and Institute for Systems Research
University of Maryland at College Park, MD 20742, USA.

ABSTRACT

Data mining for information gathering has become a critical part in decision-making. Manufacturers today are competing in a global market. Staying in competition calls for the fullest use of information related to design and manufacturing to assure products with high quality, low cost and on-time delivery. In this chapter, a business strategy is set forth to relocate a manufacturing site that is closer to customers and a low-cost labor market. The study presented focuses on technical challenges related to the relocating process. To secure a smooth and productive transition, efforts have been made to review the current manufacturing process and establish a new production plan system that will be in place for controlling the production of new products. Case studies are presented in the area of route sheet preparation, shop floor layout design, animation based assembly training and parametric design assistance, demonstrating the impact of data mining on the production realization process.

D. Braha (ed.), Data Mining for Design and Manufacturing, 465–486.
© 2001 *Kluwer Academic Publishers. Printed in the Netherlands.*

INTRODUCTION

As the information technology is expanding its horizon and penetrating into every sector of industry, data mining has been a common practice in engineering for information gathering so as to speed up the product realization process. The new product design arrived will bring about a significant cost reduction with quality assurance.

In this chapter, case studies are presented to demonstrate the relationship between data mining and engineering practice under a manufacturing environment. These case studies are centered at a business decision of relocating a manufacturing facility where injection-molding machines are manufactured. Starting in 1990s, the manufacturing industry of injection molding machines has enjoyed an era, in which orders of injection molding machines are increasing continuously and associated profit margins are increasing as the balance between demand and supply provides favorable ticket prices for those machines. However, things are changing, not because of the demand in the market but because of the competition among the manufacturers. There are more manufacturers, which are capable of producing injection-molding machines with quality. As a result, manufacturers are now expanding their capability while eyeing low labor cost and collaborations with vendors, who are able to provide components or sub-assemblies. One of the strategic business practices is to establish and/or relocate the manufacturing site to a new location where a low labor cost can be expected, and products can be delivered to customers on time.

The role played by the data mining process is the impact of data and information gathering on the engineering practice. Four case studies are presented. The first case is gathering the massive information provided and hidden in route sheet documents for establishing a new production system. The second case is the evaluation and validation of a manufacturing operation, demonstrating the impact of data mining on improving the manufacturing operation. The third case is the computer animation using the data and information gathered for training new employees working in assembly. The fourth case is the computer database supported tool for design assistance. In the conclusion part, a comprehensive understanding of the importance of transforming the traditional manufacturing practice to a new e-manufacturing system is emphasized. It points out the process of data mining is critical to succeed in this endeavor.

ASPECTS OF THE DATA MINING PROCESS

Utilization Of The Current And Existing Information

Figure 1 presents the current production hierarchy of manufacturing injection-molding machines. The key processes include sales & marketing, design, production planning and inspection. As illustrated, the department of marketing issues a six-month sales forecast to initiate the production planning. As a result, the industrial engineering department generates the Bill of Materials (BOM) and option codes for the products ordered. Upon receiving the BOM and option codes, the department of production is responsible for implementing the production plan [1]. The plan consists of two components. The first component is the purchase of the standard parts and signing contracts with vendors on those parts, for which the part cost from purchase would lower the production cost by its own. The second component is the master production schedule to organize the production on the shop floor. The schedule begins with the material requirement planning, such as raw materials, equipment, tooling, fixtures, etc. The second part is the production plan based on the route sheets prepared. Finally, all the standard parts and manufactured parts are assembled in the process of building injection-molding machines. Figure 1 depicts the information flow in production planning.

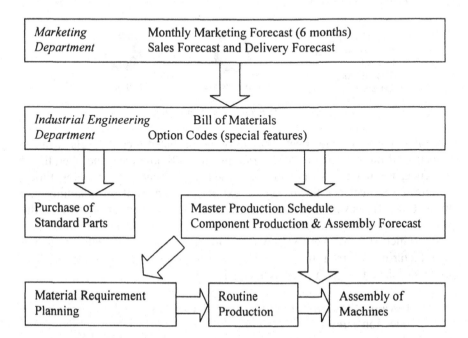

Figure 1. Flow Chart Showing Information Flow in a Manufacturing Setup.

The data mining process presented in this paper focuses on the utilization of the information massively buried in the current route sheet. The example is the route sheet used in the production of an injection screw, a key component of any injection-molding machine. The production of injection screws is closely related to sustaining the profit margin in manufacturing. The objective of the data mining process is to synthesize the knowledge and utilize it to improve the current production so as to move the production of injection screws to a new and higher level after the relocation.

Figure 2 presents a typical injection screw [16]. It is a shaft with a helical slot around it. When the shaft rotates during an injection process, it advances the material being processed to form a plastic component or components. As illustrated, the shaft consists of three parts: feed section, transition section and metering section to carry out their designated functions. Its production involves a series of manufacturing operations.

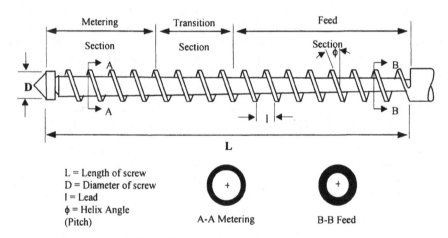

Figure 2. Detail of an Injection Screw

Table 1 lists a route sheet. The information listed merely deals with the sequence of the manufacturing operations currently involved, the operational sequence, the machines or equipment used, setup times and running times. Certainly the listed information is extremely important in planning the production. However, the route sheet presents little information on how the production is organized and no information on the layout of the production cells, which is the key information in the process of relocating the manufacturing site from the current production place to a new location. The process of data mining in relocation consists of

- Review the current production documents used for producing injection screws.

- Establish a new database managing
  - o Information on facilities that have to be shipped to, or purchased at, the new manufacturing site.
  - o Information on the tools, fixtures, gages, and measurement instruments, which have to be packed and shipped to, or purchased at, the new manufacturing site.
  - o Information on the facilities, tools, gages and measurement instruments, which should be purchased at the new manufacturing site.
- Make a decision on layout designs of facilities at the new manufacturing site.
- Prepare a set of new or revised documents to guide the production of injection screws at the new manufacturing site.

Table 1   Details of Initial Route Sheet of Injection Screw.

Seq. No.	Operations	Machine No.	Setup Time	Run Time
10	Preparation of Raw material	Machine 1	0.0000	0.0000
20	Rough Turning	Machine 2	0.3333	0.4167
30	Spiral Milling	Machine 3	0.5000	0.5833
40	Finish Turning	Machine 2	0.5000	0.5833
50	Keyway Milling	Machine 4	0.1670	0.1000
60	Spiral Polishing	Machine 5	0.0000	0.3333
70	Rough Grinding	Machine 6	0.3333	0.3333
80	Nitriding	Machine 7	0.0000	0.0000
90	Spiral Polishing	Machine 5	0.0000	0.3333
100	Finish Grinding	Machine 6	0.3333	0.4167
110	Inventory and Storage	Warehouse	0.0000	0.0000

Decision Of A Cellular Layout Design

Figure 3 presents a flow chart to illustrate the revised production plan of injection screws based on the review of the current route sheet and production documents. Each operation is confined with a block, which will be used in the process of designing cellular layouts. Figure 4 is the suggested cellular layout design based on the flow chart shown in Fig. 3 [9]. As shown in Fig. 4, the key equipment for manufacturing the injection screw includes two machines. One is the spiral-milling machine and the other is the spiral-polishing machine.

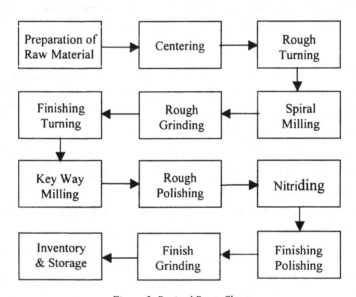

Figure 3. Revised Route Sheet

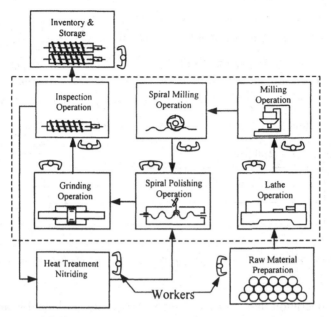

Figure 4. A Cellular Layout for the Production of Injection of Injection Screws

ESTABLISHMENT OF A NEW DATABASE TO AUTOMATE THE GENERATION OF ROUTE SHEETS

It is critical that the preparation of route sheets can be automated based on the information collected during the process of data mining [2]. The structure of the relational database is illustrated in Figure 5. A representative route sheet for preparing raw material, generated using the database, is shown in Figure 6.

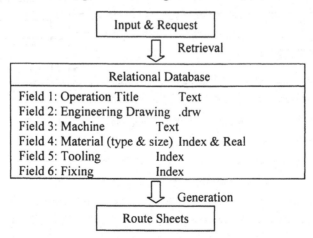

Figure 5. Relational Database Established for Managing the Process of Data Mining

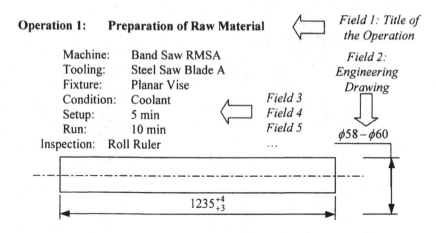

Figure 6. Suggested Format Showing Details of Ancillary Equipment Required.

For an operation that consists of several steps, an additional chart will be added and a route sheet will be prepared for each step in the operation, as shown in Figure 7.

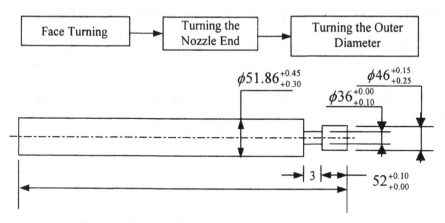

Figure 7. Additional Chart and Diagram for Multiple Operations.

Figure 8 is a chart tracking the in-process production. The chart is also generated using the established relational database. On the chart, the production engineer will be able to check the production implementation on an operation-based scale.

Operation No.	Description of process	Time
Operation 1	Preparation of Raw Material	15 min
Operation 2	Centering Operation	10 min
Operation 3	Rough Turning Operation	35 min
Operation 4	Spiral Milling Operation	45 min
Operation 5	Rough Grinding Operation	30 min
Operation 6	Finish Turning Operation	28 min
Operation 7	Milling Operation for Making the Key Way	25 min
Operation 8	Polishing Operation for Finishing the Spiral Slot	45 min
Operation 9	Nitriding Operation for the Threading Part	
Operation 10	Polishing Operation for Finishing the Spiral Slot	45 min
Operation 11	Finishing Grinding Operation for Tolerance Control	45 min
Operation 12	Final Inspection and Storage	

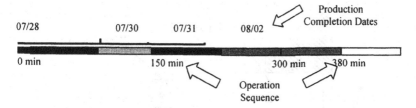

Figure 8. Chart to Track the Production of Individual Components.

EVALUATION AND VALIDATION OF A MANUFACTURING OPERATION

The availability of the relational database provides an opportunity of generating information for production management when it is integrated with other software systems, such as computer-aided design & manufacturing systems. The example presented here is related to machining operations where the Pro/Manufacture system is use to evaluate the machining time needed based on the feed, depth of cut and cutting speed specified. As illustrated in Fig. 9, the travelling time needed from the home position of the cutting tool to the start point:

$$\text{Travel\_Time} = \frac{\text{Distance (mm)}}{\text{feedrate\_for\_position (mm / rev)} \times \text{rotation speed (rev / min)}}$$

The machining time needed to cover the entire surface:

$$\text{Machining\_Time} = \frac{\text{Length} \times \text{number of travel passes}}{\text{feedrate\_for\_machining (mm / rev)} \times \text{rotation speed (rev / min)}}$$

The time needed to return from the ending position to the home position:

$$\text{Non - cutting Machining\_Time} = \frac{\text{Distance (mm)}}{\text{returning feedrate\_for\_position (mm / rev)} \times \text{rotation speed (rev / min)}}$$

$$\text{Returning\_Time} = \frac{\text{Distance}}{\text{feedrate\_for\_position}}$$

As a result, the time needed to carry out the lathe operation can be evaluated as a sum of three time intervals.

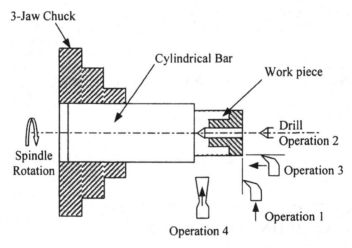

Figure 9. Sample Machining Operation for Process Time Estimation.

When such information is used to estimate the related production cost, the impact of utilizing the process of data mining is evident. To demonstrate such an application, a change of cutting speed from a low level to a higher level for cost reduction is shown below. The machining time is a function of cutting speed. When cutting speed is high, a short machining time is expected, thus improve the productivity. However, the tool life is inversely proportional to the increase of cutting speed. A cutting tool can be easily worn out when machining at high speed. As a result, the tooling cost will be increased as cutting speed increases. Besides, a frequent replacement of a worn tool takes extra time. It is not uncommon to observe that operators use moderate cutting speeds when machining. Figure 10 illustrates such a trade-off analysis in terms of the total unit cost, which is the sum of the unit variable cost and the unit fixed cost.

It is likely that a relocating process could offer an opportunity of lowering tooling cost because of rich sources in tooling material and low labor cost of fabricating tool inserts at the new location. As a result, the total unit cost curve tends moving towards to a high cutting speed region. The following calculation presents a quantitative analysis on the cost reduction through a change of cutting speed from a low to a high level as a result of the tooling cost.

Reduction of the Machining Cost of Lathe Operations by Increasing Cutting Speed

(1) Calculation of Lathe Operation Time at the present level of cutting speed: 80 meter/min and tooling cost of $2

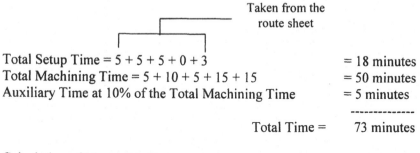

Taken from the route sheet	
Total Setup Time = 5 + 5 + 5 + 0 + 3	= 18 minutes
Total Machining Time = 5 + 10 + 5 + 15 + 15	= 50 minutes
Auxiliary Time at 10% of the Total Machining Time	= 5 minutes
Total Time =	73 minutes

Calculation of Machining Cost
    Direct labor: $120/hour x (73/60) hour   = $146
    Labor overhead at 120%: $146x 120% = $175
    Equipment overhead cost: $100/hour x (73/60) hour = $122
                             Total Cost = $443

Calculation of Tooling Cost
        Weighted tooling price: $2/minute
        Total tooling cost: $2 x 73 = $146

Variable Cost = Machining Cost + Tooling Cost
               = $443 + $146 = $589

Fixed Cost = Depreciation of Lathe and measurement instruments +
                    Administrative Support + Others (building, taxes, etc)
                = $200 per month
        Assume that 10 injection screws are manufactured on a monthly
basis. The total cost per injection screw is given by

Total Cost per Injection Screw= Variable Cost + Shared Fixed Cost
                = $589 + ($200/10) = $609

(2)   Calculation of Lathe Operation Time at the present level of cutting
        speed: 90 meter/min and tooling cost of $1.8/min

Total Setup Time = 18 minutes (unchanged)
Total Machining Time = 40 minutes (reduced)
Auxiliary Time at 10% of the Total Machining Time = 4 minutes (reduced)

                                Total Time =   62 minutes

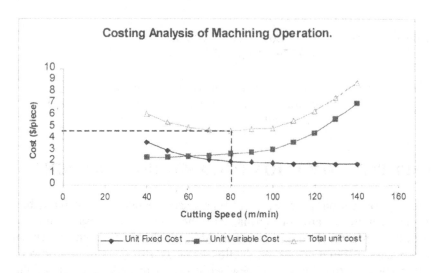

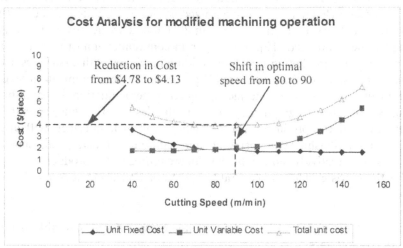

Figure 10. Improvement in cost effectiveness of modified machining operation.

Calculation of Machining Cost
  Direct labor: $120/hour x (62/60) hour = $124
  Labor overhead at 120%: $120x 120% = $149
  Equipment overhead cost: $100/hour x (62/60) hour = $103
        Total Cost = $376

Calculation of Tooling Cost
    Weighted tooling price: $1.8/min
    Total tooling cost: $1.8 x 62 = $112

Variable Cost   = Machining Cost + Tooling Cost
                = \$376 + \$112 = \$488

Fixed Cost = Depreciation of Lathe and measurement instruments +
                Administrative Support + Others (building, taxes, etc)
              = \$200 per month

COMPUTER ANIMATION FOR ASSEMBLY TRAINING

During the relocation process, recruiting local operators is unavoidable. These new employees are not familiar with the assembly process required in the process of manufacturing injection-molding machines. Training is critical in the process of building the production capacity [6]. For the purpose of quality assurance, the training of new workers has to make full use of the accumulated knowledge and be standardized so that the training process can be effective and variation in the content that needs to be covered during training can be minimized [19]. Such a gap in communication of information is often seen when marketing injection molding machines with customers and purchasing components from vendors [13]. To raise the communication level in training and marketing, an interface using computer animation for assembly training is developed to graphically simulate the design of product assembly and operation of the products. The availability of such interface will also improve the relationship with customers and with the members of the supply chain, so as to effectively reduce the development time of the product.

The three basic steps in generation of animation sequences are listed below:

- Developing geometric models of the components of assembly using CAD software.
- Assembling the CAD models of the components in an assembly. This is done under a CAD system (Pro/ENGINEER used in this study) by placing the components properly relative to each other. Various constraints regarding the positions of components are defined to finalize the assembly.
- Using the animation package, define key frame sequences that describe the position and orientation of parts and assemblies at specified times. Key frames are created by getting snap shots of the current positions and orientations of the models and interpolation between the key frames produces a smooth animation.

In this study, animation sequences using Pro/ENGINEER design animation module are developed for two types of assemblies. The first type is for assembling relatively large components. The developed animations guide the workers to follow the safety regulations, optimally utilize the resources, such as fixtures and lifting equipment, available for assembly and to accomplish the work within shortest possible time.

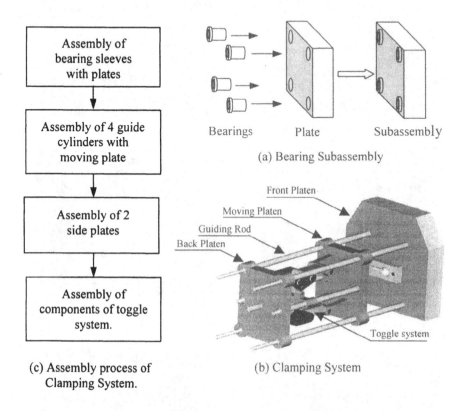

(c) Assembly process of Clamping System.

(b) Clamping System

Figure 11 Assembly Procedure for the Clamping System.

The second type of animations are for assembling small components like bearings, bolts etc. The importance of such animations is to ensure that the workers do not erroneously forget to assemble any components and also to teach them about the correct order of assembly of those components. The absence of few of such small components may normally go unnoticed during the assembly. But this later on leads to maintenance problems which are expensive and difficult to take care of. Figure 11(a) illustrates an example of an animation developed for assembly of small components. It shows the assembly of individual bearings into the respective holes of the plate. Figure 11(b) illustrates an example of a clamping system where heavy components and subassemblies are involved. Figure 12 shows the intermediate stages of the assembly animations developed for the assembly of the platens and guide rods.

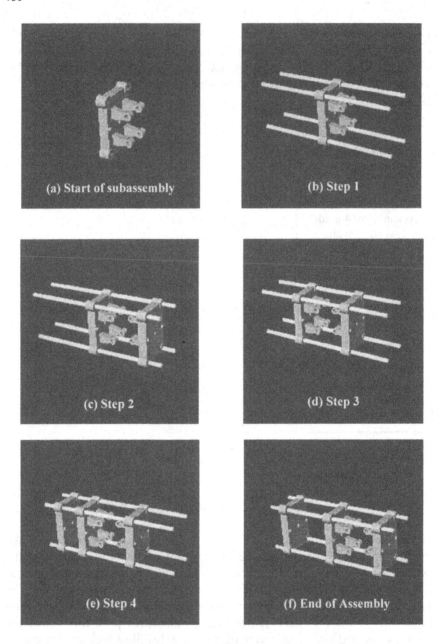

Figure 12 Intermediate Stages in Animation of Clamping System.

Figure 13 shows the process of assembling 10 components, which are presented to the assembly team before the assembly process. At the end of the assembly if there were any components that are not assembled, it will be immediately noticed since those components will still remain with the assembly team.

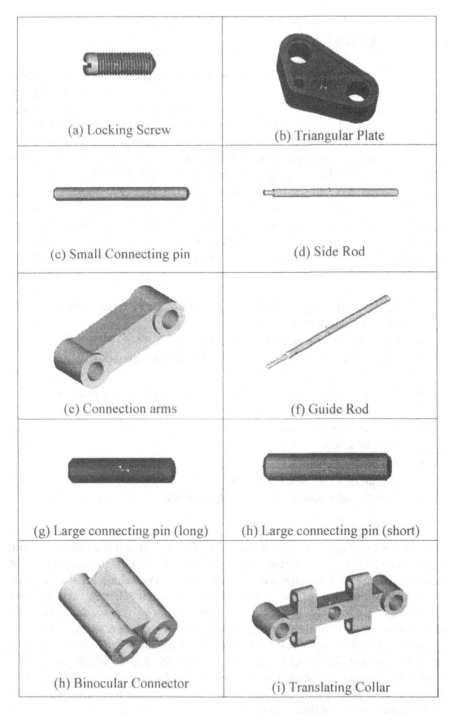

Figure 13 Components Included in the Assembly 'Case' of Toggle Mechanism.

DESIGN ASSISTANT

To demonstrate the impact of using the data mining process on the product realization process, we present a design tool that generate variants of a given design to meet the customer needs. We call such a tool the Design Assistant. The following is a brief description of the functionality of such a design tool.

The design of injection screws can vary significantly based on customer needs. However, the design of injection screws share commonality no matter what a variant is. This is because the features depend on the functionality of the screw and the physical constraints on size that are imposed due to different customer specifications. To start the design of a screw from scratch, every time there is a change in the design, would obviously be time consuming and also error prone. In the process of data mining, a parametric approach is highly recommended. For example, the injection screw consists of the following key parameters:

- Type of feeding system (reciprocating screw type, plunger type, rotary screw type etc.)
- Material used for manufacturing the screws.
- Temperature required for melting the molding material.
- Overall length of the screw.
- Relative positions and lengths of various sections of the screw.
- Variation of depth of flight along the length of screw.
- Pitch of helix of screw.
- Lead of screw.
- Radial flight height.
- Design of front nose of the screw.
- Length and diameter of shank of screw etc.

Figure 14a illustrates a template of the screw design assistant. A user can retrieve the existing CAD file of the screw design through the directory path. Figure 14b is the user interface where values of the design parameters are input by the user. Once the user has given the new values for dimensions, these can be applied to the CAD model by simple clicking on the "Set Dimensions" button. Even though the dimensions of the model have been changed, there will not be a corresponding change in the graphical output. To actually 'see' the changes that have taken place, the user needs to click on the "Repaint" button. This will present the model with the modified dimensions. Under the design assistant environment, a variant of the injection screw design can be generated with quality and the effort that has been minimized through the process of data mining.

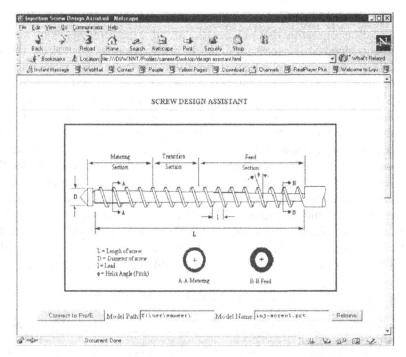

Figure 14(a) Design Assistant Template for the Reciprocating Screw

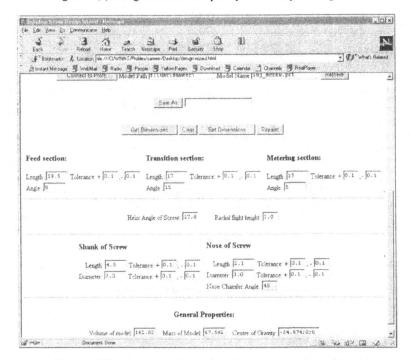

Figure 14(b) Interface for Viewing and Setting Dimensions of Screw.

Other design parameters, such as material properties like Young's modulus, Poisson ratio, thermal coefficient of expansion etc. can also be viewed and changed if necessary. Similarly dimensional tolerances, surface finishes, etc can be easily validated from the interface. Body properties like volume, mass, surface area, center of gravity etc can be verified by making use of the Pro/Web.Link API. Standard features like turning the model to different views (side view, top view), viewing cross sections, highlighting certain features of the body are also available in the developed web interface.

As a safeguard, a warning mechanism is built in. For example, there is a limit on the total length of an injection screw. It is not uncommon to see that the user input exceeds limits. Under those circumstances, an alerting signal will be shown on the design screen to inform the user that he/she is violating the design constraint(s). Figure 15 shows a warning message indicating that the total length of the screw needs to have a certain minimum ratio with the diameter of the screw. This (L/D) ratio generally is kept below 20 so that there are no significant problems of bend stability and heat dissipation.

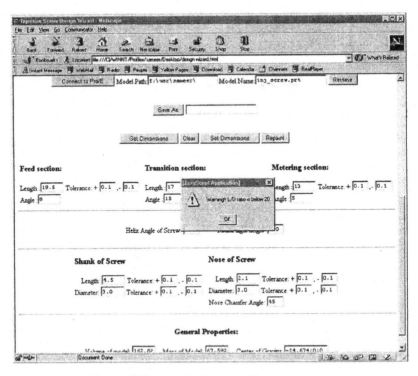

Figure 15 Warning Mechanism in Design Assistant.

CONCLUSIONS

The work presented in this chapter is related to technical challenges in the process of relocating a manufacturing facility. To succeed in the relocating process, examining the current engineering practice is the starting point. Information gathering and extraction is critical to establish a new production system that is computer data based. The data mining process has to be integrated with engineering software systems for effectiveness and applicability. It is the engineering software systems, which carry out the design and manufacturing activities, and the process of data mining serves as the manager, who controls and coordinates the routine operations based on the previous experience and directing the future action using the artificial intelligence that is built in the process of decision making.

ACKNOWLEDGEMENTS

This research has been supported by funds received from the National Science Foundation under the ECSEL program, and the Department of Mechanical Engineering at the University of Maryland at College Park. Dr. Chiang, Chen of Chen Hsong Group deserves special recognition for his strategic decision on relocation. Special thanks are due to Professor Stephen Lu at the University of Southern California for his support and encouragement.

REFERENCES

Amirouche, Farid M.L., Computer-Aided Design and Manufacturing, Prentice Hall Publication, 1994.

Thuraisingham, Bhavani M., Data mining: technologies, techniques, tools and trends, Boca Raton: CRC Press, 1999.

Michalski, Ryszard S., Machine Learning and Data mining: methods and applications, New York, J. Wiley, 1998.

Beranek, John M. et al., "Automatic Generation of Assembly Sequences for Polyhedral Assemblies", ASME Conference on Flexible Assembly, pp. 31-40, 1991.

Chang, Tien-Chien, Richard A. Wysk and Hsu-Pin Wang. Computer Aided Manufacturing, Prentice Hall International Series in Industrial and Systems Engineering, 1991.

Davis, James R. and Adelaide B. Davis, Effective training strategies: A comprehensive guide to maximizing learning in organizations, Berrett-Koehler Publishers, 1998.

Delchambre, A., CAD method for Industrial Assembly, John Wiley and Sons Publishing, 1990.

Famili, A. (Fazel), Dana S. Nau and Steven H. Kim, Artificial Intelligence applications in Manufacturing, AAAI Press/ MIT Press Publishing, 1994.

Ireson, William Grant, Factoy planning and plant layout, Prentice-Hall 1982.

Jakupec, Viktor, John Garrick, Flexible learning, human resource and organizational development: putting theory to work, Routledge, London, 2000.

Juran, J.M. and Frank M. Gryna, Quality Planning and Analysis, Mc-Graw Hill Inc.

Kalpakjian, Serope, Manufacturing Processes for engineering materials, Addison-Wesley Publishing, 1996.

Marquardt, Michael J. and greg Kearsley, Technology-based learning: maximizing human performance and corporate success, St. Lucie Press, 1999.

Rehg, James A., Computer Integrated Manufacturing, Prentice Hall Career and Technology Publishing, 1991.

Tlusty, George, Manufacturing Processes and Equipment, Prentice Hall Publishing, 1994.

Rosato, Dominic V. and Donald V. Rosato, Injection Molding Handbook, Van Nostrand Reinhold Publishing, 1986.

Senker, Peter, Towards the automatic factory: the need for training, IFS Publications, 1986.

Yokota, K. and D.R. Brough, Assembly/Disassembly Sequence Planning, Flexible Automation, pp. 31-38,Vol.12, 1993.

Zhang, G.M., Quality Management in Systems, The Commercial Press, Beijing, People's Republic of China, 1998.

CHAPTER 21

Using Imprecise Analogical Reasoning to Refine the Query Answers for Heterogeneous Multidatabase Systems in Virtual Enterprises

Z. M. Ma
maz@engr.USask.Ca
Department of Mechanical Engineering, University of Saskatchewan, 57 Campus drive, Saskatoon, SK S7N 5A9, Canada

W. J. Zhang
zhangc@engr.USask.Ca
Department of Mechanical Engineering, University of Saskatchewan, 57 Campus drive, Saskatoon, SK S7N 5A9, Canada

W. Y. Ma
mewma@cityu.edu.hk
Department of Manufacturing Engineering and Engineering Management, City University of Hong Kong, 83 Tat Chee Avenue, Kowloon, Hong Kong SAR, P. R. China

ABSTRACT

Virtual enterprise is typically one kind of information–based enterprise. The features of information system in virtual enterprise can be generalized as heterogeneous and distributed. Its organization and production management put an essential requirement on information integration. A core problem in the integration of heterogeneous information sources is the resolution of a great deal of conflicts, which result in the situation of imprecise information. Therefore, it is necessary for the production activities of virtual enterprise to acquire the more informative information from the information source with imprecision based on database systems.

In this article, we briefly review the approaches to the integration of multiple databases in virtual enterprise and the forms of conflict in the integration firstly. The imprecise information is thus introduced into the integrated database systems in virtual enterprise to resolve the conflicts. On

D. Braha (ed.), Data Mining for Design and Manufacturing, 487–503.
© 2001 *Kluwer Academic Publishers. Printed in the Netherlands.*

the basis of that, we introduce the notion of equivalence degree of imprecise data, by which we can mine the rules of functional dependencies from the integrated database systems with imprecise information. Following the rules of functional dependencies, we can apply the frame of analogical reasoning to refine the imprecise query answer for the database systems to support the organization and management of virtual enterprise.

INTRODUCTION

Current manufacturing industries face the buyers' market. In order to increase the competitiveness of products and enterprises, manufacturing enterprises try to make their product with low price, high quality and short product delivery time on one hand. On the other hand, the change from sellers' market to buyers' market results in steadily decreasing product life–cycles and appearing the production activities of individual product or small batches made to order. All those require manufacturing enterprises can quickly respond to the market's changes. The organizational principle of virtual enterprise makes the aims above possible.

Virtual enterprise, according to the definition of Rolstadas (1995), consists of a number of units geographically dispersed but managed as one total unit, although the sub–units may be under separate management. In a virtual enterprise, there is a master company, which may not have its own fabrication facilities and develops products by relying on other manufacturing companies, called partner companies or partners. Moreover, the organizational structure of a virtual enterprise is product–oriented, which implies that the relationship between master and partners is not permanent. In virtual enterprise, the product or configuration design and manufacturing are essentially the procedure that master company selects or syntheses the partner companies and monitors their production activities by deriving information concerned. So it can be concluded that, from the viewpoint of information processing, virtual enterprise is information–based enterprise.

Being typically information–based enterprise, virtual enterprises generally employ multiple information systems, which are independently developed, locally administrated, and different in logical or physical designs. Therefore, some new challenges have been proposed in the information management of virtual enterprise, in which the modeling of information (Zhang and Li, 1999) and the sharing of information for master company across organizational, geographical, or structural boundaries (Hardwick, Spooner, Rando and Morris, 1996; Cheung and Hsu, 1996) are the key problems.

According to the distributed and heterogeneous features of the information systems in virtual enterprises, an approach for integrate multidatabase systems based on query language and view frame has been proposed (Ma, Zhang and Li, 1999) to share manufacturing information in virtual enterprises. As shown in (Ma, Zhang and Li, 1999), however, there are a large number of semantic conflicts of information to resolve, which exist in different partner companies, and the resolutions of these conflicts result in information imprecision in the integrated database systems.

Let us look at an example. Two database relations belonging respectively to two partner companies in different sites are shown in Figure 1.1. Following the strategies and approaches in (Ma, Zhang and Li, 1999), we have the integrated result relation shown in Figure 1.2, where [54.50, 55.49] and [15, 21] stand for intervals of values, namely, partial values, and φ stands for applicable missing data, namely, one type of null value. Partial values and null values are called imprecise values. Since there are imprecise values in the relation of Figure 1.2, the answers of queries for "the price of the part 10011" and "the lead–time of the part 10012" are [54.50, 55.49] and φ, respectively, which are imprecise. However, if we have the knowledge that the lead–time and the price of a part is closely related to its material, in other words, the material of a part can determine its lead–time and its price in a way, we can refine the imprecise answers above using analogical reasoning. We can conclude that the price of the part 10011 may be 54.8 because its material is the same as one of the part 10012 and the price of the part 10012 is 54.8. Based on the same consideration, the lead–time of the part 10012 may be [15, 21]. Note that the refined query answers are the approximate answers, not meaning that they should or must be such results. The conclusions above may be incorrect due to other factors that affect the lead–time and the price of a part. Nevertheless, the more refined approximate answer is meaningful for the production activities and decision–making of virtual enterprises. It is especially true for the situation that the query answer is null.

Site 1: Partner relation r   Site 2: Partner relation s

Part ID	Material	Lead-time (Weeks)	Price (Dollars)	Part ID	Material	Cost (Dollars)
10010	Fe	2	40	10010	Fe	40.30
10011	Cu	3	55	10012	Cu	54.80
10013	Alloy	4	67	10013	Alloy	67.47

Figure 1.1 Database relations of partner companies

Part ID	Material	Price (Dollars)	Lead–time (Days)
10010	Fe	40.30	[8, 14]
10011	Cu	[54.50, 55.49]	[15, 21]
10012	Cu	54.8	φ
10013	Alloy	67.47	[22, 28]

Figure 1.2 Integrated database relation with imprecise information

The relationship between the attributes *material* and *lead–time/price* above is considered as one kind of association rule, called rule of functional dependencies in viewpoint of database systems. In this article, we concentrate on two issues. One is how to find the rules of functional dependencies from the integrated database systems in the virtual enterprises. The other one is how to apply such rules to refine the query answers.

The remainder of the article is organized as follows. In Section 2, we briefly review the related work and give the basic knowledge. The notion of equivalence degree of data is introduced in Section 3 to describe the proximity degree of two data, which may be partial values, null values, or crisp values. In Section 4, imprecise functional dependencies are defined. The approach to refining imprecise query answers using imprecise analogical reasoning is presented in Section 5. Section 6 concludes this article and points out the future work.

RELATED WORK AND BASIC KNOWLEDGE

Database Schema Integration and Conflict Resolution

As mentioned above, the master company should access and derive data from multiple independent databases in partner companies in order to organize production activities effectively. However, the data sources in partner companies are created and developed independently, so the databases of partner companies are distributed and heterogeneous. The problem of database integration in virtual enterprise management is involved. In database schema integration, a core problem is to resolve a large number of incompatibilities that exist in different databases for semantically related data.

The approach that provides users a multidatabase query language is adopted by Ma, Zhang and Li (1999) to integrate multidatabase systems in virtual enterprises. Being not the same as the global schema integration and the federated database systems, this approach does not require a global schema for all integrating databases or a local schema describing data that the application may be accessing. This approach is suitable for the database integration under virtual enterprise environment due to the possibility of unpredictable partner companies joining the virtual enterprise.

Several types of important conflict problems and their resolutions are generalized as follows (DeMichiel, 1989; Tseng et al, 1993; Ma, Zhang and Ma, 1999a).

(a) Naming conflicts. This type of conflict can be divided two aspects. One is semantically related data items are named differently and the other is semantically unrelated data items are named equivalently, which can be resolved by mapping semantically related data items and semantically unrelated data items to a common virtual attribute and different attribute with one–to–one mapping, respectively.

(b) Data representation conflicts. This occurs when semantically related data items are represented in different data types, which can be resolved by mapping semantically related data items to a common virtual attribute with one–to–one mapping.

(c) Data scaling conflicts. This occurs when semantically related data items are represented in different databases using different units of measure, which can be resolved by mapping semantically related data items to a common virtual attribute with either many–to–one mapping or one–to–many mapping, depending on the concrete situation.

(d) Missing data. This occurs when the schemas of the integrating databases have different attribute sets. This kind of conflicts can be resolved by outerunion operation or outerjoin operation.

For the one–to–many mapping, the integrated result is to produce partial values, one special kind of value, on the virtual attribute. A partial value is an interval of values on continua attribute domain or a set of values on discrete attribute domain, in which exactly one of the values is true value. For outerunion or outerjoin operation, the integrated result is to produce null values on the virtual attribute, in which exactly one of the values is true value.

An example for multidatabase integration in virtual enterprises is illustrated in Figure 1.1 and 1.2, in which the conflicts as well as their resolution and imprecise data are shown. The detailed issues on multidatabase integration can be found in (Tseng et al, 1993; Ma, Zhang and Li, 1999; Ma, Zhang and Ma, 1999a)

Imprecise Data and Extended Data Model

Partial values and null values have been introduced into the integrated database systems in order to resolve the semantic conflicts in multidatabase systems of virtual enterprises. Partial values (Grant, 1979; DeMichiel, 1989) and null values (Codd, 1986; Codd, 1987) are applied to model information imprecision in databases. Null values with the semantics of "applicable" can be considered as the special type of partial values that the true value can be any one value in the corresponding domain, namely, an applicable null value corresponds to the whole domain. Following the issues in (DeMichiel, 1989; Ma, Zhang and Li, 1999), the notion of a partial value is formally illustrated as follows.

Definition 2.1: A partial value on a universe of discourse U corresponds to a finite set of possible values in which exactly one of the values in the set is the true value, denoted by $\{a_1, a_2, ..., a_m\}$ for discrete U or $[a_1, a_n]$ for continua

U, in which $\{a_1, a_2, ..., a_m\} \subseteq U$ or $[a_1, a_n] \subseteq U$. Let η be a partial value, then sub (η) and sup (η) are used to represent the minimum and maximum in the set.

Definition 2.2: For a partial value $\eta = \{a_1, a_2, ..., a_m\}$, the number of possible values in η is called the modular of η whereas the length of the interval for a partial value $\eta = [a_1, a_n]$ is called the modular of η. The modular of a partial value η is denoted by $\|\eta\|$.

Note that crisp data and null values can be viewed as the special cases of partial values. A crisp data on discrete universe of discourse could be represented as the form of $\{p\}$, and a crisp data on continua universe of discourse could be represented as the form of $[p, p]$. The true value of a null value can be any one value in the corresponding domain, i.e., a null value corresponds to the whole domain. Moreover, the partial value without containing any element is called an *empty partial value*, denoted by $\perp$. In fact, the symbol $\perp$ means an inapplicable missing data (Codd, 1986; Codd, 1987). Null values, partial values, and crisp values are thus represented with a uniform format.

Considering crisp value and empty partial value, we define the modular of a partial value.

Definition 2.3: Let η be a partial value in the universe of discourse U. Then
(a) $\|\eta\| = 0$ if $\eta = \perp$, i.e., η is an empty partial value.
(b) $\|\eta\| = 1$ if $\eta = \{a\}$ when U is a discrete universe of discourse, or $\|\eta\| = \delta$ if $\eta = [a, a]$ when U is a continua universe of discourse, where δ is a very small number compared with $\|U\|$. Generally, δ is viewed to be close to 0.
(c) $\|\eta\| = b - a$ if $\eta = [a, b]$ and $b \neq a$, or $\|\eta\|$ is equal to the number of possible values in η if η is the partial value in discrete universe of discourse U.

The partial values in discrete universe of discourse whose modular are greater than 1, and the partial values in continues universe of discourse whose modular are neither 0 nor δ are called *proper partial values* on this article.

For a proper partial value $\eta = \{a_1, a_2, ..., a_m\}$ with its modular $\|\eta\| = m$, assume m elements in η can be sorted according to the ascending order. Then two functions for processing partial values are defined: *Lstring* (η, i) and *Rstring* (η, j). The result of *Lstring* (η, i) is a partial value consisting of the first i elements in η, and the result of *Rstring* (η, j) is a partial value consisting of the last j elements in η, where $0 \leq i, j \leq m$. If a proper partial value $\eta = [a_1, a_n]$ and $\|\eta\| = n$, *Lstring* $(\eta, i) = [a_1, a_1 + i]$ and *Rstring* $(\eta, j) = [a_n - j, a_n]$, where $0 \leq i, j \leq n$.

Example 2.1: Let $\eta_1 = \{a_1, a_2, a_3, a_4, a_5, a_6\}$ and $\eta_2 = [24.5, 44.5]$, then *Lstring* $(\eta_1, 4) = \{a_1, a_2, a_3, a_4\}$, *Rstring* $(\eta_1, 3) = \{a_4, a_5, a_6\}$, *Lstring* $(\eta_2, 4) = [24.5, 28.5]$, and *Rstring* $(\eta_2, 3) = [41.5, 44.5]$.

When partial values and null values are introduced into relational databases, an attribute value in a relation may be a precise atomic value, an imprecise partial value, or an imprecise null value. Traditional relational data model has no ability to represent imprecise data. In order to resolve the information conflicts in schema integration and implement the information integration in virtual enterprises, relational data model must be extended.

In classical relational databases, relational schema R is a finite set {A1, A2, ..., An}, denoted R = {A1, A2, ..., An} or R (A1, A2, ..., An), in which Ai ($1\leq$ i $\leq$n) is called attribute. Each attribute Ai associates with a domain, denoted Dom (Ai) or Di, which is a set of possible values for that attribute. It is evident that the domain of attribute is a finite set of values and can be discrete or continua. A relation r on schema R (A1, A2, ..., An) is a set of tuples and each tuple consists of attribute values, which come from corresponding attribute domains respectively. Each attribute value must be atomic under classical relational database environment.

Let 2^{Di} denote the power set of the set Di. Then instead of Di, 2^{Di} will be a new domain of attribute Ai in the extended relational data model, called extended domain of attribute and denoted by Dom* (Ai) to differentiate the traditional domain of attribute Dom (Ai). Based on the extended relational model, a relation r on relational schema R (A1, A2, ..., An) is defined as a subset of the Cartesian product of Dom* (A1) × Dom* (A2) × ... × Dom* (An). Because primary key is taken as identifier of tuples in the relation, imprecise data is not permitted as the attribute value on primary key.

Imprecise Analogical Reasoning

Analogy is an important inference tool in human cognition and is a powerful computation tool for general inference in artificial intelligence and decision-making. Approximate reasoning refers to such an inferring process that an object S (the source) has properties P and Q, respectively, and another object T (the target) shares the property P, then we can infer that T may also possess the property Q. In other words, S and T have the same property Q if S and T are matched on P to each other. This process can be expressed by the following logical rule.

IF P (S) and Q (S) and P (T) **THEN** Q (T)

Note that the properties P and Q must have an associated relationship in applying the approximate reasoning above. Two products A and B, for example, are all made of the material Cu, the cost of A are HK\$ 25, and the length of A is 25 cm. It is clear that the material of a product determines its cost in a way. Therefore, we can infer that the cost of B may also be HK\$ 25 if we do not know what is the cost of B. Of course, this approximate conclusion may be incorrect due to other factors affecting the cost of the product. Nevertheless, such conclusion is more meaningful during the lack of information. On the other hand, if we do not know what is the length of B, we can not infer that the length B is also 25 cm because the material of a product is generally irrelevant for drawing a conclusion about the value of the length.

It is not difficult to see that there are two issues involved in approximate reasoning. One is that which property P can determine the property Q. Another is that which object S has the same property as the object T on property P such that T has the same property as S on property Q.

Viewed from relational databases, the relationship between properties P and Q in approximate reasoning is essentially the constraints of functional dependency between P and Q, where P and Q can be seen as attributes. According to semantics of functional dependency, two tuples, i.e., objects S and T, must have the same attribute values on Q if they have the same attribute values on P. Therefore, in relational databases, approximate reasoning can be emulated by functional dependency.

Imprecise data may appear in relational databases due to data unavailability or data and schema incompatibilities in multidatabases. As discussed above, partial values and null values have been used to model imprecise data in virtual enterprises to resolve semantic conflicts of multidatabase integration. Therefore, the query answers for such multidatabases may be imprecise or even null. Applying approximate reasoning, it is clear that we can refine the imprecise query answer of tuple T on attribute Q when there exists a more precise attribute value of tuple S on attribute Q. The refinement result is that the value of T on Q is the same as the value of S on Q.

Of course, under imprecise information environment, the associated relationship between P and Q may be imprecise. On the other hand, the match between S and T on P is generally imprecise. It is necessary to quantitatively describe the degree of imprecise match to choose the object S when employing approximate reasoning under such environment. The approximate reasoning under such conditions is called *imprecise approximate reasoning*.

When applying imprecise approximate reasoning, we should know which property P imprecisely determines the property Q and which object S has the highest match degree with the object T on property P such that T has the same property Q as S. It is clear that in imprecise relational databases, imprecise functional dependency is used to emulate imprecise approximate reasoning.

EQUIVALENCE DEGREE OF DATA

From the discussion above, it can be seen that measuring the equivalence degree of imprecise data is the key for implementing analogical reasoning. In

(Ma, Zhang and Li, 1999), three notions for describing the semantic relationship between imprecise data are proposed.

Definition 3.1: Let t and s be two tuples of a relation r (R) and $A \in R$ be an attribute. Then

(a) t [A] and s [A] are equivalent, denoted by t [A] $\equiv$ s [A], if t [A] $=$ s [A]; i.e., they are the same two crisp values,

(b) t [X] and s [X] are inclusive, denoted by t [X] $\approx$ s [X], if t [X] $\subseteq$ s [X] or t [X] $\supseteq$ s [X], and

(c) t [A] and s [A] are interrelated, denoted by t [A] $\sim$ s [A], if t [A] $\cap$ s [A] $\neq$ Φ.

The above three semantic relationships between two data items are called equivalence, inclusion, and interrelation, respectively. It is not difficult to see that the equivalence relationship is reflexivity, symmetry, and transitivity, the inclusion relationship is reflexivity and transitivity, and the interrelation relationship is reflexivity and symmetry. For two imprecise data, say η_1 and η_2, $\eta_1 \equiv \eta_2 \Rightarrow \eta_1 \approx \eta_2$, and $\eta_1 \approx \eta_2 \Rightarrow \eta_1 \sim \eta_2$, where $\Rightarrow$ indicates implication. The direction of these inferences is irreversible.

The notions of equivalence, inclusion, and interrelation are actually utilized to differentiate the degrees of information "similarity". Let $\eta_1 = [11, 13]$, $\eta_2 = [10, 15]$, and $\eta_3 = [12, 18]$ be three partial values. According to the Definition 3.1, $\eta_1 \approx \eta_2$, $\eta_1 \sim \eta_3$, and $\eta_2 \sim \eta_3$. Intuitively, η_1 is more similar to η_2 than to η_3. But what are their similar degrees? Especially, η_1 and η_2 are all similar to η_3, but which one is more similar to η_3? The notions above cannot answer these questions because they only qualitatively describe the semantic relationship between data items. We define a novel notion *equivalence degree* to measure the semantic relationship between imprecise data in this article, which stems from the idea of the nearness measure and semantic proximity in fuzzy relational data model (Rundensteiner et al, 1989; Shenoi and Melton, 1989; Liao, Wang and Liu, 1999; Ma, Zhang and Ma, 1999b).

Definition 3.2: Let η_1, η_2 be two partial values over the attribute A. The equivalence degree of η_1 and η_2, denoted by ED (η_1, η_2), is expressed as follows.

$$ED\ (\eta_1, \eta_2) = \|\eta_1 \cap \eta_2\| / \|\eta_1 \cup \eta_2\| - \|\eta_1 \cap \eta_2\| / \|Dom\ (A)\|$$

Here Dom (A) and $\|\eta\|$ denote the traditional attribute domain and the modular of a partial value η, respectively.

Example 3.1: Let the domain of attribute A be $\{a_1, a_2, ..., a_{20}\}$, namely, Dom (A) $= \{a_1, a_2, ..., a_{20}\}$ and $\|Dom\ (A)\| = 20$.

(a) Suppose $\eta_1 = \{a_{11}\}$ and $\eta_2 = \{a_{11}\}$. Then ED (η_1, η_2) = 1/1 - 1/20 = 1.

(b) Suppose $\eta_1 = \{a_{12}, a_{13}, a_{14}, a_{15}\}$ and $\eta_2 = \{a_{16}, a_{17}\}$. Then ED (η_1, η_2) = 0/6 - 0/20 = 0.

(c) Suppose $\eta_1 = \{a_{14}, a_{15}, a_{16}\}$, $\eta_2 = \{a_{14}, a_{15}, a_{16}\}$, $\eta_3 = \{a_{14}, a_{15}, a_{16}, a_{17}, a_{18}\}$, and $\eta_4 = \{a_{14}, a_{15}, a_{16}, a_{17}, a_{18}\}$. Then $ED(\eta_1, \eta_2) = 3/3 - 3/20 = 0.85$ and $ED(\eta_3, \eta_4) = 5/5 - 5/20 = 0.75$. We have $ED(\eta_1, \eta_2) > ED(\eta_1, \eta_3)$.

(d) Suppose $\eta_1 = \{a_{11}, a_{12}, a_{13}\}$, $\eta_2 = \{a_{10}, a_{11}, a_{12}, a_{13}, a_{14}, a_{15}\}$, and $\eta_3 = \{a_{12}, a_{13}, a_{14}, a_{15}, a_{16}, a_{17}, a_{18}\}$. Then $ED(\eta_1, \eta_2) = 3/6 - 3/20 = 0.35$ and $ED(\eta_1, \eta_3) = 2/8 - 2/20 = 0.15$. We have $ED(\eta_1, \eta_2) > ED(\eta_1, \eta_3)$. η_1 is more similar to η_2 than to η_3.

Intuitively, the more two partial values overlap, the more similar they are. Besides, the semantic equivalence degree of two partial values is also relevant to their modular. Case (c) in the above example can illustrate this situation well.

The notion of semantic equivalence degree of attribute values can be extended to tuples. Let t_i and t_j be two tuples in $r(R)$, where $R = \{A_1, A_2, ..., A_n\}$. The semantic equivalence degree of tuples t_i and t_j is defined as $ED(t_i, t_j) = \min \{ED(t_i[A_1], t_j[A_1]), ED(t_i[A_2], t_j[A_2]), ..., ED(t_i[A_n], t_j[A_n])\}$.

IMPRECISE FUNCTIONAL DEPENDENCIES IN DATABASE SYSTEMS

Functional dependencies (*FDs*), being one of the most important data dependencies in relational databases, express the dependency relationships among attribute values in relation. In classical relational databases, functional dependencies can be defined as follows.

Definition 4.1: For a classical relation $r(R)$, in which R denotes the set of attributes, we say r satisfies the functional dependency $FD: X \rightarrow Y$ where $XY \subseteq R$ if

$$(\forall t \in r)(\forall s \in r)(t[X] = s[X] \Rightarrow t[Y] = s[Y]).$$

For a $FD: X \rightarrow Y$, $t[X] = s[X]$ implies $t[Y] = s[Y]$. Such knowledge exists in relational databases and can easily be mined from precise databases when we do not know it. However, how to discover such knowledge in imprecise relational databases is difficult. In the following, we give the definition of imprecise functional dependencies based on the equivalence degree.

Definition 4.2: For a relation instance $r(R)$, where R denotes the schema, we say r satisfies the *imprecise functional dependency IFD*: $X \bullet Y$, where $XY \subseteq R$, if

$$(\forall t \in r)(\forall s \in r)(ED(t[X], s[X]) \leq ED(t[Y], s[Y]))$$

Example 4.1: Consider the relation r in Figure 4.1. Assume that Dom (X) = $\{a_1, a_2, ..., a_{20}\}$, Dom (Y) = $\{b_1, b_2, ..., b_{20}\}$, and $\delta = 20/1000$ on attributes X and Y. Then $ED(u[X], v[X]) = 0.4$, and $ED(u[Y], v[Y]) = 0.6$. So *IFD*: $X \bullet Y$ holds in r.

	K	X	Y
u	1003	$[a_7, a_8, a_9]$	$[b_{12}, b_{13}, b_{14}, b_{16}]$
v	1004	$[a_8, a_9, a_{10}]$	$[b_{13}, b_{14}, b_{16}]$

Figure 4.1 Integrated relation r with partial values

Imprecise functional dependencies have the following properties.

Proposition 4.1: A classical functional dependency *FD* satisfies the definition of *IFD*.

Proof: Let *FD*: $X \rightarrow Y$ hold in $r (R)$. Then for $\forall t \in r$ and $\forall s \in r, t [X] = s [X]$ $\Rightarrow t [Y] = s [Y]$, and $ED (t [X], s [X]) = ED (t [Y], s [Y]) = 1$.

REFINING IMPRECISE QUERY ANSWERS

In this section we propose a framework for refining the imprecise query answers for the database relation with imprecise information using imprecise approximate reasoning. Let an integrated database relation with imprecise data be r $(A_1, A_2, ..., A_n)$ which consists of tuples $t_1, t_2, ..., t_m$. Assume that we have an imprecise query answer $t_m [A_n]$ and we would like to refine this query answer using approximate reasoning.

As mentioned above, we have two issues to be involved in imprecise approximate reasoning. The first issue is which attribute A_i $(1 \leq i \leq n - 1)$ can determine the attribute A_n under imprecise information environment. In other words, we want to mine such a rule of imprecise functional dependency: A_i •A_n $(1 \leq i \leq n - 1)$. For the found attribute A_i, the other issue is which tuple t_j $(1 \leq j \leq m - 1)$ has the highest equivalent degree with the tuple t_m on attribute A_i. On the basis of that, we can replace $t_m [A_n]$ with $t_j [A_n]$.

Mining Imprecise Functional Dependency

According to the definition of imprecise functional dependency, the requirement $ED (t_j [A_i], t_k [A_i]) \leq ED (t_j [A_n], t_k [A_n])$ $(1 \leq j, k \leq m - 1)$ must be satisfied in a relation r $(A_1, A_2, ..., A_n)$ with imprecise information if *IFD*: A_i •A_n. In order to discover such attribute A_i from the relation r, we give the algorithm as follows.

    For i := 1 to n – 1 **do**
        For j := 1 to m – 1 **do**
            For k := j + 1 to m – 1 **do**
                If $ED (t_j [A_i], t_k [A_i]) > ED (t_j [A_n], t_k [A_n]))$ **then** loop next i;
            Next k;
        Next j;

Output A_i;
Next i.
For any one output attribute A_i, IFD: $A_i \bullet A_n$ must hold.

Measuring Equivalence Degree Between Attribute Values

Assume that IFD: $A_i \bullet A_n$ holds. Then we should find a tuple t_j ($1 \leq j \leq m - 1$) from r such that for any tuple t_k ($1 \leq k \leq m - 1$ and $k \neq j$) in r, ED (t_j $[A_i]$, t_m $[A_i]$) $\geq ED$ (t_k $[A_i]$, t_m $[A_i]$). In the following we give the corresponding algorithm.

$t := t_l$;
$x := ED$ (t_l $[A_i]$, t_m $[A_i]$);
For j := 2 to m – 1 **do**
    If ED (t_j $[A_i]$, t_m $[A_i]$) > x **then**
        {
        x := ED (t_j $[A_i]$, t_m $[A_i]$);
        $t := t_j$
        };
Next j.

From the algorithm above, it can be seen that tuple t has the highest equivalence degree with tuple t_m on attribute A_i.

Refining Imprecise Query Answer

Following the idea of analogical reasoning discussed above, we can infer that t_m $[A_n]$ should be equivalent to t $[A_n]$ because IFD: $A_i \bullet A_n$ holds and t_m $[A_i]$ is equivalent to t $[A_i]$ with the highest equivalence degree. If t $[A_n]$ is more precise data than t_m $[A_n]$, we can replace query answer t_m $[A_n]$ with t $[A_n]$ as an approximate answer. Viewed from the result, the query answer is refined.

In order to refine the imprecise query answer, the simplest approach is to replace t_m $[A_n]$ with t $[A_n]$. However the imprecise query answer t_m $[A_n]$ can generally be compressed to be a more precise value than t $[A_n]$ according to the constraints of imprecise functional dependencies. In the following, we give the strategies for compressing imprecise value by using imprecise functional dependencies. Assume the imprecise values are intervals of values on the continua domain of attribute. With respect to the partial values on the discrete domain of attribute, the similar method can be employed.

Let t $[A_i] = f1$, t_m $[A_i] = f2$, t $[A_n] = g1$, and t_m $[A_n] = g2$. Moreover, $|f1| = p$, $|f2| = q$, $|g1| = u$, $|g2| = v$, $|g1 \cap g2| = k$, and $|f1 \cap f2| = k'$, then $|g1 \cup g2| = u + v - k$ and $|f1 \cup f2| = p + q - k'$. Assume ED ($f1, f2$) > ED ($g1, g2$). It is clear that IFD: $A_i \bullet A_n$ does not hold and we should make ED ($f1, f2$) equal ED ($g1$, $g2$) after compression processing.

(a) If $0 < ED$ ($g1, g2$) < ED ($f1, f2$) = 1, g2 can be refined to make g1 and g2 closer in semantics. At this moment, g2 is compressed into $g2' = g1 \cap g2$.

(b) If $0 < ED (g1, g2) < ED (f1, f2) < 1$, we can compress $g2$ to satisfy $IFD: A_i$ •A_n.

- $sub (g1) \geq sub (g2)$ and $sup (g1) \geq sup (g2)$

  Let $g2' = Rstring (g2, x)$ such that $ED (g1, g2') = ED (f1, f2)$, where $|g1|$ = u, $|g2'| = x$, $|g1 \cap g2'| = k$, and $|g1 \cup g2'| = u + x - k$. Hence x can be obtained uniquely. If $x \leq k$, let $x = k$. When $sub (g1) \leq sub (g2)$ and $sup (g1) \leq sup (g2)$, we can apply the similar method to gain the unique x.

- $sub (g1) \geq sub (g2)$ and $sup (g1) \leq sup (g2)$

  Let $g2' = Rstring (Lstring (g2, x), y)$ such that $ED (g1, g2') = ED (f1, f2)$ and$(n–i)/(i–j) = (sup (g2) – sup (g1))/(sub (g1) – sub (g2))$, where $|g1|$ = u, $|g2'| = y$, $|g1 \cap g2'| = u$, and $|g1 \cup g2'| = y$. So x and y can be obtained uniquely. If $x < y \leq k$, let $x = y = k$. When sub (g1) $\leq$ sub (g2) and sup (g1) $\geq$ sup (g2), we cannot refine $g2$.

Note that there may be another attribute A_j such that A_j •A_n ($1 \leq j \leq n - 1$). Under the situation that multiple attributes functionally determine A_n, we can repeat the processing procedure above to obtain the more precise query answers.

Example

In this section, we present a simple example to illustrate some of the ideas discussed. Assume that we have the database relation with imprecise information in Figure 5.1, which is created by integrating the multidatabases in virtual enterprises.

In this database relation, let Dom (Material) = {Fe, Cu, Ni, Cr, Zn, Mn, Alloy}, $\delta_M = 7/100$, Dom (Price) = [30, 50], $\delta_p = 20/1000$, Dom (Lead–time) = [5, 30], and $\delta_L = 25/1000$. When we query "What are the lead–time of part 10015 and the price of part 10010", we obtain two imprecise answers "φ" and "[51.50, 54.50]", respectively. Now we would like to refine these two imprecise query answers using the framework of analogical reasoning developed above. First of all, we refine the query answer "φ".

Part ID	Material	Price (Dollars)	Lead–time (Days)
10010	{Ni, Cr, Zn}	[51.50, 54.50]	[8, 14]
10011	{Ni, Cr}	[53.50, 55.50]	[8, 14]
10012	{Mn}	[54.20, 54.20]	[10, 10]
10013	{Mn}	[54.20, 54.20]	[10,10]
10014	{Fe, Cu, Alloy}	[63.50, 65.50]	[22, 28]
10015	{Fe, Alloy}	[64.50, 65.50]	φ

Figure 5.1 Integrated database relation r with imprecise information

(a) Because ED ({Ni, Cr, Zn}, {Ni, Cr}) = 2/3 – 2/7 = 0.38, ED ([8, 14], [8, 14]) = 6/6– 6/25 = 0.76, ED ({Mn}, {Mn}) = δ_M/δ_M – δ_M /7 $\approx$ 1, and ED ([10, 10], [10, 10]) = δ_L/δ_L – δ_L/25 $\approx$ 1, IFD: $Material$ •$Lead$–$time$ holds in r.

(b) ED ({Ni, Cr, Zn}, {Fe, Alloy}) = ED ({Ni, Cr}, {Fe, Alloy}) = ED ({Mn}, {Fe, Alloy}) = 0, and ED ({Fe, Cu, Alloy}, {Fe, Alloy}) = 2/3 – 2/7 = 0.38, therefore, part 10014 has the highest equivalence degree with part 10015 on attribute Material.

(c) According to the results in previous two steps, instead of "φ", we infer that the lead–time of part 10015 may also be [22, 28] days. It is evident that the query answer is refined.

Now let us look at the refinement of imprecise query answer [51.50, 54.50]. Based on the same ideas above, we can conclude that IFD: $Material$ •$Price$ holds in r and part 10011 has the highest equivalence degree with part 10010 on attribute Material. So we refine [51.50, 54.50] with [53.50, 55.50] by using the method defined above.

Let [51.50, 54.50] be refined as [54.50 – x, 54.50]. Then we have ED (54.50 – x, 54.50], [53.50, 55.50]) = ED ({Ni, Cr, Zn}, {Ni, Cr}), namely, 1/(2 + x – 1) – 1/20 = 2/3 – 2/7. Hence, x = 1.33 and [51.50, 54.50] is refined to be [53.50, 54.50].

SUMMARY AND FUTURE WORK

Imprecise data may appear in databases due to data unavailability or data and schema incompatibilities in multidatabase systems. Partial values and null values have been used to model imprecise data in virtual enterprises to resolve semantic conflicts of multidatabase integration. Functional dependencies, which are one kind of integrity constraints in relational databases, not only play a critical role in a logical database design but also can be employ to reasoning, being knowledge.

In this article, we discuss the issues on refining imprecise query answers for heterogeneous multidatabase systems in virtual enterprises by using analogical reasoning. In order to do that, we introduce the notion of equivalence degree of imprecise data, by which we define the imprecise functional dependencies in the integrated database systems. On the basis of that, we give the approach to mining the rules of imprecise functional dependencies and apply the frame of analogical reasoning to refine the imprecise query answer.

Virtual enterprises are important organization forms of current manufacturing enterprises. With characteristics of agile manufacturing, virtual enterprises can quickly respond to the market's changes and its products have effective competitiveness. However, virtual enterprises are the information–

based enterprises and their implementations have put some new requirements on the modeling of complex data and information management. In this article, we have put our focus on multidatabases integration in virtual enterprises and the refinement of imprecise query answers using analogical reasoning based on imprecise functional dependencies. In the future work, we will concentrate on mining the other association rules from engineering database systems and applying the other reasoning forms to effectively support the design of intelligent manufacturing systems.

ACKNOWLEDGEMENTS

The authors want to thank the support for this study from the NSERC and AECL, Canada and the City University of Hong Kong.

REFERENCES

Atzeni, P. and Morfuni, N. M., "Functional Dependencies and Constraints on Null Values in Database Relations", *Information and Control*, 70 (1): 1–31, 1986.

Cheung, W. M. and Hsu, C., "The Model–Assisted Global Query System for Multiple Databases in Distributed Enterprises", *ACM Transactions on Information Systems*, 14 (4): 421–470, 1996.

Codd, E. F., "Extending the Database Relational Model to Capture More Meaning", *ACM Transactions on Database Systems*, 4 (4): 397–434, 1979.

Codd, E. F., "Missing Information (Applicable and Inapplicable) in Relational Databases", *SIGMOD Record*, 15 (4): 53–78, 1986.

Codd, E. F., "More Commentary on Missing Information in Relational Databases (Applicable and Inapplicable Information)", *SIGMOD Record*, 16 (1): 42–50, 1987.

DeMichiel, L. G., "Resolving Database Incompatibility: An Approach to Performing Relational Operations over Mismatched Domains", *IEEE Transactions on Knowledge and Data Engineering*, 1 (4): 485–493, 1989.

Dutta, S., "Approximate Reasoning by Analogy to Null Queries", *International Journal of Approximate Reasoning*, 5: 373–398, 1991.

Grahne, G., "Dependency Satisfaction in Databases with Incomplete Information", in *Proceedings of 10th International Conference on Very Large Data Bases Conference*, Singapore, August, pp. 37–45, 1984.

Grant, J. and Minker, J., "Answering Queries in Indefinite Databases and the Null Value Problem", in *Advances in Computing Research*, (3), *The Theory of Databases*, JAI Press, pp. 247–267, 1986.

Grant, J., "Incomplete Information in a Relational Database", *Fundamenta Informaticae*, 3 (3): 363–378, 1980.

Grant, J., "Partial Values in a Tabular Database Model", *Information Processing Letters*, 9 (2): 97–99, 1979.

Hale, J. and Shenoi, S., "Analyzing FD Inference in Relational Databases", Data and Knowledge Engineering, 18: 167–183, 1996.

Hardwick, M., Spooner, D. L., Rando, T. and Morris, K. C., "Sharing Manufacturing Information in Virtual Enterprises", *Communications of the ACM*, 39 (2): 46–54, 1996.

Hathaway, R. H., Bezdek, J. C. and Pedrycz, W., "A Parametric Model for Fusing Heterogeneous Fuzzy Data", *IEEE Transactions on Fuzzy Systems*, 4 (3): 270–281, 1996.

He, Jichao, "Extending Relational Model to Deal with Null Value", *Chinese Journal of Computers*, 10 (8): 449–459, 1987.

Imielinski, T. and Lipski, W., "Incomplete Information and Dependencies in Relational Databases", in *Proceedings of ACM SIGMOD International Conference on Management of Data*, pp. 178–184, 1983.

Imielinski, T. and Vadaparty, K., "Complexy of Query Processing in Databases with OR-Objects", in *Proceedings of ACM SIGMOD–SIGACT–SIGART Symposium on Principles of Database Systems*, pp. 51–65, 1989.

Liao, S. Y., Wang, H. Q. and Liu, W. Y., "Functional Dependencies with Null Values, Fuzzy Values, and Crisp Values", IEEE Transactions on Fuzzy Systems, 7 (1): 97–103, 1999.

Lipski, W., "On Semantic Issues Connected with Incomplete Information Databases", *ACM Transactions on Database Systems*, 4 (3): 262–296, 1979.

Ma, Z. M., Zhang, W. J. and Li, Q., "Extending Relational Data Model to Resolve the Conflicts in Schema Integration of Multiple Databases in Virtual Enterprise", in *Proceedings of the 1999 ASME Design Engineering Technical Conferences*, 1999.

Ma, Z. M., Zhang, W. J. and Ma, W. Y., "View Relation for Schema Integration on Multiple Databases and Data Dependencies", in *Proceedings of 9th International Database Conference on Heterogeneous and Internet Database*, pp. 278–289, 1999a.

Ma, Z. M., Zhang, W. J. and Ma, W. Y., "Assessment of Data Redundancy in Fuzzy Relational Databases Based on semantic Inclusion Degree", *Information Processing Letters*, 72 (1–2): 25–29, 1999b.

Minker, J., "On Indefinite Databases and the Closed World Assumption", in *Lecture Notes in Computer Science*, *N238* (Springer Verlag, New York), pp. 292–308, 1982.

Rolstadas, A., "Enterprise Modeling for Competitive Manufacturing", *International Journal of Control Engineering Practice*, 3 (1): 43–50, 1995.

Rundensteiner, E. A., Hawkes, L. W. and Bandler, W., "On Nearness Measures in Fuzzy Relational Data Models", *International Journal of Approximate Reasoning*, 3: 267–98, 1989.

Shenoi, S. and Melton, A., "Proximity Relations in the Fuzzy Relational Databases", *Fuzzy Sets and Systems*, 31)3): 285–296, 1989.

Tseng, F. S. C., Chen, A. L. P. and Yang, W. P., "Refining Imprecise Data by Integrity Constraints", *Data and Knowledge Engineering*, 11 (3): 299–316, 1993.
Zaniolo, C., "Database Systems with Null Values", *Journal of Computer and System Sciences*, 28: 142– 166, 1984.

Zhang, W. J. and Li, Q., "Information modeling for made–to–order virtual enterprise manufacturing systems", *Journal of Computer–Aided Design*, 31 (10): 611–619, 1999.

CHAPTER 22

The Use of Process Capability Data
in Design

Anna Thornton
Athornton@analytics-consulting.com
Analytics Operations Engineering, 1 State Street Suite 300, Boston, MA
02109

ABSTRACT

When a design is created, the designer specifies the geometry, dimensions, and material of the parts. This information is used to tell manufacturing how to produce the parts. In addition, the designer specifies tolerances, which are used to guide the process selection and parts acceptance. Acceptability can be determined in a number of ways, including process validation and inspection.

Traditionally, the process of tolerance specification has been uni-directional; i.e., the design specifies the tolerances, and manufacturing is expected to meet them. However, in the new paradigm of design for manufacturing and concurrent engineering, communication with manufacturing is necessary to ensure that the tolerances specified are consistent with manufacturing's capability. By ensuring producibility before a design is released to manufacturing, expensive redesign, processes, rework, and customer dissatisfaction can be avoided.

This chapter was designed to help planning a process capability database implementation. It is based the author's experience with implementing process capability databases in a number of large manufacturing/design organizations. The chapter starts by reviewing the typical data content and structures. Next, the chapter reviews the issues of using surrogate data. The final section discusses implementation issues.

D. Braha (ed.), Data Mining for Design and Manufacturing, 505–518.

INTRODUCTION

When a design is created, the designer specifies the geometry, dimensions, and material of the parts. This information is used to tell manufacturing how to produce the parts. In addition, the designer specifies tolerances, which are used to guide the process selection and parts acceptance. Acceptability can be determined in a number of ways, including process validation and inspection.

Traditionally, the process of tolerance specification has been uni-directional; i.e., the design specifies the tolerances, and manufacturing is expected to meet them. However, in the new paradigm of design for manufacturing and concurrent engineering, communication with manufacturing is necessary to ensure that the tolerances specified are consistent with manufacturing's capability. By ensuring producibility before a design is released to manufacturing, expensive redesign, processes, rework, and customer dissatisfaction can be avoided.

A significant amount of literature stresses the importance of selecting tolerances appropriately. Several articles (Chase et al. 1996; DeGarmo et al. 1997; Parkinson et al. 1993) stress the importance of tolerance allocation. If tolerances are too tight, unnecessarily expensive operations are used; if tolerances are too loose, the part may not function properly. Tolerances should be optimized to reduce mechanical errors (Lee et al. 1993; Lin et al. 1997; Zhang and Wang 1998), minimize assembly problems (Ting and Long 1996), and improve product performance (Michelena, 1994; Wang, 1993). Setting tolerances to match process capability and design intent is also the subject of a significant amount of literature (Gao et al. 1998; Liu et al. 1996; Srinivasan et al. 1996).

To achieve this, organizations must develop several capabilities. First, the manufacturing organization must have an accurate understanding of its process capabilities. This capability is becoming more widespread through the extensive use of statistical process control (SPC), automatic measurement systems, and electronic data-collection methods. Second, the manufacturing organization must be able to communicate the data to the engineering community. Third, the engineering community must know what data to ask for and how to ask for it. Finally, the engineering community must know how to use the data. The focus of this chapter is on communicating the existing process capability data between manufacturing and design.

Typically, one of two methods is used. The first method depends on one-to-one communication between design and manufacturing. In this method, engineering typically asks manufacturing, "on this design, can you achieve this tolerance?" Manufacturing then uses its own internal data resources and internal expertise to answer the question. The second method depends on a database that contains information on manufacturing capability, and is accessible to the entire community. The database is queried using a set of criteria, and the appropriate data are returned. Each method has its benefits and shortcomings. The first is very time-consuming but typically is accurate. The second is, theoretically, less time-consuming but has many technical issues surrounding the implementation and use of the database.

A limited number of journal articles address the implementation of process-capability databases. Several articles address the specific problem of using process capability in electronic-systems design (Lucca *et al.* 1995; Nagler 1996). Campbell and Bernie (1996) discuss requirements for a formalized rapid prototyping database. Perzyk and Meftah (1998) describe a process-selection system that includes general data on process capabilities. While many papers discuss the possibility of using an automated checking system when applying process capability in design, none describe the implementation of a working system.

To understand how companies use process-capability data, a survey of more than twenty large design and manufacturing firms was undertaken (Tata and Thornton 1999). In general, all but one of the companies had spent significant resources to implement process-capability databases. However, none of the companies felt it was making efficient use of its data.

The survey identified both organizational and technical barriers to using process-capability data. Organizational barriers include a lack of systematic procedures and policies for using the data (e.g., incentive structures and standard design practices), and a lack of good communication between design and manufacturing (e.g., teams didn't trust each other, access to databases was limited, and needs were communicated poorly). Technical barriers include poor indexing schemes, lack of statistical validity, excessive time required to access the data, and poor user interfaces. In general, the technical barriers prevent designers from getting the right data quickly.

In general, when industries were interviewed about implementing process-capability databases, their first reaction was, "this should be easy." They believed that if data was simply put online in a structured format, engineering will be able to access and use the data, and ensure producible designs However, this idealistic view of the problem is very far from the truth of what is involved in creating a useable system. There are many issues around indexing methods, the use of surrogate data, user interfaces, usability, and data control.

This chapter was designed to help planning a process capability database implementation. It is based the author's experience with implementing process capability databases in a number of large manufacturing/design organizations. The chapter starts by reviewing the typical data content and structures. Next, the chapter reviews the issues of using surrogate data. The final section discusses implementation issues.

DATA STRUCTURE

The ability to find and use process-capability depends highly on the data's structure. Interestingly, the databases of the surveyed companies all had similar structures and data content. The database is typically made up of the raw data (the actual measurements), the descriptors (the method by which the data is indexed) and the aggregated data (the means and standard deviations).

Raw data

In a typical scenario, the factory floor first manufactures a part or assembly, then takes a measurement of a key variable. This measurement has been determined to be either a Key Product Characteristic or a Key Process Characteristic (Thornton 1999) by the engineering community or by customers. The measurement is recorded along with a set of descriptors that describe who builds what part. This data can either be plotted in an SPC format or analyzed as runs (described below) to calculate measures of process variation.

The raw data can have a number of flaws. Errors in measurement and data entering can corrupt the measurements. In addition, selective entering of data can skew the true process capability. Selective data entering can happen when a person chooses only the portion of the data that shows stable performance and throws away the portions of the data taken when the process was out of control.

Descriptors

Once the data is entered, the designer then needs a way to access the data. In older versions of the database, the data was often indexed based on the drawing number. In order to look up a set of data, the person would need to know what drawing had a similar part. This made it difficult for someone to find data. In order to alleviate this problem, several companies started to index the data based on descriptors.

When a team asks the question, *What is the standard deviation and mean shift for drilled hole?* The answer they often receive is, *It depends.* The search for the correct capability is non-trivial because many factors other than the feature type can influence the final process variation, and these factors need to be recorded alongside the raw data. A number of different descriptors can be included (the example of a hole is used to illustrate each descriptor):

- **Process:** Holes can be created using a variety of processes. Each process will introduce different amounts of variation.

- **Material:** Given the same process, some materials will introduce more variation than others.
- **Feature geometry:** The process variation for hole-drilling will depend on the aspect of the feature being measured. For example, the hole depth may have less variation than the diameter.
- **Feature size:** The process variation in holes with larger diameters may be larger than that for holes with small diameters.
- **Machine:** Many designers will set tolerances assuming the best process capability. However, in many cases, the best machine may not have the capacity to produce all parts. Consequently, overflow parts may be made on older, less capable machines.
- **Tolerance:** The tolerance specified on the part drawing may change the process used and hence the variation introduced. For example, for tight tolerance surfaces, NC path planners may cut at slower speeds. However, tighter tolerances typically come at a cost: larger labor content, processing cost, and/or throughput time.
- **Operator:** Machine operators can influence process variability. Some operators are better than others.
- **Feature/part characteristics:** A feature's location in a part can influence process variation. For example, more variation will be introduced when drilling a hole in a thin part than in a thick part. Fixtures and post-processing will also influence the final variation.
- **Part:** The part in which the feature is created will have a major impact on the
- **Number of parts:** In low-volume production, parts may be made by specialists in a job-shop environment; this may or may not increase quality. The use of automation in high-volume production may standardize processes, but the job shop may be more capable of controlling variation.

Several of these descriptors are numeric (e.g., tolerance, number of parts), but others are based on a set of possible values (i.e., material, process, feature, machine). Still others are too general to be recorded in a structured format (i.e., feature/part characteristics). There is always a tradeoff between specificity of a descriptor set and the overhead involved in managing the data. It will never be possible to capture completely all the factors that influence the capability in a structured data set; at some level, the manufacturing experts will need to explain the process capability.

To facilitate a consistent and systematic process for managing the descriptors, a number of companies use an indexing scheme. An indexing scheme is a systematic method for describing typical part/process/material combinations. The standard features, materials, and processes are given unique codes. A combination of a feature, material, and process uniquely identifies each data record. Typically, each indexing code has a hierarchical structure used to facilitate coding and retrieval. Table 1 shows a sample coding from an existing indexing system. If a machinist measures the length of a hole made using a handheld drill in a Titanium extrusion 6AL6V, the measurement is recorded using the code M4.1.1. P7.1.1 F3.2

Table 1: Sample Feature Codes

Material	Process	Feature
4. Titanium	7. Hole Preparation	3. Hole
4.1. Extrusion	7.1. Drill	3.1. Location
4.1.1 6AL6V	7.1.1. Handheld	3.2. Length
4.1.2 6AL6V2SN	7.1.2. Semi-Automatic	3.3. Diameter
4.1.3 6AL4V	7.1.3. Numerical Control	3.4. True Position
4.2. Plate, Sheet	7.2. Countersink	3.5. Perpendicularity
4.2.1 6AL6V		
4.2.2 6AL6V2SN		
4.2.3 6AL4V		

Using a feature-indexing scheme enables a team to find data without having to know the drawing numbering system. The common classification scheme also facilitates information sharing among manufacturing groups. Indexing is not without problems, however. One indexing scheme we saw could encode 52 million possible combinations, but only 50,000 of the combinations (0.1%) were feasible, and many feasible index combinations were not populated. An indexing scheme must be developed with an appropriate specificity of coding. Codes that are too highly detailed will result in a sparse dataset; codes that are too general may be difficult to interpret.

Process Capability

Raw data that share a common set of descriptors are combined into "runs" to determine process capability. Process capability is the amount of variation introduced when a feature is created by a process. The standard deviation and bias are used to characterize the process[1]. In addition, product-development teams often use Cpk and Cp to evaluate producibility. Typically, a team aims for tolerances produced with a Cpk of 1.33.

The standard deviation and bias for a process are estimated from the individual data points. The run's bias, b, is a function of n individual data points, x_i, and the target value, m.

---

[1] In most cases, the variation is assumed to have a Gaussian distribution.

$$\mu = \sum_{i=1}^{n} \frac{x_i}{n} \quad and \ b = \mu - m \tag{1}$$

The run's standard deviation is a function of the individual points, x_i, and the mean μ.

$$\sigma = \sqrt{\sum_{i=1}^{n} \frac{(x_i - \mu)^2}{n-1}} \tag{2}$$

Cp and Cpk are a function of the upper and lower tolerance limits $[LL, UL]$.

$$Cp = \frac{UL - LL}{6\sigma} \quad and \ Cpk = \frac{\min(UL - \mu, \mu - LL)}{3\sigma} \tag{3}$$

Data Structure Summary

The following example summarizes a typical data structure. Individual measurements are collected and recorded on the plant floor. Table 2 shows an example of the raw data, taken at machine 1031, of a hole location, using a numerically controlled machine in a titanium extrusion. The tolerance is [.0666,.0966] with a target value of .0766. The plant floor took a measurement of .0756. Typically, the date and operator information also is included.

Table 2: Sample Record

Part	Material	Process	Feature	Target	UL	LL	Mach.	Measurement
3461	4.1.1	7.1.3	3.1	0.0766	0.0666	0.0966	1031	0.0756

After a number of parts are built, it is possible to calculate the process capability for a run in which a hole position in a titanium plate for machine 1031 in part 3461.

Table 3: Sample Run

Part	Feature Code	Target	LL	UL	Mach.	Bias	Std Dev	n	Cp	Cpk
3461	M1.1 P7.1.3 F3.1	0.0766	.0666	0.0966	1031	0.0001	0.0047	136	1.07	0.72

Now that these data are available, the designers can access them when designing a similar plate. However, as we keep saying, *it isn't that easy.*

PROBLEMS WITH SURROGATE DATA

The ability to use historical data as a predictor for the future relies on several assumptions. It will never be possible to exactly predict future process capability from historical data. Even if an organization produces the exact same part, there will be differences. It is necessary to understand and communicate these differences to prevent invalid process certification.

Differences Between Future and Past Capability

Even assuming that a team wishes to build the same part on the same machine with the same operator, historical data may not be a good indicator of future capability. First, capability can degrade with age due to machine wear or lack of maintenance. Second, capability can get better over time. Process improvement, regular maintenance, standard operating procedures, statistical process control, etc., can result in an improved process capability. If historical process capability is used without understanding these trends, the surrogate data may over or underestimate capability.

Complete Set of Descriptors

The ability to locate correct surrogate data depends on having a set of descriptors that captures possible effects. If the set of descriptors is not thorough enough, incorrect surrogate data may be selected. For example, if the machine number is not included and there are multiple machines with various capabilities, the wrong data may be used. In addition, the user, typically a designer, may be unable to specify or control some of the descriptors. For example, he or she may not know, or have control over, what machine the part is built on, or what operator is used. This is potentially dangerous, because a designer may be tempted to choose the best process capability with which to validate his or her designs, but ultimately, the part may not be built on that machine.

Existence of Surrogate Data

When a user selects descriptors and requests a process capability, one of three events will happen: no records, a single record, or multiple records will be returned. When no records are returned, it is impossible for the database to predict the capability. However, it may be possible to select a "surrogate" capability based on similar parts. For example, a designer needs information about a ½-inch hole in a titanium extrusion but there are only data for ½-inch holes in an aluminum plate and ¼-inch holes in a titanium plate. In this case, the system needs to be able to know which point is a better surrogate for the other.

There are a number of ways to handle this in the database system. The first is to create the correlations and similarities through committee. This is very labor-intensive and subjective, and may change with time. The second option is to correlate the data statistically by basing the value on a linear fit through the nearest neighbors. Where there are non-linearities in the relationships or where there is co-variance among the axes, a more complex estimation may be needed. The success of this approach depends on the existence of enough data in the area of the unpopulated point to be able to accurately estimate the new value.

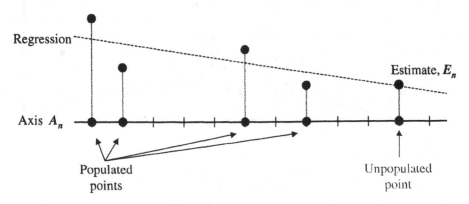

Figure 1: Estimation of empty record

Multiple Data Points

In many cases, the search for a surrogate data point may return multiple runs. This may occur when a frequently used process/feature/material combination is requested (e.g., locations of drilled holes in an aluminum extrusion in aerospace manufacturing). Because of differences in machines, operators, part characteristics, machine settings etc., the data returned may indicate a wide range of process capabilities. When this occurs, it is necessary to differentiate possible patterns or trends in the data. It is not valid to average the runs together, as the data comes from different populations.

Thornton and Tata (2000) proposed an iterative approach to process-capability analysis that uses a visualization method to facilitate the search.

The graph (Figure 2) shows the data points from a sample set of runs sharing the same index. Each run is plotted as a single point. The bias for each run is plotted on the X-axis and the standard deviation on the Y-axis. The acceptability limits for *Cpk* = 1.33 and *Cpk* = 1.00 are shown as lines on the graph for a tolerance band of [-0.01,+0.02]. The top line represents *Cpk* = 1.0 and the bottom line, *Cpk* = 1.33. Any run falling below both lines will have a *Cpk* of greater than 1.33; any falling above both lines will have a *Cpk* of less than 1.0.

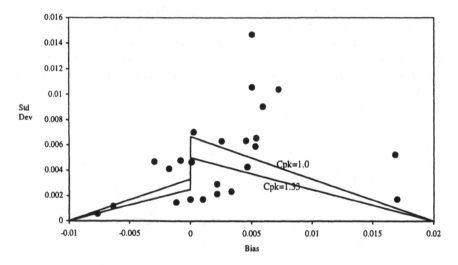

Figure 2: *Cpk* Graph

The graph helps a user to assess the range of possible *Cpk* values. If all of the points fall above the line, there is a good chance the tolerance can not be achieved. Conversely, if all of the points fall below the line there is a good chance the tolerance can be achieved. However, if some fall below and some fall above the acceptability limit, the user must also be able to assess quickly which runs are relevant. The difficulty is in identifying which descriptors are relevant and how they impact capability. Relevance is a function of the process characteristics that a new design shares with the surrogate run. To facilitate the discovery of patterns, secondary descriptors can encoded visually (Thornton and Tata 2000).

However, because more than two process characteristics typically influence the standard deviation and bias, the simplest approach is to allow a user to select the encoding scheme. This can be time-consuming for the user because he or she must test multiple encodings until a pattern emerges. The user needs help to determine the most useful refinement parameters. In order to help a designer to locate trends; the two most significant contributors are identified and encoded automatically. The most relevant factors are automatically encoded using a multivariate linear regression analysis to identify the most significant factors.

Summary

In summary, there are many possible problems with using process-capability data during design. Many companies we spoke with are under the false impression that the use of process-capability data will replace the process engineer in the early stages of the product development process. Given the number of possible pitfalls associated with using surrogate data, it is critical that the process engineer be involved with any data analysis. It is also necessary to communicate to the user what is the range of possible errors in the data. These can be estimated statistically.

IMPLEMENTATION ISSUES

In addition to the fundamental problems inherent in using surrogate data, there are a few practical issues associated with implementing a process-capability database system.

Determining the indexing scheme: The indexing scheme is critical to identifying the correct surrogate data. The following are steps that can be taken to help generate the indexing scheme:

- Review drawings of typical parts and collect a set of common features, processes, and materials; and

- Review the processes and collect the processes and materials used; and
- When designing the indexing scheme, use a standard 1., 1.1., 1.1.1, numbering scheme. This will allow for later expansion. In addition, a review board and review process should be created to approve any modifications to the indexing scheme.

Determine relevant data: In most manufacturing organizations, many sites are involved in production. Once a part is designed, production may occur at a range of sites, and the designer may not know which site will build it. Therefore, a survey of all manufacturing sites should be conducted, and the data-collection systems, existing data, and database systems catalogued for each. In addition, the review board should determine the data to include in the centralized system, taking into account the organization's manufacturing strategy. For example, if manufacturing plant A has antiquated equipment that will not be used for future production, it should not be included in the database.

Select a database approach: A number of approaches exist for implementing a process-capability database that spans multiple sites. A number of questions must be asked before making a decision:

- Is it necessary to have real-time data? Does the database need to reflect the parts being built today, or can older data be used?

- How error-prone are the data? If the data are error-prone, using a real time system may be difficult, and the data may need to be cleaned up before being posted it to a central system.

- How common are the data formats and systems across the organization? If the data formats and systems are diverse and incompatible, integrating the systems in a real time fashion may be difficult.

There are two approaches to integrating the data across the organization:

1) **Real time system:** In this system, multiple data sources from multiple sites can be accessed in real time. Users will have access to the most up-to-date data. However, implementing such a system will require significant time and information technology skills. Each site will need to add an indexing system to its database to ensure conformity of data. This approach will require automatic data checkers to remove any errors in the data.

2) **Meta-database system:** In the meta-database system, data from each site is uploaded regularly to a centralized system. During the upload process, the data is cleaned up, indexed, and put into a common format. Data in the meta-database will lag behind the real-time data by three-to-six months. If this lag is not a problem, the implementation should be significantly easier.

Accessing supplier databases: In situations where the organization designs the parts that are being outsourced, a supplier's capability needs to be understood. In general, suppliers are wary of sharing this data. We recommend working with key suppliers with whom the organization has a long relationship in order to access capability data. The meta-database approach is more compatible with including supplier data than is the real-time system.

Handling non-machine based processes: Not all processes will fit into the material/feature/process scheme described above. For example, serial processes, selective assembly, shimming, assembly, and fixtures have different descriptors than traditional machining processes, and may require unique data structures.

Integration with design process: The organization needs require the use of the process capability databases during the design process. Also, it is important to ensure that the database does not become the master source for all information relating to process capability. The database should be used to complement the process engineer's role, not replace it. The team should focus on working on the difficult 10% and use the database to support the validation of the easy 90%.

Additional data: The organization may wish to add additional data to help the design community assess issues around cost and capacity. For example, the database may also have standard costs and capacity added to the machine information. This will allow the designer to determine if tightening or loosening a tolerance will have any cost impact.

Design user-friendly interfaces: The user interface is critical to the usability of the system. The system design should be designed to be intuitive, allow the designers to specify the descriptors they know, and return the data in a way that can be interpreted easily (Thornton and Tata 2000).

CONCLUSION

In summary, the use of historical process capability data in design will improve the producibility and quality of new designs. However, the implementation of such a system is non-trivial. This chapter has described several of the technical challenges relating to both the use of surrogate data and the implementation of a multi-site process capability database.

REFERENCES

Chase, K. W., J. S. Gao, S. P. Magleby and C. D. Sorensen (1996) "Including Geometric Feature Variations in Tolerance Analysis of Mechanical Assemblies." *IIE Transactions* **28**(10), pp. 795-807.

Gao, J., K. W. Chase and S. P. Magleby (1998) "Generalized 3-D Tolerance Analysis of Mechanical Assemblies with Small Kinematic Adjustments." *IIE Transactions* **30**, pp. 367-377.

DeGarmo, A., E. Paul, J. T. Black and R. A. Kohser (1997) *Materials and Processes in Manufacturing.* Prentice Hall, Upper Saddle River, NJ.

Lee, S. J., B. J. Gilmore and M. M. Ogot (1993) "Dimensional Tolerance Allocation of Stochastic Dynamic Mechanical Systems Through Performance and Sensitivity Analysis." *Journal of Mechanical Design, Transactions of the ASME* **115**(3), pp. 392-402.

Lin, C. Y., W. H. Huang, M. C. Jeng and J. L. Doong (1997) "Study of an Assembly Tolerance Allocation Model Based Monte Carlo Simulation." *Journal of Materials Processing Technology* **70**(1-3), pp. 9-16.

Liu, C. S., S. J. Hu and T. C. Woo (1996) "Tolerance Analysis for Sheet Metal Assemblies." *Journal of Mechanical Design, Transactions of the ASME* **118**(1), pp. 62-67.

Michelena, N. F. and A. M. Agogino (1994) "Formal Solution of N-Type Taguchi Parameter Design Problems with Stochastic Noise Factors." *Journal of Mechanical Design, Transactions of the ASME* **116**(2), pp. 501-507.

Parkinson, A., C. Sorensen and N. Pourhassan (1993) "A General Approach for Robust Optimal Design." *Journal of Mechanical Design, Transactions of the ASME* **115**(1), pp. 75-81.

Srinivasan, R. S., K. L. Wood and D. A. McAdams (1996) "Functional Tolerancing: A Design for Manufacturing Methodology." *Research in Engineering Design - Theory Applications and Concurrent Engineering* **8**(2), pp. 99-115.

Tata, M. and A. Thornton (1999) "Process Capability Database Usage In Industry: Myth vs. Reality." *Design for Manufacturing Conference, ASME Design Technical Conferences*, Las Vegas, NV.

Ting, K.-L. and Y. Long (1996) "Performance Quality and Tolerance Sensitivity of Mechanisms." *Journal of Mechanical Design, Transactions of the ASME* **118**(1), pp. 144-150.

Thornton, A. C. and M. Tata (2000) "Use of graphic displays of process capability data to facilitate producibility analyses." *Artificial Intelligence for Engineering Design, Analysis, and Manufacturing* **14**, pp. 181-192.

Thornton, A. C. (1999) "A Mathematical Framework for the Key Characteristics Process," *Research in Engineering Design*, 11:145-147.

Wang, N. and T. M. Ozsoy (1993) "Automatic Generation of Tolerance Chains from Mating Relations Represented in Assembly Models." *Journal of Mechanical Design, Transactions of the ASME* **115**(4), pp. 757-761.

Zhang, C. C. and H. P. Wang (1998) "Robust Design of Assembly and Machining Tolerance Allocations." *IIE Transactions* **30**(1), pp. 17-29

INDEX